Henry Stevenson, Thomas Southwell

The Birds of Norfolk

With remarks on their habits, migration, and local distribution. Vol. 3

Henry Stevenson, Thomas Southwell

The Birds of Norfolk
With remarks on their habits, migration, and local distribution. Vol. 3

ISBN/EAN: 9783744780216

Printed in Europe, USA, Canada, Australia, Japan

Cover: Foto ©berggeist007 / pixelio.de

More available books at **www.hansebooks.com**

THE
BIRDS OF NORFOLK,

WITH

REMARKS ON THEIR HABITS, MIGRATION,
AND LOCAL DISTRIBUTION:

BY

HENRY STEVENSON, F.L.S.

MEMBER OF THE BRITISH ORNITHOLOGISTS' UNION.

CONTINUED BY

THOMAS SOUTHWELL, F.Z.S.

MEMBER OF THE BRITISH ORNITHOLOGISTS' UNION.

IN THREE VOLUMES.

VOL. III.

"The wisdom of God receives small honour from those vulgar heads that rudely stare about, and with a gross rusticity admire his works. Those highly magnifying him, whose judicious inquiry into his acts, and deliberate research into his creatures, return the duty of a devout and learned admiration."—Sir THOMAS BROWNE.
"Religio Medici," Wilkin's ed., 1835, vol. ii., p. 19.

LONDON:
GURNEY AND JACKSON, 1, PATERNOSTER ROW
(SUCCESSORS TO MR. VAN VOORST).

NORWICH:
NORFOLK CHRONICLE CO., LIMITED.
1890.

INTRODUCTION.

In taking up the work laid down by Mr. Stevenson in 1877, never, as it proved, to be resumed, I feel that some explanation is due from one whom nothing but the force of circumstances would have induced to undertake the task. I was painfully conscious of the difficulties I had to encounter in following one so gifted with the innate faculty of observation and possessed of so pleasing a style of imparting the great store of information accumulated thereby. Nothing but the reproach that such a work should remain unfinished, and his expressed wish conveyed to me by his family, and above all the generous offers of assistance from friends, many of whom had already contributed largely to the preceding volumes, would have induced me to make the attempt of which I now present the result.

I may at once say that I have neither leisure for the necessary outdoor observation, nor the literary ability to enable me to continue the charming life-histories with which Mr. Stevenson has in so many cases furnished us. I have, therefore, been much less ambitious in my aim, and have confined myself as strictly as possible to the story of each species as a Norfolk bird; in some instances, where they are rare or local, or possess particular interest from other causes, I have entered rather more fully into detail, especially where the bird is or has been in danger of becoming lost to us as a resident, but in most cases I have studied to set forth the facts in as brief a form as possible, avoiding descriptions of plumage, except where a few remarkable varieties are concerned, such information now being within the reach of all in one or other of the excellent modern works on the subject.

I have followed the same general arrangement and nomenclature as adopted by Mr. Stevenson, that of the third edition of Yarrell's "British Birds;" but, like him, I have made such changes in the scientific names as seemed desirable.

The lapse of time since the appearance of the first and second volumes has rendered the addition of such an Appendix as Mr. Stevenson had intended all the more necessary. In this I have noticed all the species which during that period have been for the first time recognised as occurring in the county, followed by brief remarks on a few which have become better known to us than when Mr. Stevenson wrote; and, finally, have enumerated a few others which were included in the "Birds of Norfolk" on what is now considered insufficient authority.

It remains only for me to express my sincere thanks for the kindly help which I have received in the prosecution of my task, and to acknowledge the sources from which I have received information; the latter I have always endeavoured to do in the text, even at the risk of repetition, but the former is a much more difficult duty, as it would require me to give a list of Norfolk observers of portentous length, and in the words of Simon Wilkin, in his preface to Sir

Thomas Browne's works, "to enumerate them were it possible to do so without omission, might have rather the appearance of parade than of gratitude: while a solitary omission would expose me to the mortifying and undeserved imputation of ingratitude;" I must, therefore, deny myself the pleasure of thanking individually those who have so kindly assisted me. My special thanks, however, are due to the late Mr. J. H. Gurney—whose death before the completion of a work in which he took so great an interest we have all to deplore—also to his son, Mr. J. H. Gurney, for much valuable help in all ways; while to Professor Newton, I have no hesitation in saying, the completion of this work is mainly due; for, in the first place, without his advice and encouragement it would never have been undertaken, and without his valuable assistance it would as surely never have been finished. I am also much indebted to the various bird preservers of the county for notices of scarce species and frequent opportunities of examining rare birds in the flesh; of these my thanks are especially due to Mr. Cole and Mr. Roberts, of Norwich; Mr. Lowne and Mr. Smith, of Yarmouth; Mr. Dack, of Holt; Mr. Pashley, of Cley; and Mr. Newby, of Thetford; I am, also, much indebted to the published notes of Mr. T. E. Gunn, which I have acknowledged in the text. For the rest mine has been a labour of love, and I trust that the contrast between the commencement and the concluding portion of the volume, however evident, will be regarded with indulgence.

NORWICH. T. S.
September, 1890.

MEMOIR OF THE AUTHOR,

HENRY STEVENSON, F.L.S.

In writing a memoir of the Rev. Richard Lubbock my late friend, Henry Stevenson, truly remarked that "many authors have acquired an established reputation not merely by the excellence but by the multiplicity of their published works, whilst others have attained a not less imperishable fame by one effort of genius —a single contribution to some particular class of literature, with which, for all time, they are personally identified." How literally true this is with regard to himself will be cheerfully admitted by all readers of his "Birds of Norfolk"—one of the earliest of the recent county faunas which have since become so numerous. It is as pre-eminent, so far as its author completed his work, for the exact acquaintance with his subject it displays, as for the skilful way in which the biography of each species is just sufficiently treated, and the pleasing style in which the results of his observations are recorded. Of Mr. Stevenson it may truly be said that this single contribution to the ornithology of his county has established for him a reputation which will last so long as the study endures.

The family of Stevenson settled in Norwich in 1785, their first representative here being William Stevenson, F.S.A., a miniature painter, and pupil of Sir Joshua Reynolds, whose father, the Rev. Seth Ellis Stevenson, was rector of Treswell in Nottinghamshire. William Stevenson entered into partnership with the proprietor of the "Norfolk Chronicle," an old-established county Tory journal, which remained the property of the family until it was formed into a Joint Stock Company in 1886.

Seth William Stevenson, son of the above and father of the author of the "Birds of Norfolk," like his predecessor, was a man of distinguished literary talent, and published two books on continental travels, which, early in the present century, possessed

an importance hardly reached in the present day by narratives of much more extended wanderings; but his chief work, which, though nearly completed in manuscript, was at the time of his death only about half through the press, is a "Dictionary of Roman Coins," since completed.

The descendant of these two men, who inherited all the literary tastes, "integrity, and goodness of heart," as well as "the retiring modesty of disposition," attributed to his predecessors, was Henry Stevenson, who was born in Surrey Street, Norwich, on the 30th March, 1833. He was educated at King's College School, London, and early displayed that love for natural science which in after years became to him almost a passion. In 1855, at the age of twenty-two, he was elected honorary secretary to the Norfolk and Norwich Museum, a post which he held to the time of his death. In 1864 Mr. Stevenson was elected a member of the British Ornithologists' Union, being one of the first additions made to the original number of the members of that society.

On the 3rd November of the same year he was elected a Fellow of the Linnean Society, his certificate bearing the signatures of Professor Newton, Mr. P. L. Sclater, and the late Professor George Busk. At the first meeting of the Norfolk and Norwich Naturalists' Society, of which he was one of the founders, on March 30th, 1869, he was chosen a vice-president of that society, and filled the office of president in 1871-72, evincing a keen interest in its proceedings till the last.

Mr. Stevenson was not only an observant naturalist but was a keen sportsman also, and his remarks on covert shooting and partridge driving will be found quoted with approval in Lord Walsingham's volume in the Badminton Library on "Field and Covert Shooting." His chief delight was snipe shooting in the marshes bordering the broads, where the variety of the bag and the constant hope of meeting with or observing the habits of rare and interesting birds had an irresistible attraction for him. But it was in summer that the broads presented to him their greatest attraction, and the charming sketches entitled "A Summer's Night on the Broads" (vol. i., p. 120) and "A Summer's Day on the Broads" (i., p. 188), are but faint reflections of the many delightful hours spent in the contemplation of nature in these once wild and still fascinating scenes; while his description of "A Sunset on the Broads," vol. ii., p. 163, shows how capable he was of appreciating the poetic beauty of the scenes so frequently presented by nature in her varying moods. The present writer first made Mr. Stevenson's personal acquaintance on coming to reside in Norwich in 1867, but a community of tastes soon established a friendship which led to many such pleasant excursions as have just been referred to, and those who enjoyed his society on such occasions can testify to the kindly, genial nature, and unfailing flow of humour, which, added to his intuitive powers of observation, rendered him such a delightful companion, and endeared him to all who enjoyed his friendship.

A series of domestic bereavements, commencing in 1862, added to acute personal suffering, clouded the latter years of his life, and rather suddenly on the 18th of August, 1888, he passed away, regretted by all who knew him.

Although Mr. Stevenson showed great interest in the charities of the city, and served the office of sheriff in 1875, he never took an active part in municipal affairs, and all public acts were repugnant to his retiring disposition.* His chief pleasure was derived from his aviary and collection of local birds, and, from his large and wide-spread correspondence, he possessed unusual opportunities of acquiring rare specimens, which from time to time occurred. He thus formed such a series of Norfolk birds,—chosen with the greatest judgment to illustrate the various states of plumage incidental to age, sex, and season,—as probably was not exceeded, viewed as a working collection, by any in the county. Added to this, he possessed an excellent library of modern ornithological works, and these, both birds and books, as well as his varied and extensive experience, it at all times afforded him the greatest pleasure to place at the disposal of any real student of ornithology who sought his ready aid.

On September 12th, 1887, on leaving the house in Unthank's Road, Norwich, which he had so long occupied, the greater part both of his birds and books were disposed of by auction, many of the principal rarities being acquired for the Norfolk and Norwich Museum ; and on March 21st and 22nd, 1889, after his death, the remainder of the collection was dispersed by the same means.

Of the large number of contributions to the ornithology of his county from Mr. Stevenson's pen the greater number appeared in the "Zoologist," his first communication, dated June 25th, 1853, being on some " Unknown Eggs lately found in the neighbourhood of Norwich," which appeared in the volume of that journal for 1853 (p. 3981); from that time he kept up an almost unbroken record till September, 1885, when his last " Ornithological Notes

---

* Professor Newton reminds me, in proof of this, that he could not even be induced to read his own admirable paper on the " Bustard in Norfolk," in Section D, at the meeting of the British Association at Norwich ; but when necessity required it he could speak fluently, and with telling effect, an instance of which will be found in a speech made in the Council Chamber on his retiring from the office of Sheriff, on November 9th, 1876.

from Norfolk," being those for the year 1883 appeared. During the whole of this long period his communications to this journal form a most valuable history of the ornithology of the county, and many were the rare species which it fell to his lot to chronicle.

On the establishment of the Norfolk and Norwich Naturalists' Society he became a frequent contributor to its "Transactions," the most important of which, in addition to the annual papers on the ornithological occurrences, were a paper on the "Meres of Wretham Heath," in August, 1869; "On the Abundance of Little Gulls on the Norfolk Coast, in the winter of 1869-70," in December, 1870; his Presidential address to that society, and an admirable paper on "Scoulton Gullery," in 1871; some valuable remarks on the "Wild Birds Protection Act of 1872," a subject in which he took the greatest interest, and concerning which he gave evidence before a select committee of the House of Commons, in 1873; a memoir of his late friend the Rev. Richard Lubbock, in 1877; a paper "On the abundance of Pomatorhine and smaller Skuas on the Norfolk Coast, in October and November, 1879," in April, 1880; a valuable paper "On the Plumage of the Waxwing," in November, 1881; "On the occurrence of the Dusky Petrel in Norfolk in 1858," read in November, 1882; his last contributions to the society's "Transactions" being in March, 1888, when he read a paper on the "Vocal and other Sounds emitted by the Common Snipe;" and a note on the "Tenacity of life in young House Martins," read in the same year.

Mr. Stevenson was also the author of the "List of Norfolk Birds," in White's "History and Directory of the County of Norfolk," in the editions of 1864 and 1883, as well as of many communications to the "Field" and "Land and Water," but the work which established his fame as a field ornithologist was his "Birds of Norfolk" of which I have now to speak.

The original plan of the "Birds of Norfolk" seems to have been on a much more modest scale than that of the finished work. A short account of each species was written in a series of note books which from internal evidence I should imagine were finished about the year 1863; these formed the basis of the articles, but as they finally appeared they had been greatly expanded, and in most cases entirely re-written, the information being brought fully up to the time when each successive portion went to press, and so great was the author's desire for accuracy and completeness that in some cases even sheets which had been printed off were cancelled in favour of more correct or more recent information. I believe it was impossible for any book to have been written with a greater regard to absolute truth or more conscientiously than the " Birds of Norfolk." Added to this, Mr. Stevenson, although the result was as a rule charming, was by no means a rapid writer, and his habit of verifying every fact for himself, or obtaining his information where possible at first hand, rendered his progress slow, and his first volume was so much delayed that its preface was not dated till December, 1866. The second volume appeared in September, 1870, the article on the bustard alone having cost an amount of labour and research which is almost incredible to those who have not engaged in similar investigations. From that time the work proceeded still more slowly, owing to ill health and other causes, which often necessitated its being laid aside for considerable periods, but the articles lost none of their value from delay, and I need but point to the account of the grey lag goose, in fact those on all the geese, the mute swan, the sheld drake, and the shoveler duck, in the third volume, as models of what such writing should be. The last article written by Mr. Stevenson was on the gadwall, which, although finished and in type, was never wholly printed off; the last portion, I believe, going to press in 1877, and from that time the work was never resumed.

Mr. Stevenson left a mass of correspondence, and a large number of note books commencing in 1850, and becoming about 1865 very voluminous. The last entry is dated June the 25th, 1888, within a few weeks of his death.

I was abroad when I heard of that event, and therefore to my lasting regret unable to join in paying a final tribute of respect to one whose friendship I valued so highly, and whose kindly nature, genial disposition, and modesty that was almost excessive—added to rare talents and a constant readiness to impart his rich stores of information to fellow-workers—endeared him not only to a large circle of personal friends, but also to many others who knew him only as a valued correspondent, or as the author of the " Birds of Norfolk."

ERRATA.

PAGE	LINE			
14 note	3 from bottom,	for	"to small"	read "too small."
65	1	„ top	„ "it sample"	read "its ample."
92	9	„ bottom	„ "*linguæ*"	read "*linguæ*."
94	20	„ top	„ "at what may be termed the elbow-joint" read "at the carpel-joint."
113	5	„ bottom	„ "1874"	read "1854."
122	12 and 17 from bottom,	for	"Willoughby"	read "Willughby."
123	9	„ „	„ "Merritt"	read "Merrett."
153	10	„ „	„ "Widgeon"	read "Wigeon."
153 Head line			„ "Sheld Drake"	read "The Shoveler."

329 end of first paragraph, *dele* "unless, indeed, they be included by the term 'whyt-plover.'"

372 line 11 from bottom, *dele* the comma after "Hempton Green."

ADDENDUM.

Muscicapa parva, Bechstein. RED-BREASTED FLYCATCHER. An example of this pretty little flycatcher, which proved to be an immature female, was shot by Mr. F. M. Ogilvie, at Cley-next-the-Sea, on the 13th September, 1890. Although too late for insertion in its proper place, a notice of this first occurrence of the species in Norfolk is of too much interest to be omitted. Seven other British examples are recorded, three of which were killed in Scilly.

THE BIRDS OF NORFOLK.

ANSER FERUS, Steph.

GREY LAG GOOSE.*

Mr. Lubbock particularly mentions the neglect of this county by the older writers on Natural History, with the exception only of Sir Thomas Browne. Lincolnshire, Cambridgeshire, and Holderness, as he says, "were mentioned by all, but Norfolk, although perhaps richer than any of these, seemed consigned to total oblivion." Even Drayton in his *Polyolbion*, "occupies pages in the enumeration of different species of birds found in Lincolnshire, but dismisses poor Norfolk with a passing intimation that the open country around Brandon is admirably suited to hawking." It is thus that the local naturalist, though with little doubt in his own mind as to the fact, is unable to substantiate his belief that the true *Anser ferus*—the the Grey Lag or fen goose, as distinguished from the bean or stubble goose—bred as regularly in former times in the Fens of Norfolk, as it is known to have done

---

* The old name of grey lag was resumed by Yarrell for this species, in place of the inappropriate term of grey-legged goose, the word "lag," according to that author, being "a modification of the English word lake, the Latin *lacus*, or perhaps an abbreviation of the Italian *lago*." In a recent number of the "Ibis," how-in other portions of the Eastern Counties.* Sir Thomas Browne throws no light on this point, this class of wild fowl being dismissed by him with the bare enumeration of "wild geese, *Anser ferus,* Scotch goose *Anser scoticus,*" the former probably including all the grey species, and the latter the black or bernicles. There seems, however, no question that the following account of its habits in this country, as given by Pennant in 1776, applied as much to Norfolk as to Cambridgeshire and Lincolnshire, more particularly as no one county is specially referred to. "This species resides in the fens the whole year; breeds there, and hatches about eight or nine young, which are often taken, easily made tame, and esteemed most excellent meat, superior to the domestic goose. The old geese which are shot, are plucked and sold in the market, as fine tame ones, and are readily bought, the purchaser being deceived by the size, but their flesh is coarse. Towards winter they collect in flocks, *but in all seasons live and feed in the fens.*" Taking it then for granted that the grey goose, as well as the bustard, was an indigenous species in Norfolk, the question next arises as to the date when it ceased to inhabit our fens. That

---

ever, for 1870 (p. 301), the word "lag," acknowledged by the editor to be "a puzzle to most people," is thus rendered on the authority of Mr. Skeat:—"Lag," late, last, or slow, whence "laggard" and "laglast," a loiterer; "lag-man," the last man; "lag-teeth," the posterior molar or wisdom teeth, (as the last to make their appearance); and "lag-clock," a clock that is behind time. "Accordingly the grey lag goose is the grey goose, which in former days *lagged* behind the others to breed in our fens, as it now does on the Sutherland lochs, when its congeners had betaken themselves to their more northern summer quarters."

* The best informed modern ornithologists believe that the grey lag is the only species of wild goose which breeds in Scotland."

it did so till near the close of the last century* is most probable, as the banishment of so many other feathered denizens of that district commenced only with those agricultural changes which, in course of time, have altogether obliterated its natural features. There the gozzard's occupation, as we know, has been gone for many years, but tradition still refers to a period not very remote, when young geese were regularly tended after the method described by Pennant as prevailing in Lincolnshire. Mr. Alfred Newton, who in 1853, took some pains to enquire into these matters, on the spot, was informed by a man named Spencer, of Feltwell (then in his fifty-third year), that he was told by his father, who had died some fifteen years before, that when a boy he used to keep young wild geese in the fens. They were taken when very young, and were "led" out to feed in the fens until they were quite fat. Such hearsay evidence is, of course, far less satisfactory than contemporary records, but it seems more than probable that these young geese were really, as described, bred wild in the fen, and presuming Spencer's father to have been ten years of age when thus employed, it would not be far short of a century ago. Hunt in his "British Ornithology," of which the second volume was published in 1815, although making no special allusion to our Norfolk fens, says of this species, "formerly numbers of them bred and continued the whole year in the fens of Lincolnshire and other swamps *contiguous to our eastern coasts.* But the

---

* Daniel, in the third volume of his "Rural Sports," published in 1812 (p. 249), remarks in reference to this species "the compiler took two broods one season, which he turned down, after having pinioned them, with the common geese, both parties seemed shy at first, but they soon associated and remained very good friends." Unfortunately the author omits the date and locality.

labour of man by draining and cultivating these fens and morasses has greatly diminished their numbers, so much so that we are uncertain whether they remain in their old accustomed haunts, but there can be no doubt that the greater part migrate northward to breed." When speaking of it, however, in his "List" of Norfolk Birds (1829), he says, "this bird is not, as its name imports, common in the county, as from some cause or other it has become rare, being much more difficult to procure than the bean goose." All other local authors repeat only the same tale of its scarcity as compared with former days, without adding anything as to its past history. Thus Messrs. Sheppard and Whitear (1825) writing of the bean goose remark, "it is said to be more common than the grey lag goose." The Messrs. Paget (1834) having applied the specific name of *ferus* to the bean goose, have, I imagine, inadvertently reversed the position of the two species as to abundance or scarcity, as in 1845 Mr. Lubbock, speaking of the grey lag as "very rare in Britain," says, he was informed by the late Mr. Lombe "that years elapsed before he could procure a specimen." Again in 1846, Messrs. Gurney and Fisher thus follow in a similar strain, "This species is said to have formerly visited Norfolk, but we never remember to have seen a specimen taken in the county. We are, however, informed that it is still occasionally, though very rarely, met with." It must not, however, be presumed from these extracts that the rarity of this goose during the last fifty or sixty years, tends in the slightest degree to invalidate its claim to be considered as a former resident, since the avocet, the black-tailed godwit, and even the bustard itself, afford evidence that certain indigenous species when once banished from their ancestral haunts may be reckoned amongst the rarest of our migratory visitants; and though still breeding in the north of Scotland, it may possibly remain there throughout the year, as it is said to have done in the Lincolnshire fens.

Of late years the following are the only examples of the grey lag goose of which I have any record as occurring in Norfolk:—

1847. November. A male in Mr. Rising's collection, killed at Horsey, near Yarmouth, is recorded by Messrs. Gurney and Fisher in the "Zoologist" (p. 1966) as the first Norfolk specimen which had come under their notice in a recent state; and the same bird is subsequently described (p. 2017) as having "black markings about the belly and between the legs,* much resembling those found on the breast of the white fronted goose, but somewhat less decided."

1849. April. A male in Mr. Gurney's collection was recorded by that gentleman in the "Zoologist" (p. 2456) as killed on Breydon towards the end of the month, and as exhibiting the white front to a greater extent than the last, but without the black bars, adding, however, that he had "never seen a specimen of the grey lag goose exhibiting either of these characteristics to so great an extent, or so definitely marked as is the case in the adult white-fronted goose."

1854. On the 16th of October I examined a stuffed specimen in the shop of Mr. Knight, a birdstuffer in this city, which had been sent up in the flesh from Yarmouth a few weeks before, but the sex was not noted at the time. In this bird the white front was slightly visible,

---

* A female grey lag goose from India, which Mr. Gurney kept in confinement for at least fifteen years, exhibited no appearance of the white front, but on the lower part of the breast and between the legs was barred with dark stripes. A grey lag goose, which was killed in Sutherlandshire, in May, by Captain H. J. Elwes, exhibited similar black markings on the under parts.

but I did not observe any of the barred markings above mentioned.

1862. On the 8th of March Captain Longe, then residing at Yarmouth, purchased a fine specimen which had been killed in the Caister marshes on the previous day; another was said to have been shot at the same time, but was not preserved. This bird had the barred markings on the breast, but not on the vent, which is pure white. Mr. Longe described the flesh of this goose as more like wild duck than any he had eaten.

1864. A male in my own collection, which I purchased on the 5th of March, in the Norwich fishmarket, had been shot on the previous day at Ludham, and was sent up with a young white-fronted goose killed at the same time. In this bird, when recently killed, the beak was flesh-coloured, becoming purplish round the edges of the mandibles and the nail white; feet livid pink; claws slate grey. Roof of the mouth roughly serrated and light pink in colour. Eyelids, white all round; irides dark reddish hazel. No appearance whatever of the white front, and only the faintest indication of bars on the breast, although the feathers on the flanks had a barred appearance, each dark feather being conspicuously edged with a lighter colour. The stomach was filled with grasses having a strong brackish odour, with several small black pebbles, some as large as swan shot. The bird was well nourished, but without much fat, and the flesh though tough, much resembled wild-duck in flavour. It weighed seven pounds and a half.* On the 12th of the same month I purchased a second specimen for the Norwich Museum, which had been

---

* Daniel gives the ordinary weight of this goose as ten pounds, but states that, "in 1799, one was shot at Horning Ferry [Norfolk], which weighed twenty-three pounds," the weight of a highly-fatted tame goose at the present day.

killed at the same place, and two or three others were said to have been seen. This bird much resembled my own in general appearance, and was also a male. The stomach was filled with short wiry grass.*

On the 15th of December of the same year another fine specimen, now in Mr. Newcome's collection, at Feltwell, was shot at Horsey, near Yarmouth; and on the 17th of February, 1865, an adult male, preserved by Mr. Baker, of Cambridge, was killed at Welney, near Lynn, the last I have heard of in this county.

The grey lag goose, whether adult or immature, may be readily distinguished from such species as most nearly assimilate in plumage, by the bluish-grey colour of the shoulders and the lower part of the back and rump, a very marked characteristic. It should be borne in mind, also, that in the grey lag and white-fronted goose the "nail" of the beak is white, but in the bean and pink-footed goose it is black; and again in the bean and white-fronted goose the legs are orange, but in the grey lag and pink-footed goose, more or less flesh-coloured.

The grey lag is generally supposed to be the origin of our domestic goose, but Yarrell is most probably correct in supposing that one other species, the white-fronted goose *(Anser albifrons)*, "has had some share in establishing our present domestic race," as amongst them we find some with yellow legs and feet, and others with the same parts pale flesh colour, thus resembling both of the above named wild species. But although some tame geese exhibit a partial white front, this alone

---

* The measurements of these two birds, taken by myself at the time, will be found in the "Zoologist" for 1864 (p. 9119). The extent of wings, as given by Macgillivray, sixty-four inches, must be a misprint, the above examples measuring only fifty-five and fifty-six respectively from tip to tip.

would not indicate relationship to *Anser albifrons*, as I have shown already that the wild grey lag occasionally exhibits this feature in a very decided manner.* It is further stated by Yarrell, in support of his belief in their double origin, that on examining the trachea in our domestic geese he has found "the tube of the wind pipe nearly cylindrical" in some, after the manner of the grey lag goose, whilst in others "the tube has been flattened at the lower portion, a character which is constant in *Anser albifrons*." The wild grey lag has been known to pair readily with the domestic bird, and in one instance, as stated by Yarrell, a male attached himself to a domestic goose in preference to either bean or white-fronted geese associated with it. Mr. Hamond, of Westacre, has also had the young of this species, taken wild in the north of Scotland, within the last few years, some of which he presented to Colonel Petre of Westwick, and the male birds have crossed both at Bawdsey and Westacre, with domestic geese. In looking over a number of tame geese upon our village greens and commons, I have frequently remarked amongst the grey birds the marked resemblance in form and plumage—even in the grey wing-coverts, flesh-coloured bill and feet, and white nail—to the true *Anser ferus*, but the large predominance of white or partially white birds is the result no doubt of selection in breeding, the white quills and down fetching the highest price. Mr. J. H. Gurney informs me that some thirty years ago, when he had frequent occasion to travel by road in Marsh-land and South Lincolnshire, he was constantly struck with the much larger proportion of whole coloured brownish-

---

* On this point, also, Mr. Darwin remarks, "we must not overlook the law of analagous variation; that is, of one species assuming some of the characters of allied species."—"Animals and Plants under Domestication," i., p. 288.

grey geese in the flocks kept there, in comparison with those kept in the more inland and upland parts of Norfolk and the adjoining counties, which seemed to indicate a later admixture with the wild race in the Fen district than in those parts which are more distant from the fen country.

If the gozzard's occupation, however, is gone in our fens, with all the attendant barbarities of plucking the young and old indiscriminately, as described by Pennant from personal observations in Lincolnshire,* Norfolk still maintains its celebrity for poultry-rearing, and especially for turkeys and geese. Besides the large numbers reared annually in our farm-yards and fed up to a certain period on our commons and waste grounds, an immense number are reared in close vicinity to the city by Mr. Bagshaw, of Magdalen-gates, Norwich, who possesses almost a monopoly of the trade; and this not only at Christmas but throughout the year. From statistics recently supplied me, I find that this enterprising speculator in poultry supplies in twelve months from sixty thousand to seventy thousand birds, of which some thirty thousand are ducks, but though at one time he reared turkeys to a considerable extent, he has of late paid attention chiefly to the fatting of geese, the demand for them having greatly increased. Besides those bred upon his own premises, the first pre-

---

* In Gough's "Additions" to Camden's "Britannia," published in 1789 (vol. ii., p. 235), the author, after quoting from Pennant's "Tour in Scotland," to the effect that "a single man would keep a thousand old geese in the fens, each of which would rear seven, so that by the end of the season he would be master of eight thousand birds," adds, "in North Holland they pull geese thrice a year, and I have seen them on Wildmore Fen, between Tattersall and Boston, raw with the plucking, and their legs and wings frequently disjointed by rough hands, so that they fall an easy prey to the crows."

paration for Christmas business commences towards the end of October with the "buying up," when more than half the number of birds required are obtained from Holland,* and the remainder from various parts of this county. The fatting commences about the middle of November, and the largest number at one time has been twelve thousand. As before stated, this process takes place entirely on Mr. Bagshaw's premises, and the noise of this great army of cacklers, may be heard at all hours of the night, until the demands of Christmas week seal their fate, and relieve the surrounding population of, no doubt, a considerable nuisance. The food on which the geese are fatted is barley meal and brewers' grains, the former being ground by Mr. Bagshaw himself, so that he may not be exposed to the adulteration which this commodity frequently undergoes; and the quantity of food required is about ninety coombs of barley meal and sixty coombs† of grains daily. The manure from such an immense number of fowls, fed upon this description of food, is very valuable, and frequent applications for the sale of it are made; but as Mr. Bagshaw holds a farm close by his poultry yard, he prefers to make use of it himself. It takes about six days to make preparations for the market, as nearly two thousand are killed every day, and about a hundred dressers are employed in the work, but as the birds are not drawn before they are sent to market the giblets are bought with them. Of those killed for Christmas, some four thousand are sent to the Goose Clubs, and the rest are forwarded to the markets at Leadenhall

---

* As early as the 14th of June, 1870, I saw a flock of 600 young geese, all Dutch birds, being driven through the city preparatory to fattening for the Michaelmas sale.

† The term coomb, as commonly used in Norfolk, signifies four bushels, or half a quarter.

and Newgate, where they are sold on commission. No less than from seventy to eighty tons weight have been sent, during the Christmas week, by rail to London and other places, the geese averaging in weight from nine to sixteen pounds, but some as much as twenty-two pounds. The feathers obtained from twelve thousand birds amounts in round numbers to four thousand pounds, the quills and waste feathers being kept separately and sold by themselves. It is evident, therefore, from these figures that a most important business in this line is carried on between Norwich and London,* the extent of which few persons have any conception of. It should be stated, however, that although the great demand for geese in the metropolis takes place at Christmas time, the goose is in Norfolk always most in request at Michaelmas, when it is, according to the old local phrase, "a stubble goose,†"

---

* Dr. Wynter, in his "Curiosities of Civilization," amongst the statistics of "The London Commissariat," states that "the bulk of the geese, ducks, and turkeys come from Norfolk, Cambridge, and Suffolk," the Eastern Counties Railway alone having brought thence, in 1853, "twenty-two thousand four hundred and sixty-two tons of fish, fowl, and good red herrings;" an amount which has, I believe, been considerably exceeded since that time.

† "The derivation of the term 'way-goose' is from the old English word *wayz*, stubble. Bailey informs us that *wayz*-goose, or stubble-goose, is an entertainment given to journeymen at the beginning of winter. Hence a wayz-goose was the head dish at the annual feast of the printers, and is not altogether unknown as a dainty dish in these days. Moxon, in his 'Mechanick Exercise' (1683), tells us that 'it is customary for the journeymen to make every year new paper windows, whether the old ones will serve again or not; because the day they make them, the master-printer gives them a *way-goose*. * * * These *way-gooses* are always kept up about Bartholomew-tide; and till the master-printer has given their *way-goose*, the journeymen did not use to work by candle-light.' The same custom was formerly common at Coventry, where it was usual in the large manufactories of ribbons and

namely, a goose which has fattened on the grains which have shelled out in the stubbles, and of which the supply was no doubt considerably more in the old days of sickle reaping than it is under the modern system of gathering the corn crops.

The custom which prevails at the Great Hospital in St. Helen's, Norwich, of supplying the aged inmates with a dinner of roast goose on new Michaelmas-day, probably dates back to the year 1762, when, according to extracts from the records of the Institution, kindly furnished me by Mr. G. Simpson, the governor, Mr. John Spurrell "liberally contributed a yearly sum to be expended in feasts." In 1816, however, the annual dinner of roast goose having been discontinued since the death of the former benefactor, Mr. Robert Partridge signified his intention to the Hospital Committee of giving one hundred pounds as a benefaction that the gift of a Michaelmas dinner of goose to the old people of that hospital "might be revived and continued in future." Up to the year 1846, a difficult problem in carving presented itself, inasmuch as each bird had to be divided into five equal parts, but each inmate is now supplied with a quarter of a goose, an arrangement which appears to have given general satisfaction, even though this savoury dish has been described as "too much for one but not enough for two.

Prior to the opening of the railway to Norwich in 1845, immense numbers of tame geese, reared in this county, travelled annually by road to London;* as in small

---

watches, as well as among the silk-dyers, when they commence the use of candles, to have their annual *way-goose*. 'Goose-day' is now in nearly all the London houses held in May or June, instead of Michaelmas, and is quite unconnected with the lighting-up."

JOHN TIMBS.

* In Daniel's "Rural Sports" (vol. ii., p. 93), an amusing anecdote is given of a considerable bet made between Lord

quantities they are still driven up to Norwich from the surrounding country. Besides other routes, large numbers journeyed by way of Thetford and Newmarket, but of late years the rapid transit of such goods to all parts of the kingdom has done much towards developing the present enormous business in this class of poultry. Mr. Hunt quotes from the "St. James' Chronicle" of September, 1783, a notice "of a drove of about nine thousand geese which passed through Chelmsford on their way to London from Suffolk." The drivers used to be provided with a long stick, with a red rag at one end and a hook at the other; by the former they were excited forward, and by the latter stragglers were caught by the neck and kept in order, and a hospital cart attended each drove to accommodate the lame ones. They are said to have performed the journey at the rate of eight or ten miles a day, from three in the morning till nine at night.

During the late severe winter of 1870-1, when the more common species of wild geese were unusually numerous, I could not ascertain that any grey lags had been observed amongst them, but as late as the 23rd of March, a young bird was shot by the Rev. C. J. Lucas, in Burgh Fen, near Yarmouth, apparently a straggler, as no other geese appear to have been remarked in the neighbourhood at that time. This bird on dissection proved to be a female, and though in immature plumage, had the lower part of the back grey,

---

Orford and the Marquis of Queensbury, in 1740, that a drove of geese would beat an equal number of turkeys in a race from Norwich to London; the result being as Lord Orford predicted. The geese kept on the road at a steady pace, whilst the turkeys at night flew into the roadside trees to roost, and were dislodged with much difficulty in the morning by the drivers. The geese not wanting to stop and sleep, thus beat the turkeys hollow, arriving in London two days before the turkeys.

as also a considerable portion of each wing in the region of the carpal joint. It had, moreover, two dark feathers on the breast, where slight bars are observable in some older birds.

ANSER BRACHYRHYNCHUS, Baillon.

PINK-FOOTED GOOSE.

Since the specific distinctions of this short billed goose were first pointed out by M. Baillon in 1833, and subsequently by Mr. Bartlett, at a meeting of the Zoological Society in 1839, when, quite unaware that it had been previously named, he proposed for it the specific title of *phœnicopus*, it has proved to be both a constant and abundant winter visitant in Norfolk, though to a great extent confined to the western side of the county, and especially to certain localities in the neighbourhood of Holkham on the coast, and Wretham in the interior.

The earliest record of its identification in this county* is apparently the notice by Yarrell of a specimen killed at Holkham, in January, 1841, by the present Earl of Leicester out of a flock of about twenty, since which time this goose has proved to be by far the most common species that frequents the Holkham marshes. The following notes on its habits, as observed in that neigh-

---

* It is worthy of note, however, that the bird figured by Hunt in 1815, as the bean goose (*A. segetum*), and in all probability from a local specimen, represents, unquestionably, the pink-footed species, since the bill, though painted orange in the plate, is far too small for that of the bean, and the feet are coloured red as in the adult pink-footed goose. Mr. Hunt moreover adds the provincial name of "small grey goose."

bourhood have been very kindly supplied me by Lord Leicester for use in this work.

"As long as I can recollect wild geese frequented the Holkham and Burnham Marshes. Their time of appearing in this district is generally the last week of October, and their departure the end of March, varying a little according to the season. Till November they rarely alight in the marshes or elsewhere in the neighbourhood, but are seen passing to and from the sea. Where they feed in October I know not, as I have reason to believe that they do not obtain much food off the muds like the brents, but live much on grass and new sown wheat. From early in November till their time of departure for the north, the Holkham marshes have almost daily some hundreds of geese feeding on them. There are periods of a week or a fortnight when the greater portion of them go elsewhere, but rarely all go. When on the marshes they are mostly in one or two flocks, but in stormy weather, or even on certain still days, for some unaccountable reason they break up into small lots. My keepers informed me that one day last week [about the middle of November, 1870], which was perfectly calm and still, they were flying about in small lots very low, and that a great many might have been killed."

Referring also to the goose shot by himself in 1841,* and identified by Yarrell as the pink-footed, his lordship adds, "Of the many geese killed here before then, I have reason to believe from their habits they were nearly all the same as those now here—that they were the pink-footed; and of the many hundreds killed since, with the exception, I believe, of only one bean goose and a few white-fronted, they were all pink-footed. The greatest number obtained in one year was in the severe winter of

---

* See "Game Birds and Wild Fowl" by A. E. Knox, M.A., F.L.S., p. 101.

1860-1,* when one hundred and thirty-eight were killed, all pink-footed."

Mr. Dowell, who is also well acquainted with the habits of this species on the north-western side of the county, informs me that they feed in flocks of from one or two to six or seven hundred on the uplands by day, and he has known as many as twenty-seven shot in the day by sportsmen laying up for them behind gate posts in the Holkham marshes, on a gale of wind, when the geese fly low. On one occasion, when driving along the road in very snowy weather, at West Barsham, he saw a flock within twenty yards of the fence, but which, strangely enough, did not move, though he halloed to frighten them. These were probably fatigued by a long flight. He has never met with this species at any time in the salt marshes and tideway at Blakeney. In 1858 he saw a flock of fifty at South Creake as early as the 13th of October, and some were said to have been seen that year on the 1st of the month. In the winter of 1869, a flock of about five hundred geese, which were no doubt all pink-footed, frequented some barley stubbles within sight of his house at Dunton, near Fakenham. They used to arrive from the coast soon after daylight, and remain till late in the afternoon. The pink-footed, like the bean goose, also frequents the large upland fields about Anmer and Westacre, and still further inland the open country about Wretham heath. Their nocturnal movements may be inferred from the fact that, on one occasion, as Mr. Cresswell informs me, a fowler, named Charles Hornigold, took seven grey geese at a stroke in a short length of netting,† on the shores of the

---

* See the Rev. C. A. John's remarks in his "British Birds in their haunts," upon the grey geese and brents observed by him about Holkham and Thornham in that extremely severe winter.

† See vol. ii., p. 376, for the method of netting *Tringæ* and other birds on the shores of the Estuary at Lynn.

Wash, near Lynn. These birds, which were in all probability pink-footed geese, had so entangled themselves that the net had to be cut to get them out.

From the localities then most commonly frequented by this species in Norfolk, it is very natural that the majority of specimens procured by the gunners should find their way into the hands of the game dealers at Lynn, by whom, there being but little local demand for such birds, the chief portion are sent up to the London markets. From notes made at Lynn by Mr. Thomas Southwell, between the years 1848 and 1854, I find that examples of this goose were remarked by him hanging for sale every winter, between November and February, the earliest date being November 6th, 1854, and the latest February 18th, 1853. I have records also, between 1852 and 1869, of some fourteen or fifteen, from various parts of the county, that came under my own notice during that period. These, with but one exception—observed in the Norwich market on the 8th of March, 1862—were all obtained between November and February, and for the most part in hard weather.

Although so local in its habits, as a rule, this species is much more dispersed during severe frosts. In the neighbourhood of Cromer I have known two or three killed between Runton and Beeston,* and, scarce as it is in the "Broad" district, I have one in my collection shot at Ludham on the 6th of January, 1854. Mr. F. Norgate possesses one which was picked up dead on Whitwell Common, nearly in the centre of the county, and the following curious instance of nocturnal migration, which occurred on the extreme south-eastern border of the county, was communicated to me by the Rev. H. T.

---

* In the "Field" of February 4th, 1865, Mr. N. F. Hele records a few pink-footed geese, and one bean goose, as shot in January of that year, at Aldeburgh.

Frere:—"About ten years ago, in the month of February or March, a large flock of wild geese were attracted to the town of Diss by the lights on a foggy night, and from the sound of their cries, seemed to fly scarcely higher than the tops of the houses. They came about seven p.m., and, being Sunday evening, they appeared to be especially attracted by the lights in the church, and their incessant clamour not a little disturbed the congregation assembled for evening service. From that time until two a.m., when the fog cleared off and they departed, they continued to fly round and round utterly bewildered. One bird, happened to fly so low as to strike a gas lamp outside the town just as a police constable was passing by, who very properly, as the bird was making a great noise outside a public house, took it into custody, and the next day, being quite unhurt, it was sent off to Dr. Kirkman's lunatic asylum, at Melton, where it lived for some years, and may possibly be alive now. It proved to be a rather small specimen of the pink-footed goose."

The very severe winter of 1870-71, like that of 1860-61, before referred to, was remarkable for the immense number of wild geese, (almost entirely brents and pink-footed) that visited our coast, and of which, a very unusual number were procured. At Holkham, as Lord Leicester informs me, over a hundred pink-footed geese were shot between the 26th and 31st of December, but many of them left the marshes when the hard weather commenced. On the 30th of the same month Mr. H. T. Coldham, of Anmer, kindly sent me a specimen that he had shot out of a flock of eleven in that neighbourhood, which rose from a turnip field, on rather high ground. On the same day I saw two in the Norwich market, but could not ascertain on which side of the county they had been killed. On the 3rd of January (1871), I had two sent me from Lynn, in mistake for the bean goose, which were shot at North Wootton on recently reclaimed land; and on the 16th of February, Mr. A. W. Partridge met with a flock of from two hundred to two hundred and fifty wild geese at Wretham, on some young rye. They were extremely wild, but he made a long shot at a lot of five or six that separated from the rest, and bagged one which proved to be a young pink-footed, and probably most of the grey geese observed in that neighbourhood, presumed to be bean geese, are of this species.

Mr. Alfred Newton, in the "Zoologist" for 1850 (p. 2802), mentions an example of this goose, shot at Wretham, in January of that year, "remarkable for a few white feathers round the base of the bill similar to the principal characteristic of the white-fronted goose, but not extending nearly so far on the forehead and cheeks as in examples of that species." I had never till last winter (1870-1), remarked this peculiarity in the pink-footed goose; but one of the birds sent me from Lynn, on the 3rd of January, has a narrow line of white feathers all round the base of the upper mandible, and a few stray ones seem to indicate that it might, in time, have extended further. This bird is apparently fully adult,—judging from the extent of grey on the wings and back, the deep red colour of the legs and feet, and the size and colour of the bill, as also the specimen shot at Anmer, which has the very slightest indication of the same variation; but the companion bird from Lynn, whose feet and legs were much lighter, though resembling the other two in the size of the bill and in general plumage, has no indication of the white front,* nor did it occur in any other examples that I examined during the same season.

---

* These three specimens all proved to be males.

There is something peculiar as well as interesting in the fact of the grey-lag occasionally throwing out black feathers on the breast after the manner of the white-fronted, and the three other grey geese of Great Britain exhibiting, more or less, white frontal feathers like *Anser albifrons*; but, as Mr. Newton remarks, with reference to the latter variation, "it is a question which remains to be answered, whether this results from age, sex, or occasional variety." In the above instance, unquestionably, the greyest and therefore the oldest pink-footed goose in my collection of skins has the most white on the forehead.

Judging from the grey geese I have lately examined in all stages of plumage, it would seem that the white-fronted and bean geese never become so grey on the back and wings as the grey lag and the pink-footed, but in the latter, as shown by adult specimens, the extent of grey on the wings as well as on the lower part of the back resembles very closely the same portions of plumage in the grey lag.* In the young birds the brown feathers on the back and wings are rather broadly edged with a lighter shade of the same colour, but these edgings apparently become narrower and almost pure white by age, whilst in the body of the feather the grey gradually absorbs the brown by a change of colour in the feather itself, and not by a general moult.

That this species has a high northern range during the breeding season, is shown from the fact that Mr. A. Newton includes it in his list of the birds of Spitzbergen,† whence he procured both eggs and nestlings in the summer of 1864.

---

* Macgillivray says of the pink-footed goose (vol. iv., p. 602), "If the name 'grey goose' could with propriety be claimed by any one of our three grey geese, it is this species to which it ought to be given, it having more grey on the upper parts than even the thick-billed [grey-lag] goose."

† See "Ibis," 1865, pp. 209-10 and 513.

ANSER SEGETUM (Gmelin).
BEAN GOOSE.

With our present knowledge of the habits and local distribution of the pink-footed goose, as a regular and abundant visitant to this county in autumn and winter, comes the question whether the true *Anser segetum*, or Bean Goose, was not always the rarer of the two? Whether the "small grey goose," as Hunt terms it, did not in former days, as now, predominate in those immense flocks, or "gaggles" (to use the more correct term of the ancient fowler), which frequented the wide open fields on the western side of the county? Lord Leicester is inclined to believe that, prior to the identification of the goose shot by himself at Holkham, in the winter of 1841, as the pink-footed, nearly all the geese frequenting that neighbourhood, belonged (judging from their habits) to this newly established species; and since that date he can recall but one example of the bean goose, and a few white-fronted, as having occurred amongst the hundreds of pink-footed geese killed on his estate. That the pink-footed goose, moreover, is by no means confined to the vicinity of the coast, but is met with in large numbers far inland at Wretham, as well as in the once favourite haunts of the bean goose about Westacre, Walton, and Anmer, I have already shown; and it is fair to presume, therefore, that such has been always the case, and that our earlier records of wild geese in such localities may be considered as applicable to both species.

Whatever its former status, however, in Norfolk, there can be no question that the bean goose (amongst the grey geese) now ranks, in point of scarcity, next to the grey-lag,* as evidenced by the few examples

---

* Mr. J. H. Gurney informs me he has remarked in Leadenhall Market that the bean goose has been scarcer of late years, in pro-

observed in our markets of late years, even in the most severe winters; and I have experienced no little difficulty in procuring a local specimen or two, in the flesh, to compare with the pink-footed for the purposes of this work.

In the "Breck" district the bean goose has been accustomed to arrive by the middle of October, and until within a comparatively recent period, as Mr. Anthony Hamond informs me, wild geese might have been seen by hundreds on the stubbles of the Great Westacre and Walton Fields,* at the time of the Swaffham coursing meetings in the second week of November. But although little, if any, alteration has taken place within the last half century in the general aspect of that portion of the county, either by enclosure or planting, the constant disturbance occasioned by the altered system of agricultural operations may account, probably in a great degree, for the marked diminution in the number of these birds, peculiarly wary by nature and suspicious of the presence of man. In 1861 Mr. Hamond shot three bean geese at Westacre, and one at Walton, but even in the severe winter of 1870-71, he could not ascertain that any wild geese had been observed feeding on the Westacre stubbles, and believes they have ceased to do so, although seen passing over in large "skeins;" nevertheless the pink-footed goose, was remarked at Anmer, in the course of that winter. A further haunt also of this species, as well as of that last described, is the wild country about Thetford and

---

portion to all the other species, than was formerly the case. Including Dutch birds, their order as to abundance appears to be brent, bernacle, pink-footed, white-fronted, bean, and grey-lag.

* The late Rev. Robert Hamond in a letter to Selby, dated Swaffham, April 28th, 1824, speaks of his endeavours to procure the Iceland falcon to hawk bustards and wild geese in the Westacre neighbourhood.

Wretham in which neighbourhood at the beginning of the present century, the notorious George Turner used to slaughter wild geese as well as bustards by cunning devices, occasionally getting a shot at the former by "crowding" a wheelbarrow before him, covered with green boughs. At Wretham, as Mr. A. Newton was informed by the late Mr. Birch, in 1853, five wild geese had for many successive years regularly made their appearance a month earlier than the main body, and in that year they had been seen by the 7th of October, but these are quite as likely to have been pink-footed as bean geese.

Messrs. Sheppard and Whitear were evidently under a wrong impression when speaking of the bean goose in Norfolk as "met with particularly about Yarmouth," as Messrs. Paget describe it as "less frequently met with" than the grey-lag; and Mr. Lubbock, though referring to its abundance in West Norfolk, speaks of wild geese, generally, as "not very abundant on the Yarmouth side of the county." Occasionally, however, I have known this species killed in severe winters near Yarmouth, both at Hickling and Horsey, and stragglers also in spring when about to leave us for the north. From the extreme watchfulness of these birds I have rarely known any obtained, except in hard weather, when specimens have been sent to Norwich from the neighbourhood of Wells, Blakeney, and Lynn; but the winter of 1851-2, from its mildness both before and after Christmas, was remarkable for the unusual number and variety of wild geese killed in various parts of the county. At that time, on the 20th of December, I saw no less than three couples of bean geese, two and a half couples of white-fronted, and two couples of bernacle geese, a rare species with us, hanging for sale at one time, having been killed at Hickling and other localities more immediately on the coast. These birds were nearly all in adult plumage, but in poor condition, as if driven southward by stress of weather, although, as before stated, the winter with us proved unusually mild. In the last week of December, several more bean-geese, and another bernacle from Burlingham, were sent to Norwich; and in the first week of January, 1852, with scarcely any appearance of frost, a pink-footed goose from Blakeney, two white-fronted from Hickling, and several brents, kept up the exceptional character of the season. In the second week of March, 1853, with severe frost and snow, several bean-geese and brents, with a variety of wild-fowl, appeared in the market; and the severe frost in January, 1864, brought examples of nearly all the more common species. In February and March, 1855, one of the most severe seasons I remember, many bean, white-fronted, and brent-geese with whooper and Bewick's swans, and a strange variety of rare fowl and other birds, came into the hands of our poulterers and bird-stuffers, as recorded by myself at the time in the "Zoologist" (p. 4660); and since then the equally remarkable winters of 1860-61 and 1870-71, have been the only occasions on which I have known wild geese, in any number, exhibited for sale in this city. This may, however, in some degree be accounted for, during the last few years, by the extension of railway communication to nearly all parts of our coast, thus enabling local gunners to send their birds direct to the London markets.

The following are the only examples of the bean goose that have come under my notice during the last ten years, in marked contrast to the numbers of pink-footed geese recorded in my notes during the same period:—One January 10th, 1861, during a sharp frost; one November 29th, 1862, after an early fall of snow; two January 15th, 1864, during sharp weather; and one January 31st, 1867, a rather mild season. All these birds were sent to our Norwich market, but I was unable to ascertain from what part of the county. From the latter date, until the commencement of 1871, I had not met with this species either at our bird-stuffers or poulterers; but on the 11th of January, during the intense frost which prevailed at that time, Mr. H. Upcher succeeded in killing one out of a flock of three that he found feeding within shot of a frozen ditch at Blakeney; and on the 11th of February Mr. Hamond sent me a fine adult male, which had been shot at Castleacre, on the 9th, by Mr. Beverley Leeds.

According to Messrs. Gurney and Fisher, a pair of bean-geese killed at Horsey, in March, 1846, "appeared to have been inclined to nest there," but unless other indications were remarked at the time, their appearance at that date would scarcely warrant such an inference. On the 12th of June, 1865, Mr. T. W. Cremer shot a bean goose at Beeston, near Cromer, in company with two Canada geese, but whether all three had escaped from some private water, or the bean, in a wild state, had been attracted by the semi-domesticated Canadians, it is impossible to say. I may here mention, also, that in July, 1866, Mr. F. Norgate, of Sparham, saw in a pond on the premises of Mr. Wells, of Sedgeford, a bean-gander, which had been shot wild, but only winged, and this bird was said to have attached itself to some swan-geese which had nests close by. The man who had charge of the fowls—there being also sheldrakes, which nested in rabbit-holes near the pond—had no doubt from the actions of the birds that a cross had been effected, but all the eggs examined by Mr. Norgate proved to be rotten, and showed no signs of impregnation.

Pennant describes this species as arriving in Lincolnshire in autumn, and taking its departure again in May, but never remaining "to breed in the fens"; he also presumes the term "bean" goose to be derived "from the likeness of the 'nail' of the bill to a horse bean." Selby, however, is perhaps correct in attributing this trivial name to its partiality for that kind of food, as he remarks that "in the early part of spring they often alight upon the newly sown bean and pea-fields." His observations also on their habits in the north of England agree altogether with our local experience, their haunts or feeding grounds, as he states, being "more frequently in the higher districts than in the lower and marshy tracts of the country, and they give the preference to open land, or where the enclosures are very large." From their feeding much upon the young wheat, and the stubbles or layers of clover* as well as grasses, it was at one time the custom, in the western parts of this county, to employ boys after the fashion of crow-keeping to drive them off the lands, but at night they betake themselves to the low grounds or sand-banks and estuaries on the coast. Macgillivray, however, states that their habits vary according to circumstances with regard to the time of resting and feeding, and describes those frequenting the Montrose basin in winter as betaking themselves to the water by day and feeding in the fields at night, whilst in the Bay of Findhorn "multitudes are seen coming from the surrounding country to repose at night," which "early in the morning betake themselves to the fields where they feed until evening, if undisturbed."

From the want of a good series of examples, of both the bean goose and pink-footed, I have been unable, till

---

* James Cater, of Fakenham, now in his eightieth year, who was formerly gamekeeper to the Rev. R. Hamond, and well acquainted with the country around Swaffham, informs me that these birds used to frequent the new layers after harvest by hundreds, and did much damage by eating out the heart of the clover.

very recently, to come to any conclusion as to the points of specific difference raised by Mr. Arthur Strickland in his well known paper on "the British Wild Geese,"* in which he maintains that the so-called pink-footed goose is no more than the young bird of the true bean goose *(A. segetum)*, and consequently that both M. Baillon and Mr. Bartlett were wrong in making it a distinct species. During the late severe winter (1870-71), however, thanks to Mr. Hamond, Mr. Upcher, and other friends, I have had the means of forming an opinion from an examination and comparison of specimens in the flesh, and feel convinced that, whatever the Yorkshire carr-lag† as distinguished from the grey-lag may have been, Mr. Strickland's *Anser paludosus* is no other than the *A. segetum* of authors, and his *A. segetum* the adult pink-footed goose, as proved by his own descriptions. Of *A. segetum*, as defined by himself, he writes—

---

* Read before the Natural History Section of the British Association at Leeds, September 24th, 1858, and subsequently published in "The Annals and Magazine of Natural History," ser. 3, vol. iii., p. 121; and in "The Naturalist" for 1858, vol. viii., p. 271.

† Mr. Strickland, in the above mentioned paper, endeavours to establish, under the name of *Anas paludosus* or long-billed goose, a species distinct from the common bean goose, and as he supposes identical with a bird which is said to have been known, up to the close of the last century, to Yorkshire fowlers and decoymen as the carr-lag to distinguish it from the grey-lag, in company with which [but not interbreeding] it then bred in the carrs, though both have ceased to do so now, and the former has become "almost a lost species." This long-billed goose he considers has been "figured and described by Yarrell, Gould, and Morris, under the name of *segetum* or bean goose." In the "Ibis" for 1859 (p. 199), the editor, in a note on Mr. Strickland's paper, though not inclined to believe that his *Anser paludosus* "has remained so long unnamed," asks, "Is it Naumann's *Anser arvensis*, as distinguished from *A. segetum* in 'Naumannia' (iii., p. 5, pl. 4) or is it the true *A. segetum?*"

"Bill short, strong, and deep, the depth at the base being nearly two-thirds of its length; *pale red* in the middle, black at the extremities, but varies much in the proportions of these colours. Old birds nearly as large and *pale coloured* as the grey-lag."

This unquestionably is an accurate description of the adult pink-footed goose, and I have already pointed out the marked affinity between the old pink-footed goose and the grey-lag in the dove-coloured tints of the back and wings. It is impossible, however, to rely upon the measurements of the bill* for specific distinction, inasmuch as that feature varies in size, in a very remarkable manner, in different specimens.

Mr. Strickland's description also of the pink-footed goose is taken evidently from an immature bird—

"Bill, nearly the same proportions and colours as the last, *but smaller and weaker;* bird less and darker coloured."

and this (as he ignores the pink-footed goose, as a species), he considers the young of his *A. segetum*, which it undoubtedly is, though under the specific title of *brachyrhynchus*. Next, as to his supposed new or recently extinct species, *Anas paludosus*, his account is as follows :—

"Bill long and weak, being exactly twice the length of its depth at the base. This is the bean goose of Mr. Yarrell's and Mr. Gould's drawings, but not of their descriptions. The colour of the bill is like that of *segetum*, and equally various."

I have not access at the present time to Mr. Gould's

---

* On this point, in the "Proceedings" of the Zoological Society for 1861 (p. 19), Mr. Bartlett writes, " Since the 3rd of this month (January) upwards of a hundred specimens have been seen and examined by me, most of them having been killed in the Eastern Counties," and adds, " the length of the bill varies from two and a quarter inches to one and five-eighths in length; I mention this, as too much importance has been attached to this character (in the goose), which has led Mr. A. Strickland to regard and describe the old male bean goose as " a new and distinct species."

plate in his "Birds of Europe," but Yarrell's figure is not only a very accurate representation of two adult bean geese now before me, but the measurements given by him, as well as his description of the plumage and tints of the bill and feet in the bean goose, agree exactly with the specimens recently procured in Norfolk; and as Mr. Strickland lays much stress upon the bill in his *A. paludosus* being "exactly twice the length of its depth at the base," and gives the relative proportions as "two inches and three-quarters long, and one inch and three-eighths deep at the base," I will here give the corresponding measurements from the male example of *A. segetum*, killed at Castleacre, and from the male, shot at Beeston, by Mr. Cremer, (kindly lent me for this purpose), which, singularly enough, are the same in both specimens:—Total length of bill, two inches and five-eighths; depth at base, one inch and two-eighths; circumference at the base, three inches and six-eighths. In Mr. Upcher's bird, a female, the measurements are all less than in the two males, but in relative proportion. This bird, moreover, weighed only six pounds and three-quarters, whilst the Castleacre specimen weighed eight pounds and two ounces. Another marked feature, however, of Mr. Strickland's *A. paludosus* is, according to his description, "the bill strongly toothed with a strong groove running the whole length of the lower mandible," as shown also in the illustration to his paper in "The Naturalist;" but here, again, the true *A. segetum* corresponds exactly, and the same toothed edge, a little less prominent, is found in the bill of the pink-footed goose, though not perceptible at all in the two other figures given by Mr. Strickland. It seems, therefore, impossible to come to any other conclusion than that *A. paludosus* and the true *A. segetum* are one and the same.

On referring to our standard authorities on "British

Birds," one is struck with the strange difference in their descriptions of the colours of the bill, legs, and feet, as well as of the plumage, in the bean goose, arising, in most cases, I imagine, from the confusion so long existing between that species and the pink-footed, and partly, perhaps, from the want of freshly killed specimens. The feet and legs of my Castleacre bird, which were a rich orange yellow, are now, in a dried state, scarcely distinguishable from the same parts in the adult pink-footed goose. Bewick, who gives a very fair representation of the grey-lag, as the domestic goose, has figured the bean goose for the grey-lag, as shown more particularly by the form of the bill and its black nail. Selby, whose large plate I have not seen, describes most accurately the adult pink-footed goose for the bean goose. Gould ("Birds of Europe") gives the light portion of the bill as "pinky-yellow, sometimes inclining to red," * * legs and webs orange;" and Morris figures the legs and bill yellow, but describes them as red or orange-red, as in the pink-footed. Macgillivray and Yarrell alone agree in describing the colour of the bill and feet in this species as orange or orange-yellow, whilst in all other points, as to general plumage and measurements, their statements agree exactly with the specimens I have examined. According to Macgillvray, the orange colour on the bill becomes more extended in old than young birds, and he considers the three white patches at the base of the bill in some examples as a mark of immaturity. But here I may remark, that these white patches (I have seen examples with much more white) are present in the Castleacre bird, but are wanting in the one from Beeston, though in general plumage they would seem to be equally adult;*

---

* Mr. J. H. Gurney assures me that in all cases in which he has remarked, both in the grey-lag and bean goose, the appear-

and the yellow on the bill of the Beeston bird extends further towards the base. Again the nail in both these specimens is pure black, but though the claws in the Beeston bird are all black but one, in the other they are all a light horn colour, except the two middle claws, which are black margined with white; and in two of my adult pink-footed geese, at least one claw on each foot is white, and in one all the claws are parti-coloured.*
Subject, however, to such slight variations, the chief distinctions between the bill and feet in the bean and pink-footed geese are thus accurately given by Mr. Harting in his "Birds of Middlesex:"—Bean: bill, orange; nail, edges and base black; legs, orange. Pink-footed: bill, pink; nail and base black; legs, pink, tinged with vermillion, like Egyptian goose. To this may be added a further distinction pointed out by Macgillivray, derived from "the knobs on the roof of the upper mandible," the bean goose "having five series, besides the two lateral rows, separated each by a deep groove from the marginal series of lamellæ;" whereas the pink-footed "has only three series of knobs, besides the two lateral rows of shorter and more flattened knobs, separated each by a very shallow groove from the marginal lamellæ." I have found this correct as to both species, but in some examples these rows of small knobs are much less distinct than in others.

---

ance of white feathers at the base of the bill, the birds have shown evidence of maturity in their general plumage.

* Thompson ("Birds of Ireland") in describing a fine bean goose which weighed eight pounds ten ounces, says, "Upper part of nail of the bill white; a central stripe of the same colour, on the nail of the lower mandible; on part of the nails of the middle toes a whitish tinge; nails of outer and inner toes of both feet white and pale horn colour; bill and toe nails otherwise coloured as usual." He also refers in a foot note to a remark of Colonel Montagu's ("App. to Sup. of Orn. Dict."), that in the white-fronted goose the nail is occasionally black.

ANSER ALBIFRONS (Gmelin.)

WHITE-FRONTED GOOSE.

This species, which is never observed in very large flocks, can scarcely be called a regular winter visitant,[*] being rarely seen in our markets, except in severe weather. As an exception, however, to this rule, in the mild winter of 1851-2, as before stated, a very unusual number of wild geese were shot in different parts of the county; and on the 20th of December, the Norwich market exhibited the unusual appearance, amongst other fowl, of two couple and a half of white-fronted, with bean and bernacle geese from Hickling and other localities; and another white-fronted, from Blakeney, was sent up to Norwich the same day. All these birds were in perfect plumage—the white-fronted geese, from the markings on the breast, being evidently adult—but their poor condition seemed to indicate "hard times," although, the weather was then unusually mild with us, and continued so up to the following February. Only in two or three unusually severe winters have I known wild geese so plentiful as in that exceptional season, when I examined upwards of twenty, of various species, for sale in this city.

From Mr. Dowell's notes, for the same year—1851—I find that on the 18th of December he saw a flock of some twenty white-fronted geese at Holkham; and on the same day he received a fine specimen which had

---

[*] Folkhard in "the Wild-fowler" (p. 190), refers to a remark of Col. Hawker's that these birds were unknown to the gunners of the Hampshire coast till the year 1830, and that he had seen more there since, from which Folkard infers that it is only in the severest winters they visit the south coast, though, from his own experience, he states, "I have every winter met with some of these birds on the eastern coast."

been killed by a gunner at Blakeney. This goose is considered by Lord Leicester to be rare at Holkham, except in hard weather, when it commonly appears in flocks of from five to ten, and, being less shy, is easier of approach than others. Mr. T. Southwell, in the "Naturalist" for 1852, recorded several as killed in the neighbourhood of Lynn towards the end of January; and two more were shot at Hickling about the same time; and in the same Journal for 1854 (p. 88), Mr. Southwell described them as unusually plentiful at Lynn in the previous winter. The few recorded in my own note books, since that time, have been all killed during sharp frosts, between December and February, which agrees with Hunt's description of this species that "they visit the fenny parts of this county in small flocks, in severe winters." In West Norfolk, according to Mr. Lubbock, a good many white-fronted geese are sometimes observed with the bean, or, as now distinguished, more probably with the pink-footed. Blakeney and Holkham have been already mentioned as localities where it is occasionally remarked, and the brackish waters of Salthouse would seem to have attractions, as an adult bird in my own collection was killed there on the 22nd of December, 1866, and Mr. Dowell had one sent him from the same place so early as the month of October, 1850. A single bird was killed at Surlingham during a severe frost in January, 1864, being over twenty miles from the coast; and a specimen in Mr. Upcher's collection, at Sherringham, was, singularly enough, shot out of a fence by the sea, where it was discovered by a Scotch terrier. The Messrs. Paget describe them as "occasionally seen on Breydon;" and Hickling Broad appears to be a favourite resort in sharp weather.

It is remarkable that during the severe winter of 1870-1, this species, as Lord Leicester informs me, was not seen at all at Holkham, and a single adult bird which I purchased in the Norwich market, on the 18th of February, was the only example that came under my notice during that inclement season.

The majority of specimens procured in Norfolk are in immature plumage, the bars on the breast (common to both sexes in an adult state) being either wanting or only partially assumed ;* but even in those most perfect as to the pectoral bands there is, as a rule, far less grey about the feathers of the back and wings than in either the adult pink-footed or grey-lag goose.

BERNICLA LEUCOPSIS (Temminck).

BERNACLE GOOSE.

The term "not uncommon" as applied to this species by Messrs. Paget in 1834, and by Messrs. Gurney and Fisher in 1846, is certainly not applicable at the present time, nor can I give any satisfactory reason for its rarity of late years on our coast, even in the most severe winters. My own notes for the last twenty years supply but a small list of specimens observed at long and uncertain intervals, either in the Norwich market or birdstuffers' shops; nor have I any reason to suppose that it has occurred much more frequently at Lynn, the chief emporium for this class of wild fowl. Mr. Dowell

---

* The chief breeding quarters of this species in Europe are probably Iceland. In the northern parts of Scandinavia it appears not to be known except possibly as an accidental straggler, its place there being taken by the smaller species *(A. erythropus)* (Linn.), the occurrence of which in the British Islands has not yet been recorded, though from the fact that it yearly visits Germany and Holland it will very likely be found some day in this country. (See Proc. Zool. Soc., 1860, pp. 339-341.)

describes the Bernacle as only an occasional visitant at Holkham and rare at Blakeney; and a specimen sent him from Salthouse* as late as the beginning of May, 1846, was one of but three examples he had known killed on that part of the Norfolk coast. In my own notes the first record dates back to the mild winter of 1851-52, when, as before stated, wild geese of various kinds were unaccountably numerous, and amongst these, on the 20th of December, two couple of Bernacles, in full adult plumage, were sent to Norwich from Hickling; and on the 28th a single bird from Burlingham. In the winter of 1854-55 I saw but one bernacle goose, killed towards the close of that most inclement season at Kimberley, on the 18th of February; and these, with an immature bird shot at Blakeney, in November, 1860, and two in the Norwich market in the following January, another exceptionally severe winter, are all that have come under my own notice. For the last ten years† I have neither seen nor heard of a Norfolk killed specimen, nor could I ascertain that any were remarked either at Lynn or on any other part of the coast throughout the long and severe winter of 1870-71. This is the more remarkable, since in Leadenhall Market, Mr. J. H. Gurney has found the bernacle, of late years, rank next in point of numbers, to the common brent; but the majority of those exhibited for sale in London are probably Dutch birds. A hybrid in Mr. Gurney's collection, between a bernacle goose and a canada gander, was bred

---

* The occurrence of a pair of bernacles at Salthouse, on the 26th of January, 1848, is recorded by Messrs. Gurney and Fisher in the "Zoologist," p. 2027.

† Mr. Cordeaux, in his "Ornithological notes from North Lincolnshire" ("Zoologist," s. s., p. 1124), after stating that small flocks of bernacles (there known as "Spanish geese") had appeared in that neighbourhood in December, 1867, remarks, "It is many years since I noted their last appearance in this district."

at Easton, near Norwich, in 1850 ("Zoologist," 1850, p. 2969), and exhibits in a marked degree, both in form and plumage, the chief characteristics of both species. Several pure bernacles were also bred on the same water during Mr. Gurney's residence at Easton. Except as regards the origin of the terms bernacle and "tree goose," as applied to this species, it is needless here to refer to the laughable fable of ancient naturalists, who associated the origin of this species with bernacle shells and the rotten wood of trees or wrecks to which those marine molluscs attach themselves. It is satisfactory, however, to know, as pointed out by Mr. Johns, in his "British Birds in their Haunts," that such mythical fancies were ridiculed and exposed nearly two hundred years ago by Ray and Willughby.

ANSER TORQUATUS, Frisch.
BRENT GOOSE.

This small species is both a regular and abundant winter visitant to our coast, in autumn its numbers increasing with the severity of the weather, and in very hard winters is met with in immense flocks. About Yarmouth, as described by the Messrs. Paget, it is not uncommon, but its chief resorts are the flat sandy shores of the northern and western parts of the county, and so essentially is this a marine species that it is but rarely met with on the broads or other inland waters.* Mr. Dowell, writing of its regular

---

* Folkard, in his "Wild Fowler" (p. 179), states that the brent goose "never feeds on fresh water herbage, nor flies far inland, nor alights in fresh water." This is not strictly correct, although to perform any of the above acts is certainly the exception and not the rule. A specimen in the Norwich Museum was killed on Barton broad, near Yarmouth.

appearance at Blakeney towards the end of autumn, remarks, "these birds arrive about the end of October in a flock of from one to three hundred, at which time they are most excellent eating, but by the end of winter they get strong and coarse. This flock does not seem to receive any accessions except in very hard weather. They never come into the harbour except at high tide, remaining till the half ebb, when they return to the sand beach and sleep till the tide makes again. But few are killed here owing to their extremely wary habits, and to their frequenting the harbour so little, and to the punts being unable, from their small size, to follow the geese to sea. By the end of March most of these geese, like other wild fowl, have taken their departure, but in 1847 I observed that a few still frequented the harbour as late as the 3rd of May." Mr. Dowell also gives another instance of the unusually late stay of this species on that portion of our Norfolk coast, a Blakeney pilot having shot two which were very tame in the first week of May, 1852. The Rev. C. A. Johns in his "British Birds in their Haunts" (p. 477), alludes to the abundance of these geese on the Norfolk coast in the severe winter of 1860-61, at which time he remarked that the brents regularly repaired to the salt marshes adjoining Thornham harbour, which he was told was their usual place of resort. A few examples of this most common species may be met with during any winter in the Norwich market and still more at Lynn in nearer proximity to their haunts; but I never remember them so abundant as during the long and severe frosts that prevailed in the winter of 1870-71.* On the 13th of January some eight

---

* As before remarked, this severe winter was remarkable for the extreme scarcity of all other species of wild geese except the pink-footed and brent. In 1803, also, when, as stated by Montagu (" Supp. Orn. Dict.") on the authority of Mr. Boys, of Sandwich,—brents were innumerable on the Kentish coast,

or ten couples were hanging for sale in our market with other fowl, and the numbers observed at that time along the shores of the Wash, from the mouth of Lynn Estuary to Hunstanton and Holme Point, and thence by Thornham and Brancaster to Holkham bay, were described, by local gunners, as almost unprecedented.

Messrs. Sheppard and Whitear remark, "the cry of a flock of these birds very much resembles the noise of a pack of hounds, and we have twice been deceived by it." Though difficult of approach under ordinary circumstances, a long chance shot, wounding one or two, will often, as Mr. Dowell informs me, afford further sport. The flock will then settle again at no great distance, and the gunner, re-loading, works his way quietly up to them when, reluctant to leave the cripples which are slow to rise on the wing, the main body thus waiting and waiting afford at last a "lumping" shot. With the exception of a few taken of late years, in the long nets occasionally erected upon the shores of the "Wash," near Lynn, (vol. ii., p. 376), I am not aware that wild geese have ever been netted in Norfolk, although Mr. Gurney informs me that all those in Leadenhall Market, from Holland, have their necks broken.

A very great difference in size is remarkable in birds of this species, as also in the colour of the under parts of the plumage, which in some is very dark, and in others so light as to be distinguishable at a considerable distance. At Blakeney, according to Mr. Dowell, the light coloured birds, which are said not to associate with the darker ones* and to be much tamer, have acquired

---

not a bernacle, grey-lag goose, nor bean goose was seen; and yet the weather was sufficiently severe to compel the whooping swan so far south.

* This is not always the case, as Mr. J. H. Gurney, jun., when at Lynn in February, 1871, observed light and dark birds together, and with a glass could distinguish one from the other at a very considerable distance.

the local name of "stranger" brents, and three of these birds, brought to him in the flesh, in the first week of March, 1853, proved on dissection to be two males and one female, the latter the lightest he had ever seen. That this difference in plumage is not a matter of either age or sex, seems evident from an examination of light and dark varieties in the collection of Mr. J. H. Gurney, jun., selected by himself from the large numbers commonly hanging for sale during winter, in Leadenhall Market. On the 6th of February, 1870, when there were several hundred brents in the market, adult and immature, he was struck by their difference in size, and remarked that those with dark underparts were more numerous than the light coloured ones. He has a light-coloured male from Yarmouth, killed in March, and a dark-coloured male from Leadenhall Market, killed in February. He also ascertained by dissection that of two immature birds, without any white on the neck, and with white edgings to the greater wing-coverts, the one with dark underparts was a male, and the light-coloured one a female.

The young birds of this species are considered very good eating, and generally find a ready sale in our markets. In the L'Estrange Household book it is once mentioned amongst articles purchased, as "Itm a brante, — — ii$^{d.}$; but the entry, "Itm a wyld goos kylled wt ye cros bowe" most probably referred to one of the larger species.

ANSER RUFICOLLIS, Pallas.

RED-BREASTED GOOSE.

The only example of this beautiful and extremely rare species, recorded as procured in Norfolk, is the one mentioned by most of our local authorities as purchased by the late Mr. Lilly Wigg in the Yarmouth market. This specimen, however, according to contemporary evidence, was plucked and eaten and its claim to a place in the Norfolk list must rest entirely on the credibility of the statements respecting it. Mr. Lilly Wigg, the main authority, was a naturalist resident at Yarmouth early in the present century, and in Sir William Hooker's MS. notes from 1807 to 1840 "touching the Natural History of Yarmouth and its environs," I find Mr. Wigg's name on the title page, associated with that of the late Mr. Dawson Turner, and others, as joint contributors; whilst amongst the entries, signed with the initials "L.W.," is, "Red-breasted goose shot near Yarmouth." Mr. Hunt, who gives a very accurately coloured plate of this species in his "British Birds," published in 1815, and was well aware of its rarity in collections, remarks, "Mr. Wigg, of Yarmouth, informs us that he purchased a specimen in the Market-place of that borough a few years since." Again in 1824, Messrs. Sheppard and Whitear (who alone give the date and locality), state that "Mr. Wigg had a specimen of this rare bird, which was killed at Halvergate, in Norfolk, in the year 1805. He says its flesh was well flavoured;" and in 1834, the Messrs. Paget in their Yarmouth "List" wrote, "Mr. Wigg accidentally bought a specimen of this bird in the market which, to his constant regret, he plucked and cooked." It seems almost incredible that any naturalist could be guilty of such an act, considering the remarkable appearance of this species, both as to the bright tints of its plumage and its small size, when compared with other wild geese, but the following explanation, as given by Mr. Dawson Turner to Mr. J. H. Gurney, after Mr. Wigg's decease, is, I have no doubt, correct. The latter gentleman it seems had a habit, arising from a curiosity, partly scientific and partly gastronomic, of tasting every variety of bird that came in his way, and not aware at the time of its real value, he bought and ate this goose as one he had never before tasted, and was not a little chagrined afterwards to find that he had sacrificed so great a rarity. Mr. Wigg, moreover, though a talented botanist was not a collector of ornithological specimens. Mr. Gurney some years afterwards received from the late Mr. Sparshall a box containing some of the smaller feathers of this specimen, which were collected so soon as its specific value was suspected from Mr. Wigg's description of it. In further confirmation of this story —as showing the probability, notwithstanding the extreme rarity of the bird, of its having occurred in Norfolk—is the fact of a remarkably fine example having been killed out of a flock of brent geese, at Maldon, in Essex, January 6th, 1871, as recorded in the "Zoologist" for that year (p. 2513) by Mr. J. E. Harting, in whose collection I have seen it, and who gives a list of the recorded occurrences of this bird, in Great Britain, from 1776 to the present time.

ANSER ÆGYPTIACUS (Linnæus.)

EGYPTIAN GOOSE.

Both the Egyptian and the Canada goose were included by Messrs. Gurney and Fisher in their "Account of birds found in Norfolk," but though, as stated by those authors, it is probable, as regards the former, that such as "occur on our coast after high easterly winds are genuine wild examples,"* yet the transatlantic

---

* Mr. J. H. Gurney has a stuffed specimen of the Egyptian goose, shot at Yarmouth during a very strong east wind, which he is inclined to think may have been a genuine wild one; and Mr.

origin of the latter scarcely admits of its introduction otherwise than as an acclimatised species. The Egyptian goose, however, being found in Northern Africa, there appears no reason why stragglers should not reach our coast at times, and although it is impossible now to distinguish between such foreign arrivals, and those which, bred in this county and left unpinioned, are killed in their wandering flights, more particularly when frozen out during sharp weather, it may still, I think, be retained in the Norfolk list.

It would be difficult to arrive at the exact date when either Egyptian or Canada geese were first introduced into this country,* though probably, as regards the former, at a remote period, but our local records of its occurrence in a supposed wild state extend no further back than the commencement of the present century, whilst the few notices of the latter are of a much more recent date.

In Sir William Hooker's M.S. notes a pair of Egyptian geese are said to have been killed near Harleston, in December, 1808, and one at Filby, near Yarmouth, in 1815. Hunt, in 1815, figured this bird from a specimen killed some few years before at Kimberley, and states that another was shot in September, 1815, at Ormesby, near Yarmouth. Colonel Hawker mentions

---

Lubbock also remarks that he has known two or three instances in which "there could be no doubt of the specimens being really wild." But even such examples may be only stragglers from Holland, where, I believe, they are kept, as with us, in some numbers.

* In Willughby's "Ornithology," published in 1678, the Egyptian goose is figured under the name of the Gambo goose *(Anser gambensis)*, and in describing the Canada goose (p. 361) the author remarks, "We saw and described both this and the precedent [the Egyptian goose of modern authors] among the King's wild fowl in St. James's Park."

two killed in Norfolk, early in this century; and Mr. Joseph Clarke also informs me that one was shot at Yarmouth in 1833, and that the late Mr. Stephen Miller once followed a flock of several in Yarmouth roads. No mention, however, of either species is made by Messrs. Sheppard and Whitear (1825) or Messrs. Paget (1834). My own notes, of course, contain more recent instances, including undoubted stragglers from "home" waters, killed on Breydon and some of the larger broads near the coast, in May and June, as well as in mild weather during the autumn months; but these are noticeable only as showing the straying habits of this species in a semi-domesticated state.

Though included by Yarrell in his "British Birds," the CANADA GOOSE *(Anser canadensis)* should undoubtedly be classed with the swan goose,* the black swan,† the Muscovy, and Carolina ducks, and other fancy fowl, of which stragglers have been killed from time to time in this county, in localities far from their domestic haunts. As an acclimatised species, it is much more plentiful in this county than the Egyptian, and the same remarks apply as to its roving habits.‡ On Lord Suffield's estate, at Gunton, near Cromer, they are so abundant that from a hundred to a hundred and fifty may be counted at one time in the park or adjoining

---

* In the "Zoologist" for 1850 (p. 2705) Mr. Newman recorded an example of the swan goose, as taken in a Norfolk decoy, and exhibited for sale in Leadenhall Market. This species is kept in a semi-domesticated condition on several estates in this county.

† A fine black swan, apparently uninjured by shot, was picked up dead on Holkham beach, in December, 1864; and another was killed on Breydon in 1863.

‡ Major King, in his "Sportsman in Canada" (p. 196), remarks of this species, " I have myself known an instance in which half a dozen, led away by the passing over them of a flock of brent geese, deserted a farm where they had been long domesticated."

stubbles, and none of them, as I learn from the Rev. H. H. Lubbock, of Hanworth, are pinioned, unless the keeper by chance finds a nest of young ones. These birds have a regular line of flight to certain favourite localities in the neighbourhood, and as their route is pretty well known they are not unfrequently shot at by gunners, laying wait for them, when beyond their preserved boundaries. On one occasion, when driving in the vicinity of Holt, early in the morning, three Canada geese flew over my head, low enough to have been touched by the whip; this was in the height of summer, but a fine bird in my own collection was killed out of a flock of three or four at Hickling, in February, 1852, by a man "flight" shooting during very severe weather, and which had no doubt been either frozen, or starved out elsewhere.*

I may here also note the occurrence near Lynn, early in the year, 1850, of a bird supposed to be a specimen of the SPUR-WINGED GOOSE *(Plectropterus gambensis)*, but of which I can unfortunately give little further information than is contained in the following memoranda kindly sent me by the Rev. F. L. Currie, at that time residing at Clenchwarton :—" On the 14th of February, 1850," he writes, " I heard of a Spur-winged Goose having been procured in this neighbourhood (Lynn) by a poor man who sold it to a Yorkshire skipper, who, in all probability, destroyed it." In Mr. Southwell's notes also is a communication from Mr. E. L. King, of Lynn,

---

* The Rev. E. S. Dixon, late rector of Intwood with Keswick, near Norwich, in his useful little work on the "History and management of ornamental and domestic poultry," remarks of this species, that the want of a proper supply of corn in winter when the grass grows but little, is the chief cause of their restlessness. "It is no migratory impulse that sets them on the move, but over-crowding and under-feeding. They have been literally starved out."

that on the 2nd of February, 1850, Mr. P. Spurgeon had informed him that a spur-winged goose was killed near that place a few days before. The spurs on the wings appear to have been specially noticed, but it is quite possible that this bird after all was only an Egyptian goose, that species having likewise a short blunt spur near the point of each wing, which fully accounts for the figure in Willughby's plate having the term "spur-winged goose" applied to it.

CYGNUS FERUS, Leach.

WHOOPER.*

The Wild Swan or Whooper is a winter visitant to our coast, rarely altogether absent, but its numbers depending mainly upon the severity of the season. Sir Thomas Browne, with his usual accuracy of observation, remarks of this species, "In hard winters, elks, a kind of wild swan, are seen in no small numbers; * * if the winter be mild, they come no further southward than Scotland; if very hard, they go lower and seek more southern places, which is the cause that, sometimes, we see them not before Christmas or the hardest time in winter." This account agrees most accurately with our experience of its habits at the present day, since, with the exception of one shot at Blakeney by Mr. T. W. Cremer, on the 6th of November, 1871, I have no reliable record of wild swans killed before December, and then only

---

* This name being derived from the peculiar trumpeting note of the species, I have preferred to spell it as in whooping-cough, the word "hooper," as it is more commonly written, having no special signification.

through an early commencement of frost and snow; the more usual time of their appearance extending from January to March. So much, however, do their numbers depend upon the severity and duration of frosty weather, that a record of severe winters will as surely furnish a list of great swan years.

Selby, in his "British Ornithology," speaks of their abundance in 1784-85 and in 1788-89, and states that, "in the winters of 1813,* 1814, 1819, 1823, 1828, and 1829, all more or less severe, they were very commonly met with in different parts of England. Messrs. Sheppard and Whitear, specially refer to the abundance of whoopers on our eastern coast in the hard weather of 1819, and according to the Rev. W. B. Daniel ("Rural Sports," vol. iii., p. 302), "in the winter of 1803, there were numbers of wild swans in the neighbourhood of Yarmouth, in Norfolk. Seventeen were shot by a man in the course of one week." Again in the severe winter of 1838, the number of wild swans, even in the south of England, is noticed by Yarrell, and by Colonel Hawker in his "Instructions to young Sportsmen." In the winter of 1843-44, as Mr. Southwell informs me, they were pretty numerous about the shores of the "Wash" and in the same locality, in the winter of 1849-50, were so common as to be hawked about the streets of Lynn by the gunners and fishermen. From that date, my own notes supply records of but three or four winters worthy to be termed swan years, although stragglers of this species occur in most seasons.

In 1854-55, a long and hard winter, when wild fowl of all kinds were extremely abundant, I saw upwards of twenty whoopers, that had been killed on our coast or inland waters, but all of them between January

---

* In the winter of 1813, he says, on the authority of Mr. Cooke, sixty wild swans were exhibited for sale in London, on one day.

and March; and this was also the case in 1860-61, when a severe frost, lasting with little intermission from December to the end of the following February, brought great numbers of wild swans and other fowl to our shores; though from the broads and other inland waters being early frozen over, they were chiefly confined to the coast and salt-marshes, or passed on further to the south. The return of these fine birds in spring, on their passage northward, is occasionally remarked, of which an instance occurred in the first week of March, 1861, when, the weather at the time being mild and open, a "herd" of twelve were seen to alight early in the morning on the open water of "Bargate," at the entrance to Surlingham Broad; but being disturbed later in the day they again took wing and quitted the neighbourhood altogether. In January, 1864, and again in the winter of 1869-70, several were shot in this county, but for the last twenty years at least there has been no such season for whoopers as that of 1870-71, when the hard weather of that memorable winter commenced with a heavy fall of snow on the 20th of December, increasing day by day until it was over a foot deep on the level. The frost was so intense that the thermometer, even by day, registered only a few degrees above zero, and this lasted with but little abatement up to the 12th or 13th of January. A rapid thaw on the 14th, cleared the ground of most of the first fall of snow, and though frosts continued at night the weather moderated considerably up to the 28th, when the snow again fell heavily, and the broads and smaller streams were thickly ice-bound up to the first week in February. My first notice of wild swans in that season was an intimation from Mr. Anthony Hamond that in the last week of December he had seen a "herd" of forty passing along the coast at Horsey, near Yarmouth; and during the first week in January a flock of twenty-six were observed on one occasion feeding close in shore off Holme Point, near Hunstanton; and another lot of seven frequented the entrance to Heacham creek. On the 12th several appeared off the Sherringham beach, passing along the coast; and on the same day, far inland, a considerable number were both heard and seen passing over the town of Wymondham, near Norwich. Several made their appearance on the smaller streams in the immediate neighbourhood of this city; and about the same time I heard of a flight, of which sixty are said to have been counted, that passed over the town of Aylsham, some ten miles from the sea. At Yarmouth, on the 18th, a "herd" of eighteen settled on Breydon water, but escaped from the gunners and, strange to say, with the exception of a fine adult bird shot on the lake at Kimberley, on the 20th of January, none appeared for sale in the Norwich Market or in the shops of our bird-stuffers till early in the following month. From the 6th to the 10th of February the weather was warm and sunny, but only to be followed by more snow and a biting wind frost, far more trying to fowl of all kinds than anything previously experienced; whilst a terrific gale on the 10th, with blinding snow storms, strewed our eastern coast with wrecks, and compelled shore birds of all descriptions to take shelter inland. During this month several whoopers were killed on Breydon, many more along the shores of the Wash, and others inland in various localities, but as to the numbers actually procured in Norfolk during that and the preceding month, I have no means of judging accurately, since by far the larger portion were sent up to London for sale, only some half-dozen appearing at intervals in the Norwich Market. Mr. J. H. Gurney, jun., was informed by a dealer in Leadenhall Market that he had received as many as a hundred whoopers during the frost, chiefly from King's Lynn; and one poulterer at Lynn stated he had had thirty.

As a rule, however, these wild swans by no means confine themselves to the sea coast, or even to the broads and streams in close vicinity, but, following the winding course of our rivers, are almost sure to make their appearance during a prolonged frost, in certain favourite localities, even though far inland. Some forty years ago, as the late Mr. Howlett, of Bowthorpe, informed me, that portion of the Yare which lies between Cringleford and Colney was so much frequented by wild swans in hard winters as to be locally termed the "Swan river," and he once counted sixteen; but though in those days the adjoining marshes were more frequently flooded, and thus afforded the most tempting feeding grounds, yet to this day, the low meadows about Earlham, Bowthorpe, and Colney, on the above river, and Costessey on the Wensum, all within three or four miles of Norwich, are a constant resort of the whooper. In the winter of 1870-71, a flock of seven took up their quarters in that particular part of the Yare, and though constantly disturbed, and two of their number shot, the survivors were remarked from time to time, at different points of the stream, up to the end of February. Mr. Gurney informs me that many years ago, at Earlham, he flushed five adult wild swans from the river, the weather being severe at the time, with snow on the ground. He at first took them for tame swans, but on his nearer approach they gathered themselves together, then rose simultaneously, and flapping slowly along the water for some distance, launched themselves on the wing, presenting a grand sight with the bright sun of a winter's morning shining on their glistening plumage. On another occasion, a gardener, named Bright, in the employ of the late Mr. J. J. Gurney, having observed that some wild swans frequented a certain part of the Earlham river, laid up for them in a cross drain, and as the birds came swimming past, five in number, endeavoured to take their heads in line; his gun, however, missed fire, but quickly recocking it, he succeeded in killing one and winging another as they rose. The latter falling on the opposite side of the water walked off over the meadows and into a lane adjoining, where it was captured by a man who sold it in Norwich Market.

The river about Sparham and Lyng Eastaugh, as Mr. F. Norgate informs me, is another favourite locality, and through his kindness I have now before me the heads of two old and three young whoopers, killed in that neighbourhood, on the 31st of December, 1869, and the 1st of January, 1870; also the head of an adult whooper with a very long bill, shot at Sparham on the 13th of February, 1871, which weighed eighteen and a half pounds. Of the first five, most probably a family group, the old birds weighed twenty-two pounds and fourteen pounds, and the cygnets sixteen pounds, fifteen pounds, and fourteen pounds respectively. The latter Mr. Norgate describes from notes taken at the time, as " white with a few greyish-brown feathers on the wings and head, feet black, but the webs mottled nearly all over with white; bills white and black instead of yellow and black as in adults." Besides the greyish-brown feathers on the heads of these cygnets (with the least possible tinge of red here and there), their immature state is marked by the loral space between the beak and the eye, which is bare in old birds, being more or less covered with minute feathers, the same extending down the ridge of the upper mandible nearly an inch beyond the ordinary limit of the frontal feathers in adults; but it is noticeable that even in very old whoopers, the feathers of the forehead extend slightly beyond the true base* of the upper mandible. The

---

* See also remarks on an "Anatomical peculiarity of the hooper's beak," by F. Boyes and W. W. Boulton. "Zoologist," 1871, p. 2504.

light coloured portion of the bill in these young swans, along the lateral margin of the upper mandible, does not extend beyond the perforation of the nostrils. The two old birds have both a very decided rufous tinge on the upper part and sides of the head, and beside the black on the anterior portions of the bill, have also, in the dried specimens, a slight black patch at the base, a band of yellow about half an inch in depth, crossing the ridge of the upper mandible between the two.* The yellow on the sides of the bill in one extends slightly beyond, in the other level with the nostrils; but specimens vary much in this respect. The old whooper killed in February, 1871, is rufous on the upper part of the head, but has no black at the base of the bill. In this bird the bill measures four inches and a half in length, being more than half an inch longer than in either of the two other adult specimens. A remarkably fine whooper in the Norwich Museum (British Series, No. 268), which was killed at Bowthorpe, in February, 1830, and is said to have weighed twenty-six pounds,† also measures four inches and a half along the ridge of the upper mandible, but has no black at the base. In January, 1870, seven whoopers were seen to alight in a meadow at Lyng, and two or three were observed at Sparham and Witchingham about the middle of February. In all these localities, the winding course

---

* Since writing the above, I have examined the head of the adult whooper, killed at Blakeney by Mr. Cremer, in November, 1871, which has also this black patch at the base of the bill. The bill measures four inches, and the head and neck are strongly tinged with orange-red, as were also, I am told, all the under parts of the plumage.

† St. John, in his "Natural History and Sport in Moray," (p. 71) gives twenty-seven pounds as the weight of the largest whooper he had ever killed. They usually varied from fifteen to eighteen pounds.

of the stream, with its many shallows and rank growth
of weeds and grasses, has special attractions, and though
during severe frosts the shoally margins are covered
with ice and snow, it is rarely indeed that the current
ceases to flow in mid-channel. In like manner, they
are attracted by the shallow weed-choked waters of the
smaller broads,* both on the Bure and the Yare, as well
as by the presence of tame swans upon our inland pools
and lakes, some of which, never having been pinioned,
are not unfrequently shot during sharp winters in mistake for wild ones. Two whoopers were shot on the
small ponds at Hempstead, near Holt, on the 9th of
January, 1868.

The distribution of colour on the bill in this species
forms the most marked external distinction between it
and the mute or tame swan *(Cygnus olor)*, in the former the base of the bill being yellow and the extremity
black; in the latter, the base black and the extremity
flesh-coloured, or reddish orange, according to age. The
internal differences exhibited by the whooper in the
convolutions of the trachea are also very marked, as
shown by Yarrell in his anatomical illustrations, but
that these had not escaped the observation of Sir
Thomas Browne is shown by his remark (when writing
of the "elks" or wild swans), that in them, "and not
in common swans, is remarkable that strange recurva-
tion of the wind pipe through the sternum, and the
same is also noticeable in the crane." The rufous
tinge on the head and cheeks, in the wild swan as

---

* St. John, in the work before quoted, thus describes the habits
of wild swans in Scotland—"While they remain with us they frequent and feed in shallow pieces of water like Lochlee, and Loch
Spynie, &c., where the water is of so small a depth that in many
places they can reach the bottom with their long necks, and pick
up the water grasses on which they feed."

in our semi-domesticated species, is noticeable more or less in most specimens, and in a very fine bird in the possession of Mr. F. Frere, of Yarmouth, shot on Breydon in February, 1865, this ferruginous or orange red upon the tips of the feathers extends likewise to the neck, and is more vivid than in any example I have seen. To this peculiarity of plumage, however, whether to be considered as a natural effect, dependent on age or sex, or, as believed by some naturalists, a merely artificial stain, I shall have occasion to refer, more at length, in my account of the mute swan.

CYGNUS BEWICKI, Yarrell.

BEWICK'S SWAN.

When the late Mr. Yarrell, in a paper read before the Linnean Society, in January, 1830,* pointed out the specific distinctions between this smaller wild swan and the common whooper, he included amongst his specimens and anatomical preparations the sternum and trachea of one shot at Yarmouth in the winter of 1827-8, the skin of which was preserved in the collection of Mr. J. B. Baker, of Hardwicke Court. Further investigation and comparison, at the time, discovered several examples in private collections, and amongst others an adult bird in the possession of the late Mr. Lombe, of Melton,† who in some MS. notes in his copy of Montagu's Dictionary, thus identifies it with this

---

* "On a new species of wild swan taken in England and hitherto confounded with the whooper."—"Linnean Transactions," vol. xvi., p. 445.

† This fine collection, as has been elsewhere stated, is now at Wymondham in the possession of Mrs. E. P. Clarke.

county. "In addition to the common wild swan another has now been ascertained to exist; several specimens were killed in England in 1829, but prior to that I had one, now in my own collection, killed in Norfolk, and preserved by Leadbeater." Subsequent observations proved the so-called Bewick's Swan to be of frequent occurrence on our coast in sharp winters, appearing in similar localities, and under similar conditions to the whooper, though, so far as my own experience goes, in less numbers. Messrs. Gurney and Fisher, in 1846, described this species as " almost as common" as the whooper, and as " frequently occurring in milder weather," but to whatever cause the discrepancy between their experience and mine may be attributable (possibly to a change in the habits of the birds themselves), I find, on referring to my own notes for the last twenty years, no record of Bewick's swans appearing when whoopers were not also met with, nor yet one single occurrence in otherwise than severe weather, during the months of December, January, February, and March. I quite concur, however, in Mr. Gurney's opinion that the whooper is much more in the habit of coming inland than the Bewick's swan, which would account in a great degree for the larger number of whoopers actually killed.* In the Broad " district" I have known the former shot at Hickling, Horsey, and Ludham, all within a short flight from the coast, but, as exceptions, may be mentioned a pair killed by Mr. F. Howlett, on the 13th of March, 1855, on the river Yare, at Bowthorpe, within three miles of Norwich; and two near Swaffham, in the same year,

---

* According to Thompson (vol. iii., pp. 17 and 21) "although *Cygnus bewicki* is considered to visit England less commonly than *C. ferus*, it is certainly of more frequent occurrence in Ireland." Of the wild swans brought into Dublin Market, he says that four-fifths had been estimated to be of the smaller species.

on the 20th of February and 16th of March, as I learn from Mr. Thomas Southwell.

The great swan years of 1854-55, 1860-61, and 1870-71—judging by the examples of each species, in our markets and bird-stuffers' shops—brought but a small proportion of Bewick's swans in comparison with whoopers. Thus in 1855, although a "herd" of twenty-three Bewick's swans were said to have appeared in Yarmouth roads, one of which was shot;* and according to Mr. Southwell's notes two other large flocks were observed near Lynn, by the late Mr. Newcome, of Feltwell, and Mr. D. C. Burlingham, in both cases on the wing, I have records of only eight procured in that season as against twenty-six whoopers; and in 1861 and 1864, of but two or three Bewick's swans, when whoopers were exceedingly abundant. In 1871, although from so many of the wild swans shot in this county having been sent to the London markets, it is difficult to arrive at a fair estimate; yet of the few that came under my own notice (the large numbers of wild swans observed off the coast not coming into this calculation) I should put the proportion of Bewick's swans to whoopers at less than one half. None of these Bewick's swans, also, were procured, until February and March, after a long duration of severe frost, but whoopers had been plentiful from the commencement of the hard weather, in the month of December previous. Mr. J. H. Gurney, jun., who, however, did not visit Leadenhall Market until late

---

* St. John, in his "Sport in Moray" (p. 72), speaking of Bewick's swans always appearing in smaller companies than the whooper, says, "I never see above eight together, usually four to five." He also describes them as shorter and more compact in form, as swimming higher in the water, and much tamer on their first arrival, than the common wild swan, which agrees exactly with Mr. Cordeaux's experience in North Lincolnshire. See "Zoologist," 1871, p. 2472.

in the season, saw but one Bewick's swan amongst many whoopers for sale on the 24th of February.

This species, which from its smaller size—being one third less than the whooper at the same age—is more generally known to continental authors as *Cygnus minor*, exhibits also the following external differences, as given by Yarrell in the paper before referred to. "The head is shorter and the elevation of the cranium greater in proportion to the size of the head, the beak narrow at the middle and dilated towards the point. The wings when closed do not extend quite so far beyond the roots of the tail feathers; the tail itself is somewhat cuneiform, and the toes appear shorter in proportion to the length of the tarsi." To these I may add, from the examination of several specimens, both adult and immature, since the year 1855, that the proportion of yellow to black in the bill of the adult Bewick's swan is much less than in the whooper, never extending so far along the sides of the upper mandible, but rounding off behind the nostrils. The colour itself in some freshly killed birds is decidedly more of a lemon-yellow than orange. The membrane beneath the lower mandible also, which in the whooper is yellow, is black in the adult Bewick's swan and in the young light grey, a distinction apparently overlooked by Yarrell. The distribution of black and yellow on the upper mandible varies, however, in different specimens, and I am somewhat inclined to believe that the broad band of black upon the ridge of the bill, extends nearer, by age, to the forehead, as in one or two examples in pure white plumage, I have seen traces of the black extending quite up to the base of the bill, the usual yellow band across the upper part showing faint indications of black mixed with the yellow colour.* This is not the case with birds showing

---

* Thompson, in his "Birds of Ireland" (vol. iii., p. 15), states that a specimen of Bewick's swan, which had been only slightly

the slightest remains of grey in their plumage; and in such immature examples the tints of the bill, both black and yellow, are less vivid. An adult bird, purchased in Norwich Market by Mr. F. Norgate, on the 1st of February, 1865, weighed thirteen pounds, and of two killed in the winter of 1870-71, a male weighed twelve pounds and a quarter, and a female nine pounds. In many adult birds of this species that I have seen, the feathers of the upper part of the head, especially, have been more or less tinged with rust colour. Internally the convolutions of the trachea present as marked a difference between this species and the whooper, as between the latter and the domestic swan; but a reference to Yarrell's illustrations, or an examination of the two anatomical preparations, arranged with the swans in the Norwich Museum* (British Bird series), and which exhibit the differences in this respect between *Cygnus ferus* and *C. bewicki*, will render it unnecessary for me to give here any further description. The adult specimen of Bewick's swan in the Norwich Museum is one of those before referred to as killed at Bowthorpe, in 1855.

At Blakeney, Mr. Dowell informs me, the gunners have always distinguished this smaller species from the whooper, under the name of "Spanish" swan. He killed one with a small hand gun and a charge of number seven shot, at Stiffkey freshes, on the 13th of March, 1848.

---

wounded, and kept alive four years in confinement, "had the ridge in the upper mandible black *from base to point*, a small patch of pale yellow, irregular in outline, appearing on the sides only of that mandible about three lines from the base." In four specimens the yellow was differently distributed.

* These preparations were made from examples of both species, killed in this county during the winter of 1854-55, when I had the opportunity of examining a good series in the flesh.

CYGNUS OLOR (Gmelin).

MUTE SWAN.

If we regard this bird merely as an acclimatised species, still the remote period of its introduction into this country, and its independent habits, even under the protection of man, fully entitle it, as in the case of the pheasant,* to a place in the list of " British Birds;" and this, quite apart from the probability that stragglers may, at times, reach our shores from those Northern and Eastern portions of Europe in which it is said to be found in a wild state.† The fact, however,

---

* It is most probable that both the mute swan and the pheasant were introduced into this country by the Romans. That such was the case as regards the latter seems conclusively proved by the most ancient record of its occurrence in Great Britain—namely, the tract " De inventione Sanctæ Crucis Nostræ in Monte Acuto et de ductione ejusdem apud Waltham," edited from MSS. in the British Museum, by Professor Stubbs, and published in 1861. Mr. W. Boyd Dawkins, who first drew the attention of naturalists to this document in a letter to the " Ibis" for 1869, p. 358, quotes from the " bill of fare drawn up by Harold for the Canons' households of from six to seven persons, A.D. 1059, and preserved in a MS. of the date of *circa* 1177," in which the pheasant is included amongst other proscribed articles of food, and remarks, " now the point of this passage is that it shows that *Phasianus colchicus*, had become naturalized in England before the Norman invasion; and as the English and Danes were not the introducers of strange animals in any well authenticated case, it offers fair presumptive evidence that it was introduced by the Roman conquerors, who naturalized the Fallow Deer in Britain."

† Yarrell states that the mute swan " is found wild in Russia and Siberia," and that Russian naturalists " include it among the birds found in the countries between the Black and Caspian Seas." By a review of the Vienna " Jagd-Zeitung," for 1864, published in the " Ibis " (1865, p. 225), it appears that Her Johan Zelebor, a conservator of the Imperial and Royal Museum at Vienna, in an expedition to the Theiss and lower Danube, in 1863

that some of the birds, reared annually on our rivers and broads, are left wholly unpinioned, and in hard winters, when frozen out from their usual haunts, betake themselves to the coast and are there shot by the gunners, renders any proof of actual immigration impossible; more especially as the same condition of things exists in many parts of the continent.

My remarks, then, on this beautiful and most interesting species, will be directed solely to its habits, in a semi-domesticated state, whether confined within the limits of a pond or lake for ornamental purposes, or finding its own subsistence on our rivers and broads, where but for the effect of the pinioning knife it would be *feræ naturæ* in the fullest sense of the term. The number of swans kept throughout the county by private individuals and corporate bodies—for with us it is essentially a municipal as well as a Royal bird—must be very considerable, but from their general distribution it is impossible to arrive at any satisfactory estimate. So long only as water is accessible, the smallest pond* or moat contents them though the

---

found the mute swan breeding in that locality, and procured ten cygnets in their downy plumage. Mr. Gould, also ("Birds of Great Britain"), remarks, "Mr. Dresser informs me he has himself seen it in a wild state on the banks of the southern Danube, and also on the island of Bornholm, in Denmark, whence he has eggs." The latter by no means an unlikely locality from whence stragglers might reach our shores in hard winters.

* The Rev. E. S. Dixon, late rector of Intwood with Keswick, in his work on "Ornamental and Domestic Poultry" (p. 29), gives two instances in Norfolk of swans existing for some time, in the same locality, with the most limited water supply. A single bird, supposed to have escaped from Houghton, which frequented a small pond at Bircham Tofts, and "kept" himself with but little extraneous help, and a pair, which occupied for fifteen years a small piece of water in front of Mr. Burroughes's house at Long Stratton, adjoining the high road.

limited amount of weeds, under such circumstances, necessitates a liberal supply of corn as well, and it is under these conditions only, when answering to the call of the keeper, that this noble species can be termed, appropriately, the tame or domestic swan. On the other hand it would be difficult indeed to find localities so adapted to its natural habits as are presented in the sluggish winding course of our Norfolk rivers, with their shoally margins, sheltering islets, and deep sedgy "rands,"* since, as has been well remarked by Mr. Dixon, "a yard of margin is worth a mile of deep stream; one muddy Norfolk Broad, with its oozy banks, labyrinthine creeks, and its forests of rushes, reeds, and sedges is better, in this respect, than all 'the blue rushing of the arrowy Rhone,' or 'the whole azure expanse of the brilliant Lake of Geneva.'" With the rivers and broads, also, must be mentioned those wide connecting "dykes" through which the tidal current ebbs and flows and the many narrower channels that form, as well, the drainage of the marshes; for these are a favourite resort of our river swans at all seasons, where, free from the constant traffic of the main stream, the weed-choked waters afford an inexhaustible supply of food; well stocked as they are with mollusca, aquatic insects, the roots of marshy plants, fish spawn, and innumerable small fry; if the two last named really constitute a portion of their diet.†

Though on the western side of the county, the Great and Little Ouze, the Nar, and the Thet, with the

---

* This term is applied in Norfolk to the margin between the wall or bank of a river and the water; a space usually overgrown with reeds or water-plants. Pronounced rand or rond.

† Mr. Dixon, on one occasion, observed his swans greedily devouring some dead shrimps, which had by chance been thrown into the water.

Waveney on its southern boundary, have many pairs upon their tortuous streams, representing the "swan rights" of adjacent properties, still the "Broad" district must be regarded as the chief nursery of our Norfolk swans, and the Yare, within the Corporate boundaries of this city, the swannery of the Eastern Counties. Scarcely a broad, throughout the winding course of the Bure and its tributaries from Belaugh to Breydon (although the number kept on that river is far less than in former years), is wanting in that greatest ornament to those reed-locked waters—

> "The stately sailing swan,
> Who gives his snowy plumage to the gale,
> And arching proud his neck, with oary feet,
> Bears forward fierce, and guards his osier isle,
> Protective of his young."

Both on the Wensum and the Yare, above Norwich, many private owners supply their tables from the cygnets, reared amongst the carrs, reed-beds, and rushy islets of the two rivers; but it is upon the Yare, below this city, from its junction with the Wensum, at Trowse Eye, to Hardley Cross,* near Reedham (the ancient

---

* This ancient boundary mark is situated on the right bank of the river Yare, at a point where the Chet—a small stream which, under the "Chet Valley improvement Act," is being widened for wherry traffic to Loddon, some three miles and a half—empties itself into the Norwich river. The Cross, which has of late years been re-erected upon a new basement, dates back to the year 1543, when, according to Blomefield, the Norfolk historian ("Hist. Norwich," 8vo., vol. i., p. 214), " There was a new cross, with a crucifix carved on one side, and the city arms on the other, painted and carried to Hardley, and there set up in presence of the sheriffs, in the place where "the shrevys of Norwyche yerely do kepe a court." This custom still continues, since on or about the 23rd of August in each year, the Chairman of the Tonnage Committee, with the Mayor, Sheriff, and other members of the Norwich Corporation, usually accompanied by the Mayors of Yarmouth and Beccles, and the Haven and Pier Commissioners, proceed by steamboat from the

boundary of the jurisdiction of Norwich and Yarmouth), that these birds will be found most abundant, including, of course, the "Broad" waters of Surlingham and Rockland. And it is in these localities that my own observation of their habits has been chiefly made, whilst gleaning at the same time many interesting facts from the experience of two intelligent marshmen, James Rich and John Trett, of Surlingham, who, for the last half century, have gained a living upon these broads, and for almost as long a period have had the care of swans.

Of the five and twenty pairs which nest annually in the vicinity of the Yare between Thorpe and Reedham, each will be found to have its prescribed limits within which no rivals are admitted without a challenge; and, as is the case with the robins in our gardens and shrubberies, any infringement of such rights of boundary leads to instant expulsion or a fierce fight between the males. Frequent contentions also take place in spring, between the male swans, as soon as pairing commences. One bird, seizing the other by the neck with its bill, and having thus got its opponent's head into "chancery," pummels him with the stump of his pinioned wing, and as both rise and fall in their struggles, they become enveloped in the spray thrown up by the rapid beating of their quills on the water. The victor invariably celebrates his triumph by loud notes of satisfaction, and first flapping his great wings, as he poises himself for a second almost erect in the water, sails back to his mate, every feather bristling with excitement, and every action of his graceful neck

---

Foundry Bridge, Norwich, to inspect the banks and staithes within the civic boundaries; and, on arriving at the confluence of the Chet, a time-honoured proclamation is read by the Sheriff, or Sheriff elect (there being now only one), from the steps of Hardley Cross. By a recent Act the jurisdiction, at the present time, extends to the "Dickey works" at the entrance of Breydon water.

denoting a proud consciousness of superiority. It is rarely that such combats have a fatal termination through injuries received, but occasionally a male bird will be found dead amongst the reeds, which, having lost heart after defeat, has sought a quiet spot to mope and die. Should the champion of the stream, also, as I am credibly informed, chance to lose its mate, his whole nature becomes changed by his misfortune, and, failing even the courage to resent attack, is driven from place to place by every former antagonist, until he withdraws altogether from society, and pines away in solitude. Even the females seem to owe a grudge to the once despotic chief, and resent his advances, whilst a less pugnacious widower more commonly consoles himself with a fresh consort. "The peaceful monarch of the lake," then, as this swan has been termed by poetical license, is so only when separated from others of his own sex. The old swans usually commence their nests in March, but in cold backward seasons are a week or two later, and for a fortnight or three weeks before the eggs are laid, may be seen busily pulling and carrying the stuff. I cannot ascertain, however, that the hen birds, as stated by Mr. Boyes, of Beverley, in a recent letter to Mr. J. H. Gurney, jun., ever lay their first eggs on the ground,* except in cases where the nest has been destroyed, or the birds driven from their first site just as the female was ready to lay. The foundation of the nest is, in most cases, composed of dried fodder from the "rauds," provided for their use, but supplemented by reeds, rushes, and other

---

* Mr. Boyes remarks, "The mute swan is an illustration of birds laying before the nest is completed. She frequently lays the first egg on the bare ground. On visiting the place to lay the next egg she collects a few materials and shapes the nest, and on every subsequent visit to lay adds considerably to the nest till it is finished."

coarse herbage of their own collecting and added to more or less throughout the time of incubation. The interior is composed of somewhat finer materials, mixed with their own down and feathers. Though generally high enough to escape the effects of any ordinary flood, they have been known to raise them suddenly,—either collecting materials of their own accord or using such as the forethought of the marshmen may have supplied,—and thus, by a marvellous instinct, as in the case recorded by Yarrell, anticipate an extraordinary high tide.* At such times, both birds are employed in the work, the male collecting materials and its mate arranging them and shifting her eggs. The process, as observed by Rich on more than one occasion, appears to be as follows:—The fresh stuff is piled up on one side of the nest, and having been roughly laid with the bill, is flattened down with the crown of the head; the eggs are then carefully rolled on to the higher surface by means of the head and beak, the under part of the lower mandible being inserted under each, and the same course is then adopted on the other side, and lastly, having raised the centre in proportion, the eggs are returned to their proper position. The eggs, however, are not exposed during all this time, but are covered at intervals by the female to keep them warm, and this even when the waters are rising rapidly.†

---

* Waterton, in his chapter on the mute swan, in the second series of his "Essays," describes his own swans as constantly and needlessly raising their nests, which were placed "on an island quite above the reach of a flood," even on one occasion using "two huge bunches of oaten straw," supplied for the purpose. In this case, one still recognises an instinctive impulse, even though uncalled for. A "propensity," in fact, as Paley describes it, "prior to experience and independent of instruction."

† The same means is not unfrequently adopted, under similar circumstances, by coots and water-hens; and black-headed gulls are known to raise their nests at times.

The swan's nest, from it sample dimensions, is always a conspicuous object, whether placed amongst the rank herbage on the river's bank, at the mouth of a marsh drain, or on the little islands and reedy margins of the broads themselves; and from the summit of that littered mass the sitting bird commands all approaches whilst her mate keeps guard below. To my mind an old male swan never looks more beautiful than when, thus "on duty," he sails forth from the margin of the stream to meet intruders; with his head and neck thrown back between his snowy pinions,* and every feather quivering with excitement, he drives through the rippling water, contenting himself, if unmolested, with a quiet assertion of his rights, but with loud hisses and threatening actions resenting an attack. When the young, too, under the joint convoy of their parents, have taken to the water, the actions of both birds are full of grace and vigour, and the deep call notes of the old pair mingle with the soft whistlings of their downy nestlings. What prettier sight presents itself upon our inland waters than such a group desporting themselves in the bright sunshine of a summer's day, when the pure whiteness of the old birds' feathers contrasts with the green background of reeds and rushes, and the little grey cygnets on their mother's back are peeping with bright bead-like eyes from the shelter of her spotless plumes? This habit of taking the young on her back is not, as some have supposed, adopted only as a means of safety when crossing a strong current, but is a method of brooding her young on the water, very commonly practised by the female

---

* Waterton, with his accustomed power of observation, points out that "the snow white feathers in the wing receive additional beauty by the muscular power which the swan possesses of raising them without extending the wing itself."

K

swan whilst her cygnets are small, and she will sink herself low in the water that they may mount more easily. Whether at the same time she gives them a "leg up" by raising them on the broad webs of her own feet I cannot say positively, but this is not improbable, since a favourite action in swans is that of swimming with one foot resting upon the lower part of the back, the sole of the foot being uppermost. With reference also to this means of transporting the young from one spot to another, a curious fact has recently come to my knowledge, which marks as well the attachment of these birds to their accustomed nesting places. An old hen swan, one of Rich's protégées, which in autumn and winter frequents the Yare, between Thorpe and Whitlingham, regularly as the spring comes round makes her appearance, with her mate, on Surlingham Broad—a distance (allowing for the windings of the stream) of nearly six miles—and proceeds at once to collect materials for her nest in that locality. Shortly, however, after the young are hatched this same pair, the female with her little progeny on her back, may be seen passing from the broad into the river, where, turning their heads up stream with an evidently settled purpose in view, they commence the return journey to Thorpe; and from observations made by ferry keepers and others, to whom this habit is known, it is believed the whole route is accomplished without stopping to feed by the way. Two other pairs that regularly nest on the same broad return with their young during the summer to their winter quarters at Bramerton, some three miles up the river; and on one occasion a female which had been conveyed to Surlingham in the spring, and paired on the broad, returned with her cygnets to a spot near the Clarence Harbour Inn, at Carrow, distant about seven miles, where she had by chance been turned off in the previous winter.

Swans pair for life, build a fresh nest each season, and, if left unmolested, will keep pretty close to the same locality. Adult birds invariably nest earlier than young, owing chiefly, no doubt, to the advantage which an old cock swan possesses over a young one, of maintaining his right to any selected spot, whilst a young couple just "setting-up" for themselves have many drawbacks to encounter. "Might is right" in such matters, and many battles have to be fought to secure a *locus standi*, and occasionally young couples will select unfavourable sites, from which the marshmen, well knowing their eggs would be stolen, drive them off as soon as they commence building. A young male, though paired with an old female, would have an equal difficulty in holding his ground. Young hen birds[*] do not lay till their second year, some not until the third or fourth, and commence by laying from three to five eggs. A second year bird paired with a male of her own age usually lays three eggs the first season, but will probably commence with five, if paired with an old male. Commencing with five eggs, the same bird will lay from seven to nine the next season, and in the following year from ten to eleven, being then in her prime, at four years old. Hen birds which have not paired till their third or fourth year will lay from seven to nine in their first season, but, from Rich's observations, it would seem that a second year female, commencing with only three eggs, rarely, if ever, lays more than nine. I have no reason to question the accuracy of these statistics (though necessarily beyond my own cognizance), my informant having been accustomed for years, in company with the recognised swanherd, to examine the

---

[*] Rich was told by Trett's father that he once knew a year-old cygnet lay, but he has never met with such an instance himself, although adult cock swans will tread young hens only twelve-months' old.

majority of the nests in his neighbourhood whilst the birds are laying, carefully noticing the number of the eggs in each, the owners' marks on the birds, and their respective ages, and his statements may, I think, in some decree, account for the strange discrepancy in the works of British ornithologists[*] as to the number of eggs laid by the mute swan.

The wide range afforded the swans on the Yare, with the abundant supply of their natural diet, accounts to some extent, no doubt, for the very large number of cygnets reared within our civic boundaries. On this point Mr. Dixon states that in one year, on that portion of the Yare next to his late residence at Cringleford, three pairs of swans had each a brood of nine cygnets, which he considered above the average; but he had even known seven reared on a very small moat. On the Yare below Norwich, and on the adjacent broads, allowing for accidents and the difference in laying between young and old females, I believe seven to be about the average of young hatched, though many a proud mother there launches her little fleet of ten or even eleven cygnets. The following return from the swanherd's book of the produce of a single pair during eight seasons, though very exceptional, will show the effect of selection and the introduction of new blood into the local stock. The cock swan, a remarkably fine young male, was brought from some inland water,[†] and

---

[*] Yarrell states that the mute swan lays from six to seven eggs; Bewick six to eight; Mudie from five to eight; Hunt ("British Ornithology") six to eight; Pennant seven to eight; Daniel ("Rural Sports") six to ten; and Morris from two to eight or nine.

[†] I am told that, as compared with those bred in the "Broad" district, the swans of our inland waters rarely pair until the third year, and nest somewhat later in the season. The cygnets, also,

paired on Surlingham Broad with a female in her third year:—

Year.	Eggs.	Cygnets.
1866	9	8
1867	10	9
1868	11	11
1869	10	10
1870	11	11
1871	11	11
1872	11	11
1873	12	11
	85	82

I can myself speak to the number of cygnets brought off by this pair during the last four years, as the birds have regularly nested, or "timbered," as the marshmen term it, on a part of Surlingham Broad known as "King's Fleet," where I have seen them repeatedly with eleven young ones. In the summer of 1870 I saw a hen bird on Bargate, said to be eight years old, which had twelve cygnets, but in this case one egg had been added to her own from another nest.* In like manner Rich once knew of thirteen cygnets hatched by one female, but as the same bird laid only eleven eggs the next year, when he had the charge of her, the probability is that she hatched at least two foster cygnets in the previous summer. With cygnets, however, as with lambs, much depends upon a favourable season, and Mr. Dixon alludes to "a common notion in Norfolk that the cygnets cannot be hatched till a thunder storm comes

---

sent to the Norwich swan pit, from artificial ponds or lakes to be fattened, are invariably smaller; but the Blickling swans, having a wide range on the river, are always fine birds.

* It is stated by Morris that "at Beddington Park, in Norfolk, twelve eggs were deposited, and the brood all reared in 1850"; but in the first place there is no such locality in this county, and, secondly, no evidence that they were all laid by the same bird.

to break the eggs;" a superstition at least so far consistent with natural causes, that warm sultry weather is preferable for incubation to a cold backward spring. The eggs, which vary somewhat in tint, are of a greenish white, and in size proportioned to so large a bird. The appearance, however, of such addled specimens, as may be seen in Rich's cottage, occupying an ornamental position on the mantel-shelf, might puzzle many an oologist, unaware that their rich brown polished hue, like knotted oak, is the result only of the "gude wife's" patience, in first encircling them with thin layers of onion peeling, and then extracting and impressing the stain by careful boiling.

Incubation usually occupies five weeks, or about a week longer should the weather be very cold, but if the eggs prove addled the hen will continue sitting for seven or eight weeks, or till driven from her nest by the marshmen. A remarkable and fatal act of perseverance in this respect is thus recorded by Hunt in his "British Ornithology" (vol. ii., p. 194):—"A swan having built a nest on a part of Bungay Common, the season being very wet and unpropitious for incubation, the eggs became addled, nevertheless the female continued sitting upon them until she dropped dead from her nest and floated down the river." The cock bird is at this period particularly attentive to his mate, and boldly resents the intrusion of man or beast within too close proximity to the nest. Of this fact Mr. Hunt, as an eye-witness, also supplies us with a local instance which occurred in the well-known channel that connects the "Bargate" entrance to Surlingham Broad with the main river. Here, as not unfrequently happens at the present day, a swan had built her nest on a small island, near the mouth of the broad, and being at that time sitting on her eggs, her mate for a considerable time disputed the passage of a water party, "boldly attacking the boat and striking its wings with great violence against the oars, and at the persons rowing the same. At length a fine spaniel sprang from the boat into the water, and the swan immediately attacked the dog, and would have succeeded in drowning him but for the exertions of persons on board." On the same water, within the last few years, an old male swan was in the habit of attacking strangers in boats, who passed near the shore on which its nest was located. This bird was seen by Rich on one occasion to seize a man from behind as he sat in a low counter-stern boat, and gaining a foothold on the woodwork, begin striking at him with his wings, until driven off with blows from an oar. Country people were often afraid to venture across that portion of the broad for fear of this bird, unless accompanied by a marshman, when no opposition was offered; but his fierceness had no doubt been increased by the wanton provocation to which they are too often subjected. In another instance, which occurred on the river Yare, at Postwick, an old swan actually sprang from the water into a boat containing a man and a boy, and commenced pummelling the latter with its wings, and, though driven out of the boat with a stick, returned again to the charge.

From the above statements, then, it is very evident that an old male swan, in a temper (and his fierceness is not always confined to the breeding season)* is a formidable antagonist, either on land or water, but I have never been able to verify the assertion that one has been known to break a man's arm or leg by a blow from its pinioned wing. It has, I know, been so stated by Bewick and others, but no such occurrence is known, even by tradition, amongst our local swanherds, though lesser injuries have been sustained by some of them; and these men when cap-

---

* See Thompson's "Natural History of Ireland," vol. iii., p. 22.

turing the cygnets, or seizing the old swans to examine their marks, are most liable to attack. Rich once knew a man's finger dislocated when thus employed; and Trett's father had on one occasion a thumb put out of joint, and on another a very serious blow on the back from the stump of a pinioned wing. In the latter case the man was attacked by an old male swan, as he was examining the eggs in a nest, to which, being in a boggy place, he had crawled on his hands and knees. The swan coming up behind him, unperceived, struck him so violently on the back that he had difficulty in regaining his boat, where he laid for some time in great pain, and though he managed at length to pull home, he was confined to his bed for more than a week. Rich has himself been struck on the thigh, in like manner, and describes the force of the blow and the pain occasioned by it as something incredible.

To return, however, to the nesting habits of this species, the cock swan is, I find, not merely the guardian of the nest he has helped to build or to raise, but usually sits for two or three days before the hen begins to lay, thus shaping and warming the interior at the same time. Whilst the female is laying her full complement of eggs —which she does at the rate of about ten eggs in fourteen days—the cock takes charge of and broods them in her absence, often most reluctantly resigning his post on her return. The female, wherever coarse herbage is at hand —and in some cases dried litter is supplied for that purpose—covers her eggs when she goes off to feed, and her mate usually uncovers them wholly or partially before taking his turn; and as the hen invariably dries herself before resuming her sitting, her spouse awaits the completion of her toilet, and has even then not unfrequently to be shouldered off. In instances where the female has unfortunately died before her eggs have been hatched, the cock swan has been known to continue sitting, and take the sole charge of the cygnets he has succeeded in rearing, but this is exceptional. More commonly when the female dies off her nest, as the marshmen say, from fever, the male also leaves it and the eggs are spoiled. The hen bird when thus affected, which is usually after sitting about a month, quits her nest altogether, and withdraws into the thick reeds or sedges, to mope and die. The beak is kept constantly open, whilst a watery secretion appears to flow continually from the mandibles, and the bird rapidly wastes to mere skin and bone. When found in this state, Rich has occasionally saved the swan's life by taking it at once into the river, if nesting on the broad, or from one part of the river to another; thus changing the water, and probably varying the diet at the same time. Some few years back, on Surlingham Broad, a female, which only a day or two before had hatched ten cygnets, died suddenly, off the nest, when the cock bird continued to brood them, and when able to leave the nest, proudly "mothered" his numerous progeny, often carrying them on his back, and never ceased his watchful care of them until the autumnal swan "upping" relieved him of his charge. The same bird paired again the next season, but his second mate, lived only two years, and then, singularly enough, died as the former one had done, but leaving him with a sitting of ten eggs. This time, although he took to the nest as before, he only brought off two cygnets, contenting himself with those first hatched, and allowed the rest of the eggs to become cold. Swans, as a rule, have but one brood a year, not laying a second time, unless the first eggs are bad, or are otherwise destroyed.* Some

---

* A swan which had by some accident lost a sitting of ten eggs, in about three weeks laid six or seven more. Another, which had eight eggs stolen, laid seven the second time.

years back, however, a pair of swans, under the charge of Trett's father, had eight or nine cygnets, which, with one exception, crept into a large bow-net, set in the mouth of a marsh "dyke," and were drowned with the rising tide. In this most exceptional instance, the female, though still attended by her one surviving cygnet, laid again and brought off four young ones; but these, like the former clutch, came to an untimely end, trampled to death, it was supposed, by their big brother, in his efforts to secure a footing in his former nursery. As far as I can ascertain the young cygnets are never destroyed on these waters by either rats or pike, but occasionally, when very small, if they are feeding in a drain, and the water becomes low, they have much difficulty in climbing out, and become weakened, and at last drowned by falling back into the stream. The old birds, also, in their anxiety to assist them, not unfrequently trample them down, or kill one or more accidentally in their excitement with a stroke of the wing. A loss in eggs may arise from various causes besides egg stealing or a sudden flood. A young pair are sometimes driven off by an old cock swan and the eggs are spoilt; or some may roll out and get broken owing to a badly constructed nest, whilst, occasionally, a like result occurs during a fierce contest between rival males.

The down of the nestling cygnets, which is of a light grey on the head and neck, with a brownish tinge upon the back and wings, is replaced by feathers of a darker hue, approaching to a sooty grey; yet still lightest on the under parts. But, though in some a sprinkling of white feathers may be seen in their first autumn as early as the beginning of August, they do not complete their pure white plumage till the following summer, when from twelve to fourteen months old. It is, however, in this intermediate stage, the least attractive as regards form or plumage,—representing, in fact, the "ugly duckling" of Hans Andersen's charming story,—that they are in request, at the present day, for edible purposes. The assumption of the white plumage seems to be very gradual, and the change in the colour of the dark feathers at the same period is likewise remarkable, being latterly more of a light reddish brown than grey. A dozen birds in the St. Helen's Swan-pit, in February, 1873, all hatched in the previous summer, exhibited a strangely mottled appearance, no two birds being alike in their state of change. In some the white feathers were thickly sprinkled amongst the brown ones on the back, in others the brown was limited to the secondaries and upper tail coverts, and the most forward bird was already becoming pink on the upper mandible, in contrast to the other "blue beaks." In all, the breast and under parts of the plumage appeared to be pure white, their heads and necks having still a greyish tinge, darkest on the crown; but with a soiled look, as if partly due to their close confinement.

It is by no means unusual, throughout the summer months, to find a school of from ten to fourteen swans on the Yare, between Surlingham and Coldham Hall, consisting of second year birds in their white plumage, and a few older, distinguished by the colour of their bills, which may either have had their eggs destroyed, or from some cause have not paired for that season. These, though differing in age, seem to agree fairly amongst themselves, but it is amusing to see them, when drifted by the tide, in feeding, past the territory of an old cock swan attending his mate, scattered in all directions by the vigorous onset of that *pater familias*, a very host in himself, in his pride and gallantry.

There is one feature, also, in the adult plumage of the domestic swan, which it possesses in common with the two preceding species, in their wild state—

namely, the orange or ferruginous tint that pervades, more or less, the head and neck of most individuals, but is confined to the tips of the feathers only. As to the cause of this very localised colouring, naturalists have long been divided in opinion, some regarding it as a natural effect, intensified by age, others as simply an artificial stain, acquired in the act of feeding, by contact with ferruginous sands and the roots of aquatic plants similarly impregnated. My own opinion, from such observations as I have been able to make, has always inclined to the latter theory, and such is, I find, the belief of the swanherds on the Yare, who maintain that the swans in the river become red, more or less, during autumn and winter, owing to the soil washed down from the uplands into the dykes connected with the main stream. That it is not a matter of age seems evident from the fact that birds only two or three years old, brought from Blickling and other inland waters to Surlingham and Rockland have in some instances redder heads than the oldest swans on these broads, whilst the same birds are said to become gradually less rufous in their new quarters, and to lose it altogether at their autumn moult. In the spring of 1870 I saw a three year old swan on Surlingham Broad, which had decidedly more red on the head than others in the same locality, known to be from eight to ten years old. Swans kept upon private waters, or frequenting throughout the year some inland stream having a strong ferruginous deposit, would naturally appear to retain this colouring as a permanent feature, inasmuch as even during the process of moulting, the new feathers would become tinged; whilst, on the other hand, the birds that pass their nesting season on the broads, and are subjected to no such influences, become pure white at the moult, however much their heads may have been previously reddened. Perpetual contact, therefore, with ferruginous matter would account for the vivid colouring on the heads of some old swans, and the large size of the head in an old cock swan, and the breadth of surface coloured, would convey the impression, no doubt, that the hen bird was always higher coloured in this respect than the male. The feathers thus tinged, as before remarked, are only tipped with red, the posterior portions and the down remaining pure white, and only those parts of the plumage which in feeding would come in contact with the bed of the stream are thus affected. The extent of this, both on the head and neck, varies considerably. In Mr. Frere's whooper (ante, p. 53) it extends far down the neck, and though deepest, as usual, on the crown of the head, is generally distributed over the sides as well. Mr. Cremer's whooper, killed at Blakeney, in November, 1871, which had the head and neck strongly tinged with red (ante, p. 51), was said to have been similarly stained on the under parts of the plumage; but this I have never observed myself in any swan, wild or otherwise. From such facts, then, as I have here given, I had arrived at the conclusion that the colouring in question is acquired and not natural, due, in fact, merely to external causes; but it remained to prove by some chemical test the actual presence of iron in the red feathers themselves. The idea of instituting some such experiment occurred to me through reading a review, in the "Ibis" for 1862 (1st series, vol. iv., p. 182), of a paper by Herr Conservator F. W. Meves,* "On the red colouring in Gypaëtus," a raptorial genus between the vultures and eagles. By

---

* This paper was communicated to the "Summary of the Transactions of the Royal Academy of Sciences, at Stockholm, for 1860." Herr Meves is the same naturalist whose ingenious theory as to the drumming noise of the snipe is described in vol. ii. of this work, p. 316.

a simple chemical test, Herr Meves ascertained that the ferruginous tint in the plumage of these birds is owing to "a superficial deposit of oxide of iron on the feathers," a similar stain on the eggs arising from the same cause. The same peculiarity observed in some of the feathers of the crane, in Jemtland, he also found to be caused by the presence of iron, and from these data, the editor of the "Ibis" suggested that similar experiments should be made on the rufous-tinged feathers of the whooper and Bewick's swan. To this end—being desirous of establishing the point on unquestionable authority—I solicited the assistance of my friend Mr. F. Kitton, of Norwich, whose name is so well known in connection with microscopic investigations, and having placed in his hands the head of a very adult mute swan, strongly tinged with ferruginous matter, he has furnished me with the following most satisfactory result of his investigations:—"As I anticipated," he writes, "the colouring matter is iron (peroxide Fe_2O_3.) On testing some of the deeply stained feathers from the head with ferro cyanuret of potassium the characteristic deep blue colour immediately appeared (sesqui ferrocyanide of iron.) On placing white feathers from the neck in contact with some red crag debris and water, they acquired a pale buff tint, and these became blue like the red feathers of the head, when treated with the ferro cyanuret of potassium. I afterwards mounted some of the tested feathers in Canada balsam, and examined them with the micro-spectroscope, and found that the spectra of the originally and experimentally stained feathers were identical. I think you are correct in your surmise that the rufous tint is produced by contact with ferruginous sand." It remains now only to test the water and the subsoil in certain localities where swans are known to exhibit this rufous colouring most vividly, and I may here add that the delicate buff tint on the white feathers placed by Mr. Kitton in water in contact with red crag, is particularly interesting, as it corresponds exactly with the colouring so often remarked on the necks of domestic swans—just so far as they are usually submerged in feeding—occasioned more probably by the water than by actual contact with the soil.

The beak in this species varies in colour with every stage of growth from the nestling to the fully adult. The cygnets in their first autumn have this feature of a deep lead colour, the nostrils, nail, and marginal line of the upper mandible being black, as in old birds. As they assume their white plumage a lighter grey tinged with green takes the place of the lead colour, but up to the autumn of their second year they are termed "blue beaks." Before the close of that year, however, the dark tints have gradually given way to a pinkish flesh colour, and in the following spring the beak becomes orange red, the last proof of maturity, the orange predominating in the oldest birds. In the females the colours are somewhat less vivid than in the males. The development of the frontal knob or "berry" is a matter of age. This very marked feature of the mute swan is extraordinarily developed in very old males, but is always smaller in females. The difference of sex is also distinguishable in the water by the females swimming less buoyantly, by their necks being somewhat less robust, and by their more subdued, though scarcely less graceful, movements. The hen bird, I have noticed, commences the autumn moult earlier than the male, and is usually in perfect plumage before her mate has shot his old and ragged quills.

Rich had at one time a pair of swans which regularly hatched, out of their usual number, two cygnets, having the down of a decided blue tint instead of greyish brown; and these retained their distinctive colouring till "taken" up for fattening. Any such variation is rarely met with, and all my enquiries have failed to elicit a single instance of a white cygnet having been hatched with others of a normal colour. Such a variety appears to be unknown, even by tradition, to our local swanherds, a fact which seems to have an important bearing upon the question of specific difference between the mute swan and the so-called Polish or "changeless" swan, whose cygnets are said to be always white. Yarrell, however, gives the grey colour of the feet in *Cygnus immutabilis* as a specific difference, but this can scarcely hold good as a point of distinction, as the webs in the cygnets of the mute swan are frequently grey instead of black, and I have observed the same variation in birds which had acquired their full plumage. A pair of cygnets, hatched some few years back on Surlingham Broad, had each the lower mandible projecting beyond the upper, about three-quarters of an inch, but this deformity in no way interfered with their feeding, and they grew up as fine and as weighty birds as any of their companions.

Such cygnets as either elude the pursuit of the swanherds in August, or are intentionally left with their parents, associate with the old ones throughout the winter, but are invariably driven away to shift for themselves in the following February or March; and these cygnets then congregate in small parties, until their pairing time arrives in the next season. On contracted pieces of water the persecution of the young birds by the old at this season is persistent, often forcing them to take refuge on land; but for this Waterton adopted a simple and harmless remedy, viz., cutting the webs of the old swans' feet so as to enable the young to out-swim them when pursued. A very curious incident, however, in connection with this habit of the old swans, occurred in the spring of 1872, on a piece of ornamental water, at Thickthorn, near Norwich.* An old pair, having one surviving cygnet of the previous year, hatched again, when the male swan, as usual, commenced persecuting its former progeny, until the owners had to remove the poor bird altogether. On the same water were some Muscovy ducks, and the male swan next made an attack upon the drake, but found, in this case, that he had caught a tartar, for the Muscovy suddenly sprang on the swan's back, and, safely seated between its wings, pecked fiercely at his neck, in spite of the frantic efforts of the big bully to dislodge him, and the drake had at length to be driven off.

Even in their comparatively wild state, and when finding their own living, mute swans become very weighty, the males especially. A remarkably fine bird when "taken up," a few years back, on Surlingham Broad, weighed thirty-three pounds; and of an adult pair, killed on Somerton Broad in 1871, the male weighed thirty and the female eighteen pounds; but between twenty-five and thirty pounds is the ordinary weight of a full grown cock swan. We can scarcely wonder at this, or at their rapid growth, when we observe old and young alike feeding almost incessantly throughout the day, with only brief intervals of repose or of attention to the toilet, and with this characteristic comes the question of their utility or otherwise, more especially upon the shallow waters of our Norfolk Broads. "There is no bird," writes Mr. Dixon, "comparable to the swan for clearing a pond of weeds,"† and

---

* Communicated to the "Zoologist" for 1873 (p. 3413), by Miss Brightwell, through Dr. Gray.

† In Beeton's "Book of Home Pets" it is stated that two pairs of swans which, in 1796, were placed by the then Marquis of Exeter upon a sheet of water over-run with weeds, effectually cleared it in one year, and kept the weeds down afterwards, though previously, for the same purpose, three men had been employed during six months in the year.

the same remark applies to much larger pieces of water, their long necks enabling them to perform an important service to man by thus "cleansing the half stagnant water courses," and consuming the submerged vegetable refuse which would become offensive and obstructive at the same time. They have their preferences, however, in the matter of weed-diet, and on this head I cannot do better than quote Mr. Dixon's experience for the information of those who may desire to stock their ponds or lakes with the most suitable vegetation. Swans prefer, he says, "first what we call the lower forms of vegetation, the *confervæ* and the *cheraciæ*; then the *Callitriche aquatica* or water starwort, and the long list of Potamogetons or pond-weeds. The rhizomata of all sorts of reeds, rushes, arrow heads, &c., are greedily torn up and devoured," but the roots of water-lilies (white or yellow) "they scarcely ever touch except, perhaps, in a young state," though they probably devour the seeds. "The soft starchy parts of aquatic plants" are selected when in a state of freedom, as in confinement "the spare garden stuff, spinach, &c., thrown out to them, is liked all the better for having laid soaking at least twenty-four hours;" and their preference for food in a sodden state is sufficiently indicated by their peculiar method of feeding, rinsing each mouthful in water before swallowing it. Being an introduction only of late years we find no mention in this bill of fare of that greatest pest of our Broad waters and marsh drains, the American *Anacharis alsinastrum*, or "Cambridge[*] weed" as it is here commonly termed, but I have reason to believe, from enquiries recently made, that swans are of great service in keeping down, if not in extirpating, this transatlantic "difficulty." At Hoveton,

---

[*] The history of its introduction into the waters of East Anglia from that seat of learning, has a moral which should not be lost on members of Acclimatization Societies.

as I am informed by the Rev. T. J. Blofeld, it is pulled up by his swans in large quantities; and Rich informs me that at Surlingham a pair of swans, which had strayed from the Broad to a marsh-dyke near his house filled with this weed, cleared the channel very effectually in a short time. To grebes and divers, seeking their finny prey beneath the surface of the water, its endless ramifications must be particularly objectionable, but, as proved both on Hoveton and Wroxham Broads, its abundance is a great attraction to coots and to some species of wild fowl, especially widgeon and tufted ducks, and like them, the swans, no doubt, feed not only on the weed itself but on the minute mollusca that swarm on its matted fibres. Were it, however, not gathered at all by the swans as an article of food but merely torn up in the search for other roots and plants, I think even Mr. Frank Buckland might cease to denounce these birds as "spawn-eating brutes," in consideration of the eminent services they would even thus render to man. It is a disputed point I know whether swans do eat the spawn of our river fish, but though I have no direct evidence of this by dissection, the testimony of our broadmen is so far confirmatory of the watchers on the Thames,* that

---

* In "Land and Water" for November 30th, 1872, is a copy of a petition recently forwarded by the officers of the Great Marlow Thames Angling Association to the Lord Chamberlain, praying for a reduction in the number of the Queen's swans on the Thames, and similar applications have been made to the Dyers' and Vintners' Companies, on the ground of the injury done to the fishing by these birds in the consumption of fish spawn. Mr. Francis Francis supports this petition in the "Field" for November 23rd, 1872, and the same subject has been referred to in that journal on many previous occasions. At the present time the number of swans on the Thames are said to be—" Her Majesty the Queen, 397; Dyers' Company, 67; Vintners' Company, 55; total, 519. A few years ago the number belonging to the Queen was 500." At the great swannery belonging to the Earl of Ilchester, at Abbotsbury,

whilst they acquit the old swans of eating the spawn themselves, they assert that they pull up the weeds with spawn on for their young ones. Whether correct or not, however, on this point, it is undoubtedly in May and June, when the roach and bream enter the broads and dykes in shoals to deposit their spawn, that the marshmen invariably find both old and young swans collected together in those shallow waters, busily foraging amongst the herbage under the banks of the stream, where spawn had been previously noticed.

Taking the annual number of breeding swans on the Yare, between Thorpe and Hardley Cross—including, of course, the broads at Surlingham and Rockland—at twenty-five pairs, this can scarcely be deemed an excessive number to be spread over an area of from ten to twelve miles, but, inasmuch as the upper portions of the river, from the absence of "ronds," are less suitable for nesting purposes, and as from their close vicinity to the city the eggs* are more liable to be stolen, the

---

Dorsetshire, the numbers counted in 1868, as I learn from Mr. Alfred Newton, amounted to 856, of which he saw over 230 when visiting the spot in 1869.

* "In the Coke reports (see "The Swan case") the ancient penalty for taking swans' eggs is thus given under the statute of II. Hen., 7, cap. 17 :—" He who stealeth the eggs of swans out of the nest shall be imprisoned for a year and a day and fined to the King, one moiety to the King the other to the owner of the land where the eggs were taken." At the present time, by an Act passed in the reign of William IV. (1 and 2 Wm. IV., c. 32, s. 24), the taking or destroying the eggs of any bird of game, or any Swan, wild duck, teal, or widgeon, by any person "*not having the right of killing the game upon any land*, nor having permission from the person having such right," shall, on conviction, pay for every egg so taken or destroyed, "such sum of money not exceeding 5s. as may be decided by justices, together with the costs of conviction."

Mr. Cordeaux in his "Birds of the Humber District" (p. 100, note) quotes from "Thompson's History of Boston" the following Fen laws, passed at "the court view of free pledges, and court-

majority of these birds are congregated with their cygnets throughout the summer, either on the Broads themselves, or that part of the main stream which lies between the Surlingham and Buckenham Ferries, a far narrower limit of space, within which, owing to the shallowness of the Broads and their connecting channels, a taste for fish spawn can be abundantly gratified. If the over-yeared birds, also, which remain on the river, though not paired for the season, still retain a taste for spawn-diet, a school of from twelve to fourteen young swans, such as I have before referred to at Coldham Hall, with no other occupation throughout the summer months than that of seeking the choicest feeding grounds, would unquestionably be "all there" when the spawn is thickest in the dykes; a question not unworthy the consideration of our Angler's Society, though it is difficult to suggest a remedy, for if these swans were removed to any other part of the stream during the spawning season, they would be sure to find their way back.

One other charge I have heard made against the tame swan, which affects the sportsman only, is that of disturbing wild fowl on the same waters, but this, I think, is not "proven" to any appreciable extent. Occasionally a swan may be seen to peck at and chase a duck that may have approached too near in the act of feeding, but in severe winters, when the Broads and other shallow waters are covered with ice, the "wake" kept open by the constant paddling of the swans is a great attraction to the fowl, and not less so the food which in such hard times is supplied by the marshmen to the swans under

---

lect of the East, West, and North fens, with their members, held at Revesby, 19th of October, 1780," to the effect that "no person shall bring up or take any *swan's eggs* or crane's [heron's?] eggs or young birds of that kind, on pain of forfeiting for every offence three shillings and fourpence."

their care. Sometimes, however, the too close vicinity of the fowl has caused the death of a swan, some reckless sportsman bagging more than he intended.

Swans have been said to live to a very advanced age, but I have failed to obtain any confirmation from local authorities of the centenarian instances recorded by some authors.* A tradition certainly exists in these parts to the effect that a swan which formerly frequented the Lakenham river was not less than one hundred years old, but from thirty to forty years is the most that I can arrive at from the experience of the oldest swanherds living, or from the "hearsay" evidence of their predecessors. The oldest birds at the present time on the Yare are not more than from ten to fourteen years old, but a female was recently removed from the Yare, which had nested regularly for some twenty seasons. No doubt on private ponds or lakes, where they are kept more for ornament than profit, some birds may be found which are known to be of great age, but on the Yare the experience of the past few years has proved that comparatively young birds and the introduction of fresh blood now and then is most conducive to a large head of cygnets.

The term "mute," as applied to this swan, is scarcely appropriate, since, though wanting the sonorous note of the wild swan, it has various utterances, including the hiss of anger and defiance, uttered with the bill open, a strange yapping noise like the bark of a small dog when endeavouring to draw the cygnets from the

---

* Mr. Broderip in his "Zoological Recreations" quotes, at length, from the "Morning Post" of July 9th, 1840, an account of the death (accidentally), at the age of seventy years, of a venerable swan, known as "Old Jack," on the waters of St. James's Park. This bird is said to have been hatched about the year 1770, on the water attached to Old Buckingham House, and to have been a favourite with Queen Charlotte in its earlier days.

nest, the ordinary call or croak, and the more harmonious double note with which it responds to the low soft whistlings of its young in summer. The first of these, though frequently heard by day, is particularly striking on a dark or foggy night, when the presence of a family group, disturbed by some passing boat, is thus indicated. If not mute, however, when living, we have Waterton's experience to refute for ever the well known fable of its powers of song when dying. Alas! for that pretty fiction of mythology, and Tennyson's elaborately descriptive poem on the dying swan, wherein we learn that—

> "At first to the ear,
> The warble was loud and full and clear;
> And floating about the under sky.
> * * * *
> But anon her awful jubilant voice,
> With a music strange and manifold,
> Flow'd forth on a carol full and bold,
> As when a mighty people rejoice."
> * * * *

till at length the willows, mosses, and soughing reeds,

> "And the silvery marish flowers that throng,
> The desolate cricks and pools among,
> Were flooded over with eddying song."

Alas! I repeat, for the poet's too vivid imagination, the swan attended in its last moments by a matter of fact naturalist, "never even uttered" (to quote his own words) "its wonted cry, nor so much as a sound, to indicate what he felt within."

This is, moreover, one of the "vulgar errors" discussed by Sir Thomas Browne in his *Pseudodoxia epidemica*, published in 1646, and to which he sums up his objections in the following terms:—"When, therefore, we consider the dissention of authors, the falsity of relations, the indisposition of the organs [the conformation of the windpipe] and the unmusical note of all we ever beheld or heard of, if generally taken, and comprehending all swans, or of all places, we cannot assent thereto."

SWAN "UPPING" ON THE YARE.

The annual swan "upping" or "hopping," as it is variously termed, is fixed by ancient custom on the second Monday in August, and but once in the last twenty years has it been postponed through a cold backward season. Whatever may have been the pageantry of the occasion in olden times—or at least prior to the passing of the Municipal Reform Act, in 1835, and the abolition of the Guild-day and other festivities—of late years, the proceedings have had no special feature beyond the swan-hunt itself; and even the time-honoured dinner at Coldham Hall to the swanherds and other corporate officials, at which the Mayor occasionally presided, has been discontinued, I believe, since 1845. In this strictly utilitarian age, also, the office of swanherd to the corporation was not likely to pass long unchallenged, and from somewhere about the above date, no such official has been elected by the Town Council, the corporation swans having been placed entirely in the hands of Mr. Simpson, the governor of St. Helen's Hospital* in this city, who, from the Swan-pit, on those

---

* St. Helen's Hospital, or Almshouse, for aged men and women, in Bishopgate Street, known also as the Great, St. Giles', and the Old Men's Hospital, occupies the site of the dissolved hospital of St. Giles', founded by Walter Suffield alias Calthorp, Bishop of Norwich, in 1249, but which, in 1547, was granted by Edward VI., in accordance with the will of his late father, to "the mayor, sheriffs, citizens, and commonalty," with all the revenues belonging thereto, "to be henceforward a place and house for the relief of poor people, and to be called God's House, or the House of the Poor in Holmstreet."—See "Blomefield's History of Norwich," 8vo, vol. ii., pp. 376-391.

premises, supplies the tables of the Mayor and other members of the Corporation, as well as the many private owners who consign their cygnets to his care, and most effective system of fattening. In thus farming, as it were, the swan-rights of the Corporation and others, Mr. Simpson employs his own swanherd, a man named Steward acting in that capacity, who keeps a register of the eggs laid, and the ownership, by their marks, of the swans which have paired for the season. Rich, of Surlingham, who is also engaged by certain private owners to look after their swans, has a general supervision of the eggs on the broads but not on the river. The keepers of other proprietors look after their own birds, and for such attention the watchers are paid two shillings and sixpence for each cygnet. The "upping" commences with an early breakfast at Buckenham "Horse Shoes,"* a time-honoured rendezvous, where Mr. Simpson and his men are met by the keepers of such noblemen and gentlemen as have swan-rights† on the stream, each of whom takes up and marks his own cygnets, and is responsible for their safety. Two men to a boat is the usual complement—one to row and the other to seize and pinion the birds, and the first catch is generally made at Hassingham, below Buckenham, but occasionally a pair of swans have nested as far down as Cantley, and on one occasion, recently, on the banks of the Chet, at Hardley Cross.

---

* Since this was written the "Horse Shoes" has been converted by the owner, Sir Thomas Beauchamp, into cottages, and the meet is now held at the "Ferry House" on the opposite side of the river.

† Besides the Mayor and Corporation, the Bishop of Norwich, and the Trustees of St. Helen's Hospital, the Duke of Norfolk, the Earl of Rosebery, the Marquis of Lothian, Sir Thos. Beauchamp, Bart., Mr. C. E. Tuck, Mr. Pratt, of Surlingham, and Mr. Gilbert, of Cantley, have swan-rights on the Yare between Thorpe and Reedham.

Being desirous of witnessing, for once, this novel scene, I arranged with my friend Mr. Thomas Southwell, to join the "upping" party at Buckenham, in August, 1871, and, with John Trett as boatman and general referee upon all points affecting the swanherd's craft, we watched the sport from beginning to end with considerable interest. On this occasion, as no birds had located themselves below Buckenham, the small fleet of boats passed up-stream from the "Horse Shoes," and the first "take" occurred in Rockland Dyke. A stern chase is proverbially a long one, and both in the river and the adjoining dykes, when once the old birds became aware of pursuit, it usually took some time before the family group could be surrounded by the boats, or, when thus "headed," driven out on to the bank or "rond." In the water the capture of both swans and cygnets is effected by means of a long pole with a kind of iron "note of interrogation" at one end, after the fashion of a shepherd's crook, and this being passed round the neck of the bird it is, with gentle handling, drawn to the boat side and secured. Accidents, however, may happen by this process, and in unskilled hands the use of the crook has occasionally resulted in a sudden dislocation of the neck. On shore the capture of old and young is comparatively easy. The cygnets, if left quiet for a few minutes, will lie down in a group, and are then taken by hand with little trouble, and should the ownership of the old swans be at all doubtful, or the marks on their beaks require recutting, they are run down, and when boldly seized by the neck, after the swanherd's fashion, are soon *hors de combat*, though a novice, however powerful, would probably receive some hard knocks in the struggle. I must own I expected to find the old swans much more resentful at the wholesale seizure of their young, and more clamorous and persistent in following the boats than, as a rule, they proved to be. The whistling of the captive cygnets usually led the parents to follow for a time, but they soon fell behind, and later in the day were seen feeding very contentedly, and apparently quite resigned to their loss; old couples, possibly, having a dim recollection of similar bereavements. On Rockland Broad, owing to the extent of water and the shelter of the large reed beds, the "stowing up" process was much more difficult, and nearly the whole morning was spent in pursuit of some two or three families. In one case, by a little quiet manœuvring, the old swans were induced to lead their young into a short dyke, on one side of the broad, where they were soon landed, but another old pair, perhaps cunning by experience and too closely pressed at first, drew their cygnets away into the thickest part of the reeds, where, one by one, they lay up by themselves, and the parents, having accomplished this clever ruse, passed out alone into the open water. Nearly two hours had been lost in this fruitless chase, and, after all, these cygnets had to be left until the second day, as we were told they would not quit their hiding places till summoned by the old swans, when all danger of pursuit seemed over. It is from such causes and from such localities that cygnets at times escape altogether, and, being unpinioned, take flight in winter during severe weather, and are shot as strangers on other inland waters, or occasionally even on the coast. The cygnets as soon as they are taken up have their legs and feet turned over their backs and tied with soft strips of list; they are then laid at the bottom of the boats on their breasts, and on hot days should always have wet reeds or rushes to lie upon, as otherwise the heat of the sun and the privation of water for a long period is very exhausting. Indeed, considering the number of hours that elapse between the first "take" in the morning and the arrival of the Corporation birds at St. Helen's pit between nine and ten the same evening, it is somewhat remarkable that all should survive the ordeal. The old birds are caught as well as the young ones whenever the ownership of the pair is doubtful, and their marks are then examined and recut if necessary, but as the cock swans select their own partners, it commonly happens that the parent birds belong to different owners, in which case, as by law established,* the cygnets are equally divided if in even numbers, if not, the re-

---

* Folkard, in his chapter on "Swan Laws" in "The Wild Fowler" (p. 208), alludes to this point, as "decided in a very old case—Lord Strange *versus* Sir John Charleton (Year Book, 2 Richard III., p. 15)—which is referred to by Sir Matthew Arundel in his judgment on 'The Swan Case.'" To this I may append the following amusing extract from "The Case of Swans" (Coke reports, fol. 1680, part vii., p. 465):—"The Lord Strange had certain swans which were cocks, and Sir John Charleton certain swans which were hens, and they had Cignets between them, and for these Cignets the owners did join in one Action; for in such case by general custom of the Realm, which is the Common Law in such case, the Cignets do belong to both the owners in common equally, *scil.* to the owner of the Cock, and the owner of the Hen; and the Cignets shall be divided betwixt them. And the Law thereof is grounded upon a reason in nature; for the Cock Swan is an emblem or representation of an affectionate and true husband to his wife above all other Fowl; *for the Cock Swan holdeth himself to one female only;* and for this cause nature hath conferred upon him a gift beyond all others; that is, to die so joyfully, that he singeth sweetly when he dieth; upon which the Poet saith—

Dulcia defectu modulatur carmina linguæ,
Cantator, cygnus, funeris ipse sui, &c.

And therefore this case of the Swan doth differ from the case of Kine, or other bruit beasts." A curious instance this of a legal mind, led astray by mere poetical fancies.

For other points of law, with reference to the ownership, &c., of swans and swan-rights, see Folkard's admirable epitome in the chapter above referred to; and the account of the Mute Swan in the third volume of Yarrell's "British Birds."

presentatives of each proprietor toss for the odd bird. Such, at least, is the custom on the Yare, as it could never be amicably settled whether the cock or the hen* should carry the extra cygnet. A curious and most unusual circumstance, however (entirely upsetting Sir Edward Coke's theory as given in the preceding footnote), has occurred on the Yare within the last five years, two cock swans having each paired with two hens. In the one case a hen died in the first or second year of this double pairing, and the cock bird remained faithful to his surviving mate; in the other the bigamous alliance continued for four seasons with the following result as to cygnets—the above facts being vouched for both by Steward and Rich:—

    1870. Seven cygnets—one hen five, the other two.

    1871. Eight cygnets—one hen seven, the other one.

    1872. Thirteen cygnets—one hen eight, the other five.

    1873. None; the eggs dropped about and destroyed.

In the latter instance the cock bird belonged to the Corporation, the two hens to Mr. Gilbert, of Cantley, and the cygnets were divided as equally as possible.

The swan-mark, or *Cygninota*,† as Sir Edward Coke terms it, is cut with a knife in the upper surface of the beak, and has a raw, pinkish hue when freshly done, especially on the blue beaks of the birds of the year. Some swanherds are said to have rubbed in Indian ink or wet gunpowder, to render the marks more lasting,

---

    * The terms Cob and Pen in swanherds' language denoting the male and female swan, respectively, are not in use on the Yare, though they may be in other parts of the county.

    † The swan-marks, as stated by Yarrell, " consisted of annulets, chevrons, crescents, crosses, initial letters, and other devices, some of which had reference to the heraldic arms of, or the offices borne by, the swan owners." As shown by his illustrations, some are much more elaborate in design than others, but simple forms are most used on the Yare.

but this is not done at the present time, and as a rule the marks are not cut so deep as formerly, and therefore are more often renewed. Letters or other signs, however deeply incised upon the upper mandible, will wear out in time, but "nicks"* cut in the hard edge of the mandible are never lost. Other private marks are occasionally resorted to, as holes stamped through the web of one foot, or the removal of a hind claw, right or left; and, in some cases, in marking and pinioning cygnets, the pinion is removed, right or left, in accordance with the marks on the beak.

A particularly favourable season had produced at the "upping" of 1871 more cygnets than were required for fattening, and the surplus, therefore, having been marked and pinioned were turned off again with their parents. In following in the wake of the other boats, we were now and then aware of the rough surgery that had taken place by seeing the fluffy pinions of the cygnets floating down the stream, the amputation being effected at what may be termed the elbow-joint of the wing, and on subsequently watching the operation performed upon five of them reserved for fattening in a private pond, near the river, I could have wished the knife had been sharper or the hands more skilful as each bone was turned out of its socket with a most unnecessary amount of bleeding.† Cygnets when thus pinioned take much longer to fat, as they

---

* Both in Norwich and Yarmouth we have the old public-house sign of the swan with two *necks*, which, as Yarrell satisfactorily shows, is a corruption of the swan with two *nicks*, in allusion to the swan-mark of the Vintners' Company.

† Mr. Dixon remarks "a skilful operator will feel for the joint, divide the skin, and turn the bone neatly out of the socket. I will allow him to shed just one drop of blood; no more. ° ° ° Many cygnets are annually killed by the clumsy way in which the wing is lopped off. They suffer from the shock to the nervous system as much as from hemorrhage."

are liable to pine from the effects of such treatment, and so lose rather than gain flesh for a time. It is on this account that the Corporation and other cygnets under Mr. Simpson's charge have only the quill feathers of one wing clipped or plucked, which answers all the purpose for the short period that they mostly remain in the swan-pit; but in more than one instance, I believe, a cygnet has been known to escape when kept till it had re-moulted its quill feathers, and, suddenly taking wing, has exercised its new powers of flight over the barrack-yard and the adjoining river.

The first day's "upping" terminates generally between four and five in the afternoon, at Coldham Hall, where the swanherds and keepers refresh themselves before conveying the cygnets to their various destinations; the Corporation birds, especially, having still a long and weary journey to Norwich without food or water, their legs stiff and cramped as they lie bound at the bottom of the boats, writhing their long necks in fruitless restlessness, or venting their discontent in low and plaintive whistlings.

The second day's "upping" comprises the cygnets reared on Surlingham Broad and the banks of the river between Coldham and Thorpe, and late hatched birds, too young to leave their parents in August, are "taken up" later in the season, if required. For reasons, however, before stated, so prolific has been the supply of cygnets during the last few years, within the corporate boundaries, that although from one hundred to a hundred and fifty young birds have been taken annually, several broods have been left on the stream not required for culinary purposes. Of breeding swans, on the Yare, two pairs or five birds appear to be commonly allowed to each right, but the Corporation has from three to four pairs, which can scarcely be considered an excessive number.

On the Wensum, above the New Mills, usually termed the "back" river, the "upping" takes place on the last Monday in August, and many of the cygnets reared on that stream, about Hellesdon, Ringland, and Bylaugh, or on private waters in that neighbourhood, are also sent by their respective owners to be fattened in

ST. HELEN'S SWAN PIT.

This receptacle for the young birds during that brief period of their existence, in which alone they are esteemed for edible purposes, is situated at the back of the Hospital premises, and separated only by a meadow from the river from which it receives its supply of water, subject to tidal influences. The pit itself is formed of brickwork, and is about thirty-five yards long by eleven or twelve in width, with wooden feeding troughs so arranged along three sides of the water as to rise and fall with the tide. At one end a wooden staging assists the cygnets to land at pleasure, and this communicates with a small square yard, laid down in grass, where the birds retire to rest and preen themselves, and in which they are caught more easily when required; whilst a weighing machine and weights, close at hand, which complete the accessories, have a decidedly ominous and sacrificial effect. This part, however, for the sake of cleanliness, is shut off from the grass walks running the whole length of the pit from which visitors are accustomed to view the swans.

Within this tank, which has held from eighty to one hundred cygnets at a time, the fattening process is carried on which has for years, in a gastronomical sense, rendered Norwich swans almost as famous as Norfolk turkeys; and never more so than under Mr. Simpson's treatment, during the last twenty-seven years, to whom my thanks are due for much valuable information.

When first brought to the Swannery the cygnets are plentifully supplied with grass, as the best substitute for the diet they have been accustomed to on the broads and rivers, and by degrees they are led to feed on the grain placed in the troughs, with a daily supply of grass and vegetables grown specially for their use in an adjoining garden. Each cygnet consumes nearly a coomb of barley* in the process of fatting, and latterly Mr. Simpson has found Indian corn, well soaked, an excellent addition to the dietary. The cygnets, which come to the Swannery the second week in August, begin to be fat in October, and keep improving up to December, but if kept beyond that time they fall off so rapidly in condition that, according to Mr. Dixon, "a bird weighing twenty-eight pounds before Christmas has been known to shrink to seventeen or eighteen pounds by the end of January, in spite of high feeding." Male cygnets in prime condition, when thus fattened, average about two or three and twenty pounds in weight, in the feathers; and when dressed for table the largest would weigh about sixteen pounds, or from twelve to fifteen pounds on an average.† Between the males and females, when home fed, there is about the same pro-

---

* In olden times the fine for stealing a swan was paid in wheat, a custom thus referred to by Sir Edward Coke in "The Case of Swans":—"He who stealeth a swan in an open and common river, lawfully marked, the same swan (if it may be) shall be hung in a house by the beak, and he who stole it, shall, in recompence thereof, give to the owner so much wheat that may cover all the swan, by putting and turning the wheat upon the head of the swan until the head of the swan be covered with the wheat."

† The late Mr. Lombe, of Little Melton, is said to have had two swans cooked for his dinner parties, to ensure, I presume, an equal share of the prime parts to all his guests. The magnificent old cock swan in the "Lombe collection," at the Norwich Museum, which must have weighed considerably over thirty pounds, was bred at Melton, but there is no record of its age.

portionate difference in weight as in the case of the pair killed at Somerton in a semi-wild state. Various Royal and distinguished personages have from time to time received presents of cygnets from St. Helen's Swannery, amongst whom may be mentioned Queen Victoria, the Pope, and the late Emperor of the French; and a remarkably fine bird, weighing twenty-eight pounds, was sent to Sandringham, in 1873, as a present to the Prince of Wales.

Full grown cygnets at the "upping" are worth from ten to twelve shillings, and owners entrusting their birds to Mr. Simpson pay one guinea each for fatting. If purchased the price of a fat cygnet, ready for table, is two guineas.

Each bird sent out from the swannery is accompanied with the following poetical recipe* for cooking the same:—

To Roast a Swan.

Take three pounds of beef, beat fine in a mortar,
Put it into the Swan—that is, when you've caught her.
Some pepper, salt, mace, some nutmeg, an onion,
Will heighten the flavour in Gourmand's opinion.
Then tie it up tight with a small piece of tape,
That the gravy, and other things may not escape.
A meal paste (rather stiff) should be laid on the breast,
And some "whitey brown" paper should cover the rest.
Fifteen minutes at least ere the Swan you take down.
Pull the paste off the bird, that the breast may get brown.

The Gravy.

To a gravy of beef (good and strong) I opine
You'll be right if you add half a pint of port wine:
Pour this through the Swan—yes, quite through the belly,
Then serve the whole up with some hot currant jelly.

N.B.—The Swan must *not* be skinned.

---

* I only recently ascertained that these lines, which are quoted by Yarrell in his account of the Mute Swan, were written some forty years ago by a relative of my own, the Rev. J. C. Matchett, whose humorous rendering of the hitherto prosaic "instructions for cooking," was adopted by the authorities.

Mem. A swan of fifteen pounds weight requires about two hours' roasting, with a fire not too fierce.†

To this portion of the subject may appropriately be added the few references to swans in the "Northumberland Household Book," and in that of the L'Estranges of Hunstanton, from which it appears that as far back as Henry the Eighth's reign, cygnets were reckoned in season only towards the close of the year, or at feasts just before or after Christmas. In the Northumberland "Accounts" (p. 107) we find copies of the "Warraunts" to be "sewed out yerely" under the Earl's "Signet and Signe Manuell" to the keeper and under-keepers of the

---

† The following recipe for making Swan-Giblet Soup is also supplied to each purchaser:—

"Cut 6 lbs. of the knuckle of veal, and 1 lb. of lean ham in a large dish, add three onions, two turnips, one carrot, two heads of celery, a small piece of sweet basil, marjoram, thyme, parsley, and bay leaf, and a tablespoonful of salt. Butter a stew pan lightly, put in the whole of the ingredients, add five cloves, two blades of mace, and half a pint of water, stew it over a brisk fire about twenty minutes, when it becomes a nice light brown colour add eight quarts of water, directly it boils, place it at the corner of the stove, scald the giblets in boiling water, take them out and cut them into joints, the gizzard in four pieces, put them into the stock, and let them simmer gently until they are quite tender, take them out, strain the gravy through a cloth, skim off every particle of grease, put it into a clean stew pan with the giblets, and thicken it with arrowroot dissolved first in cold water, but do not make it too thick, finish by adding half a pint of sherry, the juice of half a lemon and two grains of cayenne."

With this modern recipe for cooking a swan may also be contrasted one amongst the recipes of the master cooks of Richard II., as published in Broderip's "Zoological Recreations," and termed "Swann with Chaudron." "Take the liver and the offal (that is the giblets) of the swans, put it to seethe in good broth, take it up, take out the bones, and 'hewe' the flesh small. Make a mixture of crust of bread and of the blood of the swan sodden, and put thereto powder of cloves and pepper, wine and salt, and seethe it, cast the flesh thereto 'hewed,' and 'mess it forth' with the swan."

"Carre of Arrom" to deliver to the "Countroller" of the Household and the "Clarke of the kitchinge,"

"Against the Feest of Christynmas next comynge Twenty Signetts to be taken of the breed of my Swannys within my Carre of Arrom, within my Loordeship of Lekingfeld within the Countie of Yorke."

whilst the following memorandum gives the date of issuing the warrants and the special days for serving the swans:—

"Item a Warrannt to be sewed out yerely at Michaelmas for XX Swannys for th'expencez of my Lordes hous, as too say for Christynmas Day v—Saynt Stephyns Day ij—Saynt John Day ij—Childremass Day ij—Saynt Thomas Day ij—New Yero Day iij—ande for the xijth Day of Chistynmas iiij Swannys."

The order for feeding the swans and the payment for the same is prescribed as follows (p. 103):—

"Item it is thought goode that my Lordes Swannys be taken and fedde to serve my Lordes hous and to be paid fore as thay may be bought in the Countre seynge that my Lorde hathe Swannys inew of hys owne."

In the L'Estrange Accounts, amongst the gratuities to servants for bringing presents we have two entries in the years 1519 and 1520:—

"Itm to Mr. P'Or of Castleacre svnt for bryngyng ij fatt Swannes—viij^d."

"Itm to the P'Or of Castleacre svnt in reward for bryngyng of a Cygnett—iiij^d."

Again, in 1526, amongst the articles of "gist" and "store" (paid in lieu of rent or home produce supplied to the house) we find

"It. a swanne and vij conyes of store."

and, as nearly as I can reckon, the dates from the peculiar manner in which the accounts are kept, the last gift was received in the first week of January,*

---

* Mr. Broderip, in his "Zoological Recreations" (p. 159), states that at two wedding feasts, given by Sir John Nevile, of the County of York, on the 14th of January, in the seventeenth year,

and the two former in the last week of January and last week of December respectively.

Strangely enough not one of our local historians from Blomefield (1741) downwards makes the slightest reference to St. Helen's Swan-pit, and I had almost despaired of being able to give any information as to its origin or past history when, on recently searching for some other entry in the books of the old Corporation, relating to the Hospital Trust,* Mr. Simpson discovered a minute to the effect that about May, 1793, the late Mr. Thomas Ivory *constructed a new swan yard*, and made other improvements on the premises in consideration of the Hospital committee of the Corporation having agreed to add one acre of the Hospital meadow and garden ground to the premises then in his occupation, and now known as St. Helen's House. This entry, therefore, not only marks the date of the present swan-pit, but establishes the existence of a previous one, near the same spot, before the Municipal Reform Act of 1835, up to which date, as before stated, the management of the Hospital was vested in the Corporation.

It is also, I think, most probable that a Swannery in some form or other existed in the Hospital meadows even prior to the year 1547, when, on Blomefield's authority (as referred to in a previous note), King Edward the VI. granted "to the Mayor, Sheriffs, and

---

and on the 17th of January, in the twenty-first year of the reign of Hen. VIII., the bill of fare included Swans, "2 of a dish," at a cost of £2 12s. each. And on the same authority, at the Shrievalty dinners of the same Knight, at the Lent and Lammas Assizes, though, at the former, fish in abundance but no flesh, was provided, at the latter *twenty-two swans* were served, which, if Lammas Day was then the 1st of August, must surely have been over-yeared birds.

* These books, since the appointment of the Hospital Trustees in 1835, have been transferred from the Guildhall to the Board-room of the Hospital.

Commonalty of the city and their successors for ever, "all the Site, Circuit, Compas, and Precinct of the late Hospital of St. Gyles, wythyn the Cytic of Norwych, in the Paryshe of St. Elyn, nexte Bushhope Gate there, &c." (Hist. Nor., 8vo., vol. ii., p. 390); since, indirectly, the antiquity of this Swannery may be inferred from the allusions of the same author, to local swan-rights and marks.

NORFOLK SWAN MARKS.

In his chronicle of events in this city, for the year 1482, Blomefield remarks (vol. i., 8vo., p. 170):—

"This year [22 Edw. IV. c. 6.] was the statute of qualification for swan-marks made, by which it was enacted, that no person whatever, except the King's son, should have any swan-mark or game of swans of his own, or any other to his use, except he hath freehold lands and tenements to the clear yearly value of five marks,* and all persons not so qualified shall, before Michaelmas next, sell or give away such marks and game of swans to such people as are qualified, and after that time, any person qualified may seize such game undisposed of, and he shall have half and the King the other half; upon which statute an account of all the swan-marks in this county was taken and entered in a roll, *which was renewed in the year 1598, when the order for Swans was printed; the city being then seized, according to the swan-rolls, of three swan-marks belonging to the late dissolved Hospital of St. Giles.*"

---

* Some idea of the number of tame swans kept prior to that date in this country may be formed from the fact that the famous bill of fare at the "intronization" of George Nevell, Archbishop of York, in 1464—only eighteen years before the passing of this Act, and in the reign of the same Monarch—included *four hundred swans*, with a like abundance of all kinds of provision. It appears, moreover, from the preamble to this statute that this law was passed owing to the frequent robbery of cygnets by unprincipled swan-keepers, who placed their own marks upon them to avoid detection. Nor was such robbery confined in those days to the lower orders, as, in the "Paston Letters" (Fenn's Ed., 1789, Letter xxiv.), we find Sir John Falstolf, in the twenty-ninth year of the reign of Henry VI. (only thirty-two years before the passing

On the same authority, also (Hist. Nor., 8vo., vol ii., p. 389), we learn that in 1532,

"The hospital (St. Giles') leased the site of their manor of Rokels in Trowse with the dove-house, &c., and a faldcourse in Trowse and Bixley, and three hills of bruery called Blake-hills, *with Blake's swan-mark thereto belonging.*"

To which statement is appended the following important note :—

"The city have three swan-marks on the narrow fresh water streams in Norfolk, one called Blake's mark, belonging to the manor of Rokele's in Trowse, another called Paston's, or the Hospital mark, which belonged to Margaret, widow of John Paston, Esq., daughter and heiress of John Mautby, Esq., which she gave to Edmund, her second son,* and it was then called Dawbeney's mark, and was late Robert Cutler's, clerk; and in 1503, Geffery Styward settled it on Cecily his wife, for life, and then on his eldest son, who gave it to the city. The third is called the city mark, and formerly the King's mark, and was conferred on the city by Sir John Hobard in the grand rebellion; in 1672 they had 72 swans belonging to the three marks, and the city always appointed a swanner to look after them, and paid an annual stipend to him for so doing.

---

of this Act), addressing an urgent request to his "right trusty friend and servant, Sir Thomas (Howys), parson of Castlecombe; and John Bocking, at Prince's Inn, in Norwich, or at Beccles," to attend the "Oyer and terminer," about to be held at Beccles, to prefer an indictment against "Sr John Bukk, prson of Stratford, and one John Cole, for having (to use the knight's own words) " by force, this yere and othyr yeers, taken out off my waters, at Dedham, to the nobre of xxiiij Swanny's & Signetts, & I pray you this be not forgeted."

* Mr. J. H. Gurney has kindly directed my attention to the will of the above Margaret Paston (communicated to "Norfolk Archæology," vol. iii., p. 160, by the late Mr. Dawson Turner), from which it appears that Blomefield is not quite correct on this point, since, as shown by the following extract, this lady bequeathed her swans to her grandson Robert, the son of her youngest son Edmund Paston.

"It. I geve and gaute to Robt., son of the seid Edmund, all my Swannes, merken with a merke called Dawbeney's merk, and with

For some two or three hundred years, therefore, we may trace a close connection between the swan-rights and marks of this Hospital and the Corporation, whilst a careful search amongst their archives might probably elicit some further particulars with reference to the swannery itself, in connection with swanherds' salaries, and the fatting of cygnets for our civic banquets. It is, moreover, worthy of notice that in the list of payments annually made out of the revenues of St. Helen's Hospital as quoted by Blomefield for the year 1728, and still continued, is the following item (vol. ii., p. 397) :—

"To the Chamberlain for a year's rent of the Swan-bank, £1 10s."

This is further alluded to in his description (vol. ii., pp. 102-6) of the Rectory of St. John the Evangelist, in Southgate, and the founding of "Cooke's Hospital," near the former site of St. Vedast's or St. Faith's church, in the following terms :—

"In this parish is the island in the river called the *swan-bank*, and several *bitmays* or pieces of land gained out of the river, which pay small rents to the city."

In the engraved "Plan of the City of Norwich," published by Blomefield in the second volume of his History of Norwich, several swan-marks, belonging to the City and the See, are figured with the seals and regalia of the Corporation. These, as numbered on the plan, are thus described :—

142.* The Swan Mark belonging to the See.

---

the merke late Robt. Cutler, Clerk; to have, hold, and enjoye the seid Swannes with the seid merkes to the seid Robt. and his heires for ev'more." This will was executed in 1481 and proved in 1484.

John Dawbeney, Esq., as Mr. Turner states in a foot-note, was called by the Pastons "our cousin"; the Rev. Robt. Cutler was vicar of Caister St. Edmund from 1453 to 1466, and was translated to Mautby in 1465 (vide Blomefield.)

* This is the Bishop's own mark, as figured by Yarrell (No. 10),

143. St. Bennet's Abbey Swan Mark, now the Bishop's.
144. St. Bennet's Cellarer's Mark, now the Bishop's.
195. The Swan Mark of Carrow Abbey.
196. The City Swan Mark.
197.*St. Giles' Hospital Mark, now the Mayor's.
198. Rockel's Manor Mark, late the Hospital, now the Mayor's.
199. The Hospital New Mark, now the Mayor's.
200. The Prior's Old Mark, now the Dean and Chapter.
201. The Prior's New Mark, now the Dean and Chapter.

There is little doubt, I think, that the marks thus figured by Blomefield were taken from the ancient Swan-roll of 1598, to which he alludes in his remarks on the statute of Edward IV., but as this most interesting record is at the present time missing from the muniment chest of the Corporation I am unable to compare the two, or give illustrations, as I desired, from the original. Mr. Mendham, who has filled the office of Town Clerk since 1857, assures me that he never had

---

the two other marks he holds by virtue of his office, as Abbot of St. Bennet's.

Apropos of Swans and Bishoprics, a curious passage occurs in Symmons's edition of Milton's prose works (vol. i., p. 15), under the title of "Reformation in England." After stating his opinion of what Bishops ought to be in their private lives, Milton concludes as follows:—"What a rich booty it would be, what a plump endowment to the many benefice-gaping mouth of a prelate, what a relish it would give to his canary-sucking and swan-eating palate, let old Bishop Mountain judge." Referring, no doubt, to George Mountain or Montayne, who, between the years 1617 and 1628, was successively Bishop of Lincoln, London, and Durham, and Archbishop of York. (See "Notes and Queries," 4th series, vol. xii., p. 452.)

* This mark most resembles the one now in use for the Corporation, but that figured by Yarrell (No. 9) as the Norwich Corporate mark, does not answer to any I have seen on our local swan-rolls. Yarrell's figure No. 5 is, however, identical with our present Corporate mark, though described as belonging to Sir Thomas Frowick. Possibly, through a printer's error, the two woodcuts were transposed. No separate mark is now used by the Mayor.

P

this roll in his custody, but as a copy* (though I have reason to believe not a fac simile), was executed by Mr. Ninham, of this City in 1846, and was exhibited *with the original* at a meeting of the Norwich Archæological Society, on the 22nd of October in that year (as I find by an entry in the MS. "Proceedings"† of that date), we know that its abstraction must have occurred within the last thirty years. The long period, however, that has already elapsed renders its recovery the more difficult, and, to this end I am desirous of giving every possible publicity to the fact that a roll of so much value, if only in an archæological sense, both to the city and county, is not forthcoming.

From an inspection of such swan-rolls as I have, at present, had access to, it would seem that prior to the

---

* This copy was fortunately made at the suggestion of the late Mr. Thomas Brightwell to provide against the possible loss of the original, but when completed the Town Council declined to purchase it at the sum named by Mr. Ninham, and it subsequently passed into the possession of Mr. Osborne Springfield, of Catton. This roll, which consists of several folio sheets of vellum, bound in book form, represents the heads and beaks of swans, in profile, and not, as is more usually the case, a drawing of the upper mandible only, with the marks on its surface, as in Yarrell's illustrations. Blomefield's small figures are, however, also in profile (whether taken or not from the Corporation Roll), but in his the beak of the swan is represented open, and the marks are displayed on the upper mandible only, whereas Ninham represents each swan's head with the beak closed, and, by an error in drawing, places some of the marks on the lower mandible, an impossibility, as the lower mandible shuts close into the upper one, which leads me to question this being a fac simile of the Ancient Roll. The copy contains about forty-six distinct marks, now or formerly in use on the Yare and Wensum, besides several duplicates, and from these, through the kind permission of Mr. Springfield, I purpose making selections for an additional plate, should the Roll of 1598 not be recovered by the completion of this volume.

† See also "Norfolk Archæology," vol. i., p. 371.

dissolution of the monasteries* a large proportion of the marks in use belonged to the abbeys, priories, and other religious houses, whose lands, subsequently absorbed by the crown, or apportioned amongst corporate bodies (as in the case of St. Giles' Hospital), favoured nobles, or other private individuals, carried with them the swan-rights and other privileges attached to the soil. Even at the present time a swan-mark and free fishery in the river (Blomefield " Hist. Norw.," vol. ii., p. 530) belong to the site of the Priory of Carrow, or Carrow Abbey, as it is commonly termed, " as far as the bounds of Carrow extend." The Corporation of Norwich, as before shown, possesses three marks, acquired through different properties; and a small piece of land at Surlingham, part of " Nash's estate," now in the possession of Mr. Robert Pratt, has, by ancient custom, a swan-right and mark attached.

Of the three other corporate towns in Norfolk—Yarmouth, Lynn, and Thetford—the latter, according to Martin's history of that borough (1779) possesses three swan-marks, which he figures under the following titles:—"The Prior of the Canon's mark, now Lord Petre's; Binknorth's mark, now the Corporation's; and the Prioress's mark, afterwards Sir Richard Fulmerstone's, now Henry Campion's, Esq." He also figures two distinct marks as Sir Richard Fulmerstone's

---

* On an old and very interesting swan-roll, in the possession of the Rev. T. J. Blofeld, of Hoveton, are the following marks, belonging to ecclesiastical foundations in Norfolk.—Bishop of Norwich, Prior of Norwich, Abbott of St. Bennet's, Abbott of Langley, Prior of Bramerton, Prior of Hyngham, Prior of Carrowe, Ospitall of Norwich, Convent of St. Bennett's, Celler of St. Bennett's, and Prior of St. Olaves. This roll, commencing with the King's mark and those of the Dukes of Norfolk and Suffolk, contains about thirty private swan-marks, belonging to various Halls, chiefly situated in the vicinity of the Bure and its tributaries.

and the "Channon's mark of Thetford," from a paper in his possession, which are said to "occur in an original deputation, dated May 24, 1576, made by Edward Clere, of Blickling, Esquire, chief swanner for the Queen's Majestie in Norfolk and Suffolk, to John Tirrell, Esq."

Yarmouth, so long back as 1583, was granted a swan-mark by the Crown, as will be seen by the following note, in Mr. Charles Palmer's "Continuation" of "Manship's History of Yarmouth" (p. 75), the only reference to the subject in that work:—

"In 1583 Mr. Loveday was ordered to obtain a swan-mark from Her Majesty's Swanner. In 1633 a swan-mark was granted to William Corbett, marshal, to mark cygnets for the use of the town; and in 1638 he was paid his charges for marking cygnets with the town's swan-mark. In 1641, the swanard was directed to bring, yearly, a note of all swans marked for the town, and to receive 10s. therefore."

This note is appended to a "Relation" by the bailiffs of the Cinque Ports of the events which occurred on their visit to Yarmouth in 1603, when, amongst other courtesies received by them from the bailiffs of Yarmouth, and duly acknowledged, is a present of "a fatt Swane." Mr. Palmer has never been able to meet with a drawing of the above mark, but on a curious vellum roll in the possession of Mr. Robert Fitch, of Norwich, is the swan-mark of Lord Yarmouth, with several others, all connected with this county. This mark, no doubt, belonged to the extinct Earldom of Yarmouth, as, in Manship's "Hist. of Yarm." (p. 329), and Palmer's "Continuation" (p. 251), I find that the son of Sir William Paston, of Oxnead, (who died in 1610) was created Baron Paston and Viscount Yarmouth in 1673, Lord Lieutenant and Vice-Admiral of Norfolk in 1676, and Earl of Yarmouth in 1679. He was also High Steward of that Borough in 1684, when a new charter was granted substituting a Mayor for the two former Bailiffs. His son, the second Earl, died in 1732, without male issue, and the title became extinct.* The roll on which this mark appears is in book form, with the date 1693 on one of the leaves. The front page has also an elaborate coloured drawing of a swan, wearing a collar and chain, no doubt indicative of its domestic state, above and below which are the names and dates—Robert Breese, 1728; Robert Copeman, 1758—probably former owners of the roll.

In the late Mr. Dawson Turner's illustrated copy of Blomefield, in the British Museum, comprising some sixty volumes, quarto (vol. xv., p. 145), there are also, amongst other local swan-marks, one termed Sir William Paston's mark, another "Yarmouth of Blundeston, now Sidnor;" and a third, "The Prior of Yarmouth, now Gostlin." Neither at Yarmouth or Thetford are swans kept at the present time by the Corporations.

The ancient borough of "King's Lynn," strangely enough, seems never to have possessed swan-marks amongst its other rights and privileges, as no allusion to either swans or swan-rolls appears in the late Mr. Harrod's excellent repertory or index to the Corporation records, nor in his separate index to the ancient minute books. Had there existed any such record Mr. Harrod, who was much interested in the subject of swan-marks, would certainly have alluded to it in his report. Mr. George Webster, of Lynn, who has kindly furnished me with the above information, also informs me that although a few swans are now kept by the Corporation on the Gaywood river and Kettle Mills ponds, it is a custom of late years only, and he is doubtful if the birds are marked at all. The following order of the Town Council in reference to these swans, was made as recently as the year 1870†:—"The stock of swans and cygnets on the waters in the public walks

---

* At the present time Lord Yarmouth is one of the titles of the Marquis of Hertford.

† See "Norfolk Chronicle," February 12th, 1870.

and Gaywood river, having increased to an inconvenient extent, the treasurer was authorised to sell a portion or exchange them for other aquatic birds."

In connection, however, with this part of the county are the swan-marks, five in number, of the Fincham family, figured in Mr. Blyth's* history of the village and parish of Fincham, and which are presumed to have belonged to members of that ancient family, resident at Outwell, between the years 1566 and 1624. These marks, as stated by Mr. Blyth, were taken from a MS. in the possession of the late Mr. A. H. Swatman, of Lynn, entitled "The laws, orders, and customs for swans taken forth of A. Booke which ye Lord Buckhurst deliver'd to Edward Clarke of Lyncoln's Inn to revise: Anno Elizabethæ 26°" (1584), and it appears, from the names entered in the book, that it "appertained to the district watered by the rivers Nar, Ouse, Nene, Wissey, &c."

The swan-mark of the Gurney family, as figured by Mr. D. Gurney, of Runcton, near Lynn, in his "Record of the house of Gournay," was taken, as stated by the author, from a swan-roll headed "Carolo Wyndham Equiti depinxit John Martinus, A.D. 1673," but supposed to be a copy from a more ancient swan-roll† relating

---

* "Historical notices and records of the village and parish of Fincham, in the county of Norfolk." By the Rev. William Blyth, M.A., Rector of Fincham and Rural Dean, 1863.

† This probably refers to an ancient swan-roll in the possession of Sir Thomas Hare, of Stow. Besides those previously mentioned, I have recently examined in the British Museum library a small folio volume, purchased at the late Mr. Dawson Turner's sale in 1860, in which some five different rolls of considerable antiquity have been bound up together under the title of "Swan-marks used by the Proprietors of Lands on the rivers Yare and Waveney, preceded by the order for swan Botes established by the Statutes for the Realm of England." I have also to thank the Rev. H. Evans

to Norfolk, "as it contains the mark of Carrow Abbey, dissolved in 1537, and of other religious houses." This swan-mark also appears in Mr. Blofeld's roll, as "Mr. Gurney, Barsham;" and in the ancient rolls in the British Museum, before referred to, the same mark, twice figured, is assigned to William Gurney (written Gournay in one instance), no doubt the same Mr. Gurney, of West Barsham, who had a residence, I am informed, adjoining the river at Pockthorpe, Norwich, and whose will is dated 1507.

CYGNUS IMMUTABILIS, Yarrell.

POLISH SWAN.

Since the so-called Polish Swan was distinguished by Yarrell in 1838,* from the mute swan *(Cygnus olor)* under the specific appellation of *immutabilis* or changeless,— the cygnets, as well as the old birds being described as pure white,—several specimens have, from time to time, been procured in Norfolk, and of these, most of

---

Lombe, of Great Melton, for the opportunity of inspecting two ancient rolls of Norfolk swan-marks, containing those of the Melton and Bylaugh estates, very similar in character to Mr. Blofeld's. One of these has a crown printed above the first swan's head, indicating, no doubt, the direct grant from the crown of such marks, and of which the King's swanherd was bound to keep a list. In this roll all the marks given in Mr. Blofeld's appear in almost the same order of succession, but with the addition of forty-six others. The other Melton roll commences with the date 1674, and has a memorandum at the foot, to the effect that it was given to the individual whose name is signed (now illegible) in the year 1751. but this I imagine, from the writing, is of more recent date than that first mentioned.

* "Proceedings of Zoological Society," 1838, p. 19; and "Annals Nat. Hist." for 1839, vol. ii., p. 155.

the earlier examples were identified by Yarrell himself. Whilst such occurrences, therefore, entitle it to special notice in this work, and the question whether it is or is not a good species, still remains *sub judice* with our best ornithological authorities, I do not hesitate to retain it in the position assigned it by Yarrell, whose views, as to its specific rank, were endorsed by Macgillivray from an examination and careful dissection of a pair which were kept in the Zoological Gardens at Edinburgh.

To the late Mr. William Foster, of Wisbeach, I am indebted for the information that a male specimen of this swan, in the Museum of that town, was killed at North-delph, between Upwell and Downham, in the winter of 1839, certainly the first identified on the Norfolk coast.* From that date I find no further records until 1852, when a notice appeared in the "Naturalist" for that year (p. 170), by Mr. Thomas Southwell, of three birds answering to the description of Yarrell's Polish swan, having been shot out of a flock of thirteen, at Ingoldisthorpe, near Lynn,† in December, 1851. Two of these were purchased for Mr. J. H. Gurney, and were sent up to London to be preserved by Mr. A. D. Bartlett,‡ in whose keeping they were examined by Yarrell, who, in a letter to Mr. Southwell, now before me, expresses his belief that they were both examples of *C. immutabilis*. It should be stated, however, that the swans which appeared at Ingoldisthorpe, were sup-

---

* No doubt the same bird recorded by Yarrell as killed in Cambridgeshire.

† See also a note by Mr. J. O. Harper in the same volume of the "Naturalist," p. 132.

‡ Singularly enough neither Mr. Bartlett nor Mr. Gurney, though the former remembers the fact of these swans being shown to Yarrell, can give any information as to what subsequently became of them.

posed to have escaped from Holkham Park, as about the same number were said, near that time, to have been missed from the lake, and, though I cannot vouch for the correctness of the rumour,* it is by no means improbable, that Lord Leicester may have possessed, even without being aware of it, a pair or two of Polish swans, as the Zoological and Ornithological Societies, in London, and the late Lord Derby, at Knowsley, had specimens supplied through the London dealers; and a pair of very large swans, which agree in all essential points with the reputed characteristics of the Polish swan, have been exhibited, under that name, at the Regent's Park Zoological Gardens, since the year 1871.

In the great swan year of 1855, two adult specimens were shot on Horsey Mere, on the 2nd of March, as recorded in the "Zoologist" (p. 4661). One of these remains in Mr. Rising's collection, and the other, sent to his son-in-law, in London, Mr. G. S. Frederick, was by him submitted to Yarrell, who identified it at once as *Cygnus immutabilis*. They were described as "quite alone and difficult to approach." Another example, as Mr. Rising informs me, was shot by his eldest son at Hickling, about the year 1858, and was found to correspond exactly with his preserved specimen.

On the authority of Mr. M. C. Cooke a single example is recorded in Morris's "British Birds" to have been killed in the marshes, at Horning, in this county, on the 20th of January, 1874, which specimen, according to the entry in the late Mr. G. R. Gray's "Catalogue of British Birds" (1863), should be in the British Museum, but as I learn from Mr. R. B. Sharpe, who now holds the same post in the national museum, so long occupied

---

* Mr. Southwell's notes at the time supply the fact that two out of the three birds were killed on a pond, the other on a small drain or water-course.

by Mr. Gray, this is not the case; and, notwithstanding the announcement in the "Catalogue" above referred to, and a similar entry in the "Handlist," iii., p. 78, Mr. Sharpe's private copy of the "Handlist," marked by Mr. G. R. Gray himself, intimates that this Norfolk specimen is not in the Museum.*

On the 14th of November, 1868, the Rev. T. J. Blofeld sent me a large swan, which he had just killed on Hoveton Broad, and which, from peculiar differences in the bill and feet, he rightly conjectured to be of this species. This bird had been seen for some days in the company of, though not exactly, consorting with, the mute swans, on the broad, and attention was more particularly directed to it, from its occasional flights; and it was at last shot when on the wing. It proved, on dissection, to be a female, and its general measurements were as nearly as possible those given by Macgillivray. In the left wing the first primary was short as if partially moulted, the rest perfect, but the right wing had only the two first primaries complete, the rest clipped evenly to less than half their proper length, showing that this swan was a straggler only from other waters, and not a genuine wild visitant to our coast. The feet, which together with the head, are now, through the kindness of Mr. Blofeld, in my collection, were, in a fresh state, of a light grey colour, a specific peculiarity according to Yarrell; but having remarked that some cygnets of the mute swan have light grey feet and others

---

* The only specimen of the Polish swan in the British Museum is a fine young male, shot at Nairn, N. B., September 27th, 1872, which was forwarded in the flesh to Mr. Gould by Viscount Holmesdale, who, in a letter published by Mr. Gould in the Introduction to his "Birds of Great Britain," states that it was one of five which came to a wild loch by the sea during the northerly gales prevalent at that time.

black, this distinction will not, I think, hold good. The bird weighed thirteen pounds and three-quarters, but was in poor condition,* the stomach containing only coarse sand and grit with a few fibres of vegetable matter. The trachea resembled that of the mute swan.

The Hoveton bird is the last that has come under my notice in this county, either in a semi-domesticated state on our inland waters, or apparently wild on the coast, but, considering how many of our Norfolk specimens were identified by Yarrell himself as answering in all particulars to his so-called Polish or changeless swan, I shall scarcely be travelling beyond the scope of this work if I give a brief *resumé* of its history, so far as it is at present known, and the chief points at issue with regard to its specific rank.

Yarrell's attention seems to have been first directed to it from the fact that the London dealers were in the habit of receiving from the Baltic a large swan, which they distinguished by the name of the Polish swan, and two instances came under his notice, one prior to and the other in the spring of 1836, of certain tame swans having white cygnets, and which, in the adult state, from the smallness of the black tubercle, the light grey colour of the feet and legs, and other peculiarities, appeared to him specifically distinct from the mute swan. It was not, however, until the severe winter of 1837-8 that specimens, in all respects answering to the Polish swans of the dealers, came under his notice in an undoubtedly wild state, killed in the month of January, 1838, out of large flocks observed "pursuing a southern

---

* The young male in the British Museum weighed twenty-four and a quarter pounds, but Mr. Bartlett has weighed examples at the Zoological Gardens which turned the scale at twenty-seven pounds. Of the Ingoldisthorpe birds, Mr. Southwell ascertained the weight of the largest to be twenty-three and a-half pounds, of the smallest seventeen pounds.

course along the line of our north-east coast from Scotland to the mouth of the Thames." One of four examples shot at that time on the Medway, where a flock of thirty and other smaller flocks had been seen, was exhibited by Yarrell at a meeting of the Zoological Society (Feb. 13th, 1838), and was the type on which he founded the claims of this swan to be considered specifically distinct from *Cygnus olor;* for, to use his own words ("Brit. Bds.," 2nd ed., vol. iii., p. 227):—

"The circumstance of these flocks being seen without any observable difference in the specimens obtained, all of which were distinct from our mute swan; the fact also that the cygnets, as far as observed, were of a pure white colour, like the parent birds, and did not assume, at any age, the grey colour borne for the greater part of the first two years by the young of the other species of swans, and an anatomical distinction in the form of the cranium, which was described by Mr. Pelerin in the Magazine of Natural History (1839, p. 178), induced me to consider this swan entitled to rank as a distinct species, and in reference to the unchangeable colour of the plumage, I proposed for it the name of *Cygnus immutabilis.*"

Having so far quoted from the earlier records respecting this swan, I may here introduce Yarrell's subsequent remarks in a letter to Mr. Thomas Southwell (before referred to), in which, under date of January 2nd, 1852, he expresses his belief that the two Ingoldisthorpe specimens, sent to Mr. Bartlett, belonged to the newly recognised species. Besides this result of his visit to Mr. Bartlett, he writes, "I learned something more on the subject you may like to know."

"In the sale catalogue of the collection at Knowsley, the property of the late Earl of Derby, was a pair of the *immutabilis,* which Mr. Gurney, it appears, went to buy. At the sale, also, among others was Mr. Bartlett, and a Dutch zoologist named Westerman.* The latter told Mr. Gurney that he did not

---

* Dr. Westerman, the director of the Zoological Gardens in Amsterdam.

consider *immutabilis* to be a good species, only an albino variety of the common mute swan. That of a brood of cygnets in Holland half the number were grey and the other half white, from the egg, and this opinion had so much weight with Mr. Gurney that he abandoned his intention of bidding for the living pair. As you are probably aware, the Ornithological Society of London bought, some years ago, a male, female, and a young white bird of this *immutabilis*. The female and the young bird died some time afterwards, but the old male would never breed or even associate with any of the tame female swans. Lord Derby had also a pair of this *immutabilis*, but his male bird died, and early in the year 1850, his Lordship wrote to ask me to use my influence with the Council of the Ornithological Society here, to obtain for him the old male to pair with his solitary female, his Lordship giving us other water birds in exchange. This was immediately agreed to, the male bird was sent down, and a commission sent by me produced for us at the sale this very pair of birds which Mr. Gurney declined bidding for. They are now on the water in St. James' Park. * * * * The birds have not the character which distinguish Albinos as varieties, and I have very little doubt that one of the parents of the brood Mr. Westerman referred to was a true *immutabilis*.

The albino theory thus mooted by Dr. Westerman has since been adopted by others as a satisfactory explanation of the so-called Polish swan having white cygnets; but whilst Yarrell's supposition is a fair one, that in the case cited by Westerman of a mixed brood of grey and white cygnets one of the parent birds may have been *immutabilis*, I cannot concur in the idea of albinism, from the absence of any evidence, so far as I know, of *Cygnus olor* ever having white cygnets either in a wild state, or when nesting on our rivers and broads, and in a still closer state of domestication, on ornamental and circumscribed pieces of water. No such variation, so far as I can ascertain, was ever known to our local swan-herds from their own experience or by tradition; indeed, the only exception to the usual colouring of the cygnets of the mute swan (p. 79) I have any note of was rather a melanism than otherwise.

Again, although our mute swans, when unpinioned, will betake themselves to the coast in hard weather, and are there killed, with whoopers and Bewick's swans, this in no way accounts for such large flocks of swans as appeared on our coasts in 1838, all, as Yarrell remarks, "distinct from our mute swan," and with cygnets, "so far as observed, of a pure white colour, like the parent birds." Besides the absence, however, of any proof that albinism occurs in the mute swan in its purely wild state, I know of no conclusive evidence of its occurrence on our coast, as a migrant from such parts of the Continent, as it is still known to frequent;* though the probability is great that it does so occur, only passing unnoticed from the distinctions, if any, between *C. olor* in a state of nature, and as we know it after years of domestication, being as yet undescribed. This point will, no doubt, be fully cleared up by Mr. Dresser when treating of the continental species of swans in his exhaustive work on the "Birds of Europe," and the result of such comparisons may materially assist in solving the problem, what is *Cygnus immutabilis?* whose geographical distribution† seems almost as little known at the present time as it was to Yarrell in 1838.

The most probable solution of the difficulty I have yet met with is one recently suggested by Mr. J. H. Gurney, and concurred in by no less an authority on such matters than Mr. A. D. Bartlett,—it is the

---

* See antea p. 58 for evidence of *C. olor* breeding in a wild state on the Danube and on the Danish island of Bornholm. Dr. Tristram, also, in a paper on "the Ornithology of Palestine" ("Ibis," 1868, p. 327), speaks of *C. olor* as "common in Greece and Egypt."

† Lord Lilford, in a paper on "Birds observed in the Ionian Islands, &c." ("Ibis," 1860, p. 351) describes *Cygnus immutabilis* as "not uncommon in Corfu and Epirus, in severe winters. Several were shot in the island in January, 1858."

possibility that two distinct birds have been confounded under the name of "Polish" swan, one the so-called *Cygnus immutabilis*, the race which produces white cygnets, the other the original wild race of *Cygnus olor*. Unfortunately, as Mr. Gurney remarks, "we do not know the diagnosis between these two birds accurately, and therefore cannot be sure whether the swans which occasionally occur on our coast in a wild state, belong all to the first or all to the second, or some to each." Mr. Gurney, also, reasoning from analogy in the case of the Swan-goose *(Anser cygnoides)*, so common in our parks and ornamental waters, and found wild in so many parts of Asia, suggests the probability that the tubercle on the base of the beak of our tame swans may, on comparison with examples of the wild race of *C. olor*, be found to have largely developed during years of domestication.

Pending the settlement, then, of this vexed question by competent authorities, I will conclude this notice with a few remarks on the external differences,* pointed out by Yarrell and others, between the Polish and the mute swan of our home waters, and such as have more particularly struck me in the examination of some three or four examples of this disputed species.

Besides the very important cranial differences quoted by Yarrell from Mr. Pelerin's paper in the Magazine of Natural History, and which he states were fully verified by himself, we have that author's statement, also, that an old male Polish swan, in its ninth or tenth year, had but a small tubercle at the base of

---

* The trachea of the Polish swan, according to Yarrell, resembles that of the mute swan, and though, from the examination of an adult male, Macgillivray points out some slight difference in this organ, in all other respects its internal construction seems to be the same as in *C. olor*.

the bill, thus differing even from the female mute swan at the same age. The smallness and flatness of the tubercle, or "berry," has been a particularly marked feature in all the specimens of this swan I have seen, and its upper surface has been usually studded with minute feathers, as remarked also by Mr. Southwell in the Ingoldisthorpe birds. Macgillivray points out the smallness of the eye, giving the measurement of the aperture as five-twelfths and a half. This is very noticeable in the head of the Hoveton specimen now before me, as also the fact (observable in the pair at the Zoological Gardens) that the triangular black patch at the base of the bill meets the eye anteriorly, and does not partially encircle the eye, as in adult specimens of *C. olor*, the difference being plainly shown in the heads of these two swans as figured by Yarrell. The same writer also contrasts the position of the nostrils in the Polish with those of the mute swan, stating that the nasal aperture in the former is surrounded by the orange colour of the beak, and does not, as in the latter, join the black at the base of the beak. This distinction, however, like the grey colour of the feet and legs before referred to, cannot, I think, be relied upon as a specific distinction, as I recently took some pains to examine with a glass the heads of several old as well as young mute swans on the river, and satisfied myself, in every instance, that the orange or pink colour of the beak, according to age, surrounded the nostrils, although the overhanging of the great black tubercle in adult males, at first sight, conveyed a contrary impression. A much more important difference, however, presents itself on comparing the heads of *C. immutabilis* and *C. olor*, viz., the curvature in the ridge of the upper mandible in an old mute swan, as contrasted with the comparative straightness of the same portion of the beak in the Polish swan. Judging from the specimens in the Zoological Gardens, the beak in *C. immutabilis*, is of an almost vermillion red in adult birds, and, from the predominance seemingly of red over orange, appears more vivid in colour than in the oldest mute swan I have seen alive.

The following table gives the chief measurements of the few specimens of *Cygnus immutabilis* of which I can find any such particulars, either in published records or private memoranda:—

Specimens.	Total length.	Carpal joint to end of longest quill.	Tip to tip of wings.	Tarsus.	Middle toe and claw.	Bill from base to tip.
	ft. in.	inch.	ft. in.	inch.	inch.	inch.
Gould, imm. male, Brit. Mus., Nairn	4 9	—	7 6	—	—	4
Macgillivray, male. a., Edingburgh	5 4	25	8 0	$4\frac{1}{4}$	—	$4\frac{2}{12}$
Do., female. a., Edinburgh	4 10	—	7 4	$4\frac{3}{12}$	—	$3\frac{5}{12}$
Yarrell, male?	4 9	$21\frac{1}{2}$	—	4	$5\frac{3}{4}$	—
Ingoldisthorpe largest bird	5 $4\frac{1}{2}$	$23\frac{1}{2}$	—	$4\frac{1}{4}$	$5\frac{3}{4}$	—
Do., smallest bird	5 0	23	—	4	$5\frac{1}{4}$	—
Hoveton, fem., imm.?	4 $10\frac{1}{4}$	$22\frac{1}{2}$	—	4	$5\frac{5}{8}$	$3\frac{5}{8}$

TADORNA VULPANSER (Linnæus.)

SHELD DRAKE.*

It is no little satisfaction to be still able to claim this handsome species as a resident as well as a winter visitor, though few in numbers and confined to a very limited area in comparison with former times. From local records it would seem that even so late as the commencement of the present century "Barganders," or burrow-ducks as they are called in some parts of

---

* In a small work, published by John Ray, in 1674, entitled " A collection of English words not generally used," &c., to which I

the county, nested here and there throughout the entire range of sandhills bordering the Norfolk coast, which are broken only by the lofty cliffs that extend some twenty miles, between Happisburgh and Weyborne. On the north and west of our extensive sea-board the flat shores of the Wash between Lynn and Hunstanton, and thence, in an easterly direction, to Blakeney and Cley, the "meals" and mussel "scalps," bays,* creeks, and other tidal inlets, have afforded, at all seasons, the most favourable feeding grounds.

---

have before had occasion to refer (vol. ii., p. 52), occurs the following entry amongst the "south and east country words."

"*Sheld*, flecked, party colored, Suff. *inde* Sheldrake and Sheldfowle, Suff." That this is the real derivation of a name so variously rendered and explained by authors (vide Yarrell) there can be no question, and I am indebted to Mr. J. H. Gurney, jun., for a reference to, and Canon Tristram for a sight of the scarce little volume in which it occurs. In Suffolk a cat of the colour generally called "tortoise-shell," is spoken of as a "sheld-cat." See also the reprint of Ray's 2nd. ed. of this work, 1691, by the "English Dialect Society," 1874 :—

In Ray's edition, also, of Willoughby's "Ornithology" the following passage occurs in reference to the term "sheld" in the description of this species :—"They are called by some burrow ducks, because they build in coney burroughs. By others sheldrakes, because they are parti-coloured; and by others, it should seem, burganders." For the same reason, evidently, Willoughby calls the long-tailed duck *(Harelda glacialis)*, the "swallow tailed sheldrake."

° In Forby's vocabulary of East Anglia the term "bay-duck" is applied to this species, the origin of the name being attributed to the bright colour of the bird, as in a *bay* horse. But whilst the bay or chesnut in no way predominates over the green and white of this variegated species, I am inclined to believe that its partiality for the "bays" and other indents of the coast has more to do with this provincialism than mere tints of plumage; and Holkham Bay, in the very centre of its haunts, in Norfolk; and Southwold, or Sole-Bay, on the Suffolk coast, are instances of this term being used locally in a geographical sense.

In the neighbourhood of Yarmouth the long range of sandhills, locally termed "Marrams," from the grasses which bind the loose soil together, afforded in extensive rabbit warrens every facility for the peculiar nesting habits of this species, to which its extermination on that side of the county early in the present century, if not before, is no doubt attributable; since Hunt, in his "British Ornithology," published in 1815, remarks "they were *formerly* numerous at Winterton, but being supposed to disturb the rabbits, considerable pains were taken to destroy as many as possible."* And this seems to have been done so effectually that all my enquiries have failed to identify them since, during the breeding season, with that locality.

The most remarkable circumstance, however, connected with the past history of the sheld drake in this county is contained in the following note by Sir Thomas Browne, in his "Account of Birds found in Norfolk":—

"Sheldrakes, *Sheledracus Jonstoni.*—Barganders, a noble coloured fowl *(vulpanser)*, which herd in coneyburrows about Norrold and other places;" and in a letter to Dr. Merritt, in 1688 (Wilkin's ed., vol. i., p. 402) he again refers to this species as "Burganders,† not so rare as Turn [er] makes them, common in Norfolk, so abounding in vast and spacious warrens." From the absence in either case of any allusion to the coast-warrens, one is naturally led to suppose that in his time the sheld drake nested in localities some eighteen or twenty miles from the nearest point of the coast, in fact on the extensive rabbit warrens of the south-

---

* Thompson, in his "Birds of Ireland," alleges that the destruction of rabbits on the Copeland islands led to the sheld drake also deserting those localities for nesting purposes.

† He uses the words *Bar*-gander and *Bur*-gander, presumably a contraction of burrow-gander, in reference to their nesting habits.

western part of the county; at least such is the only inference to be drawn from the first quotation, if by Norrold is meant Northwold, near Brandon, a parish to this day locally termed Norrold* by labourers and others. I have ascertained that this extract from Wilkin's edition of Sir Thomas Browne's works was faithfully copied from the original MS. in the British Museum, my friend Mr. Hampden Glasspoole having obligingly referred to the passage itself, and furnished me with a *fac simile* of Sir Thomas' writing, but whilst the spelling of some words is old and quaint, Norrold is written plainly enough, nor is there any place on or near the coast, in this or the adjoining county, at all similar in name to either Norrold or Northwold.

It is not, however, without precedent that a species commonly associated, in this and other counties in England, with the sea and its surroundings, should in former times have selected for nesting purposes such inland warrens, as, from the sandy nature of the soil and an abundance of rabbit's burrows, were suited to its habits, inasmuch as the ring dotterel, which ordinarily deposits its eggs on the beach, just above high water mark (see vol. ii., p. 84), breeds annually on the warrens near Brandon and Thetford.† But on examining the

---

* This contracted form of spelling Northwold does not occur in Blomefield's "History of Norfolk," but is mentioned as a provincialism in Chambers' history of the county (1829); and the same parish is printed Northolde on a map of Norfolk, in Speed's "Theatre of the Empire of Great Britain," 1627.

† A singular coincidence, and one I was not aware of when describing the nesting habits of the ring dotterel on our inland warrens, is the existence at the present time in those localities of certain insects, chiefly amongst the smaller *Lepidoptera*, identical with species found on the sandhills of the coast. I must refer my readers to an able paper on the subject, by Mr. C. G. Barrett (Trans. Norfolk and Norwich Nat. Society, 1870-71), in which, after

position of Norrold or Northwold on the Ordnance Map, bordering as it does immediately on the " Fen" district, and on the western limits of the parish, where it adjoins the little river Wissey, partaking much of that character of soil, it is difficult at this period of time to point to any particular spot likely, from its physical aspect, to have been a haunt of the sheld drake in Sir Thomas Browne's days, yet the expression "*about* Norrold and other places,". though apparently marking that parish as a "head centre," may still mean, to use a modern expression, "there or there abouts."* If taken, then, in this sense, there is, as I am informed by Mr. Francis Newcome, of Feltwell, a sandy tract of land either in or adjoining the parish of Northwold, on the Didlington side, where rabbits abound, whilst the next parish of Cranwich consists almost entirely of heath and warren. In the same neighbourhood, also, is Methwold,† whose warren, according to Blomefield (1739), " is large and famous to a proverb for rabbits," and though, on the same authority, "great suits have been commenced on account of their damage," "Mewell"

---

enumerating the coast insects discovered by himself and others about Brandon and Thetford, he comes to the conclusion that "the species in question [as also we may presume the ring dotterels] have occupied this suitable ground from the time of the close of the post glacial period at least," unchanged both in form and colour, and that this tract of country " was actually a range of coast sands at a comparatively recent point of the post glacial period, while the great valley of the fens was still submerged."

° That this is a correct view of Sir Thomas' meaning is, I think, proved by the many instances in which he uses the term *about*, when desirous of localising any particular species. As, for instance, he describes the common dotterel as occurring " about Thetford and the Champian," &c., &c.

† So named from its site Methelwolde, or the wold between Northwold and Hockwold, the midle wolde.

rabbits (the provincial name for Methwold) abound to this day, and are much sought for by poulterers. Below Cranwich and Mundford, again, in a south-easterly direction, to Santon, the blowing sands of that portion of the "Breck" district present even now not unfavourable conditions for the nesting of this species, though it is difficult to imagine whence food was procured for the nestlings till old enough to follow their parents; and it is to be remarked that now-a-days there is no locality known far from the sea in which this species breeds. If we accept, then, Sir Thomas Browne's statement in the sense above given, I see no reason to doubt that this inland locality is that to which he refers, and his brief record, in the absence of any local tradition, is the only evidence preserved to us—as in the case of the spoonbill and the cormorant breeding, two hundred years ago, at Claxton and Reedham—of a most interesting ornithological fact.

It would seem, however, that even of late years these birds have not entirely confined their choice of nesting places to our coast sand-hills, provided the accommodation of rabbit's burrows could be obtained elsewhere, as Mr. Robert Wells, of Heacham, informs me that some thirty years ago sheld ducks bred upon the heaths at Dersingham and Sandringham, and a pair of wild birds once hatched their young in a rabbit's burrow on his farm at Sedgeford, each of these localities being about three miles distant from the nearest point of the coast.

At the present time the few pairs that spend their summer in Norfolk are to be met with only on that portion of our coast, which extends from Holme, near

---

° Macgillivray states ("Br. Bds.," v., p. 26), that he knew of one instance of the sheld duck breeding on an island " on which there were no other quadrupeds but seals, and still the nest was in a burrow, which it must have made for itself."

Hunstanton, to the harbour at Cley. When staying at Hunstanton, in June, 1863, I saw single birds of this species on one or two occasions, near Holme point, passing from the sand-hills to their feeding grounds at low water; and in the same neighbourhood, Mr. Wells informs me, a pair still nested in 1874. At Brancaster, in 1866, Mr. F. Norgate found the remains of egg shells, at the mouth of a rabbit's burrow, apparently those of the sheld duck, in the month of July; and young birds, as I learn from Mr. Beverley Leeds, were taken there both in that and succeeding years; and so late as the year 1874 Mr. Wells knew of three pairs that frequented that part of the coast throughout the summer. In 1853 Mr. Thos. Southwell found empty egg shells of this species on the "meals" about Wells, and was informed that a pair or two nested there every season; but in August, 1872, Mr. J. H. Gurney, jun., ascertained that none had bred there that summer, but was told by a local gunner that in the previous year he had taken an old bird and eleven eggs out of one nest hole. At Blakeney, in 1872, Mr. J. H. Gurney, jun., found four young birds in the channel early in August, the remains of a much larger brood observed earlier in the season. Another pair were also said to have nested in the same locality; and in May of that year he saw a pair of old birds in the channel at Cley. As late, also, as the summer of 1874, I had reliable information that they still resort to the sand-hills on the Blakeney beach, the extreme eastern limit of the sheld duck's nesting range on the Norfolk coast, and this is the more satisfactory since, prior to the passing of the "Sea Birds' Preservation Act," in 1869, which affords them protection between the 1st of April and the 1st of August, my notes show several instances of their being killed at Blakeney and Salthouse, as well as on Breydon, in April and

May; birds just returning in the perfection of breeding plumage to their nesting quarters.*

Writing of the nesting habits of this species on the coast sand-hills of Norfolk, Messrs. Sheppard and Whitear remark that its nest "is discovered by the print of its feet on the sand, and is, therefore, most easily found in calm weather; for in windy weather the driving sand soon obliterates the impression. The old bird is sometimes taken by a snare set at the mouth of the burrow." Mr. Wells, from his own experience of late years, also informs me that their nests are most commonly found by watching the drake, when flying round and over the hole where the duck is sitting, calling her off to feed early in the morning, and having successfully reared the young birds when he resided at Sedgeford, from eggs procured on the coast in that neighbourhood, he has kindly furnished me with the following particulars:—His first sitting of eggs were hatched under a hen, and the young after two years (sheld ducks do not pair the first year) bred in some holes he had purposely made by the side of his pond to represent rabbit's burrows.† Sometimes, however, they

---

* The demand for such young ones as can be hatched out under hens, from the attractiveness of their plumage when semi-domesticated on fresh water ponds or lakes, is, of course, a strong incentive to the wholesale robbery of their nests.

† Mr. H. Durnford, in an interesting paper on the Ornithology of the "North Frisian Islands and adjacent coast" ("Ibis," 1874, p. 403) describes this species as living in a semi-domesticated state, both on the islands and mainland. "The natives make artificial burrows in the sand-hillocks, and cut a hole in the turf over the passage, covering it with a sod so as to disclose the nest when eggs are required. Several females lay indiscriminately in the nest. They are very tame, and suffer themselves to be taken by hand while sitting. Each burrow has two openings, and is made circular in shape. There are sometimes as many as a dozen or fifteen nests in a hillock within the compass of eight or nine yards. The eggs

would leave the pond, and go some distance into a pasture where they found rabbits' burrows, and they would also lay in holes and roots of hollow trees. He allowed them to hatch their own eggs, usually from eleven to thirteen, and as soon as the duck brought her young to the pond he had them fed with cockles* scalded to open them, and by throwing a few in different places, by the edge of the water, the ducklings would quickly learn to know and follow the feeder, and after a time they would learn to eat corn with the ducks. In confirmation of his statement as to their breeding formerly

---

are taken up to the 18th of June, after which they allow the birds to incubate; but they never rob a nest of all the eggs, leaving one or two to avoid driving away the birds. Each person in the village generally has a burrow, and they are scrupulously honest in not taking each other's eggs. The female always covers her eggs with down before leaving the nest."

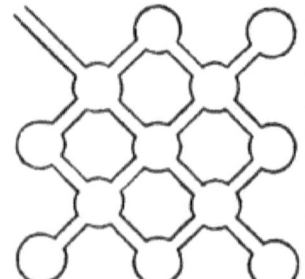

Naumann in his work "Ueber den Haushalt der nordischen Seevögel Europa's" (p. 7), gives a diagram of the ground plan of one of these artificial breeding places, showing the communication between twelve nest-chambers, all accessible by one common entrance. In his time (1819) sheld drakes were to be seen "zu Tausenden" (by thousands) around Sylt. To Mr. Alfred Newton, who, since the publication of Mr. Durnford's paper has met with a copy of this scarce publication of the great German naturalist, I am not only indebted for a sight of the work itself but for the opportunity of reproducing the diagram above mentioned, in the accompanying wood-cut.

* St. John ("Natural History and Sport in Moray," p. 73), describes the food of this species as consisting "almost wholly of small shell-fish, and more especially of cockles, which it swallows whole. It extracts these latter from the sand by paddling or stamping with both its feet. This brings the cockle quickly to the surface." He also states that he has seen tame birds of this species do the same, when impatiently waiting for their food.

at Sandringham (now the residence of the Prince of Wales), I may quote from Mr. Knox's "Game birds and Wild Fowl" the following statement as to the extraordinary docility exhibited by nestling sheld drakes, even in a state of nature. "A friend of mine told me that when at Sandringham, in Norfolk, he saw an entire family of young sheld drakes emerge from a rabbit's hole in which they had been bred, when summoned by the whistle of the gamekeeper, partake greedily of the food that was thrown to them, and return into the same retreat when the repast was finished." Mr. J. H. Gurney also informs me that he remembers in former years being told by residents in that part of the county that the sheld duck nested on Dersingham heath.

Of our "home bred" birds, as those reared on the coast sand-hills of this county may be termed, to distinguish them from migratory visitors, some fall victims to the gunners during the autumn months when dispersed along the coast, or as stragglers on inland waters, for I have known young sheld drakes, killed in the months of October and November, on Hickling Broad, as well as on the tidal backwaters of Salthouse beach. It is seldom, however, unless an unusually severe frost sets in before the close of the year, that the migratory flocks which yearly visit us accumulate on our northern shores before January and February, their numbers at such times depending much upon the severity of the season; but should a spell of hard weather set in about that time they are found by the gunners in considerable flocks, frequenting the shores of the Wash and those sandy flats and indents of the coast from Hunstanton to Blakeney, the favourite resort in summer of our native race.*

---

* Sheld drakes are occasionally taken with other shore birds in the nets placed for that purpose on the borders of the Wash, as

Mr. Wells, who has had more than thirty years experience as a shore gunner in that part of Norfolk, tells me that the boat gunners kill the largest number of these birds in February and the beginning of March; and in the severe winter of 1870-71, he saw from four to five hundred in one day, in the month of February. This agrees, also, with notes on this and other coast species received from Mr. A. W. Partridge, who was shooting in the neighbourhood of Hunstanton during that exceptionally inclement season. At such times many of these birds are exhibited for sale in the poulterers' shops at Lynn, and are thence sent to the London markets. My own notes for the last twenty years of specimens in the Norwich market, or in the hands of our bird-stuffers, are strongly confirmatory of these statements from our coast gunners, the majority of the birds entered in my journal having been seen in the months of January, February, and March. I have two or three records of sheld drakes killed in the winter months on Hickling Broad and one at Stalham; at Fritton, near Yarmouth, also, they are occasionally seen with other fowl. In February, 1869, Mr. Frere tells me he saw a small flock of six in Fritton decoy. On Breydon, as Mr. Fielding Harmer informs me, the largest number he ever saw together was a flock of nine, during hard weather, in January, 1860, smaller parties of from two to four appearing there generally every year in March, April and May. Lord Kimberley includes this species amongst the fowl occasionally

---

described in my account of the dunlin (vol. ii., p. 376, and in the Introduction to vol i., p. xlvi.)

A very intelligent keeper at the "high" lighthouse, at Lowestoft, showed me in 1865 a stuffed sheld drake, a fine old male, which had flown against the windows of the Offord Light in the previous winter, and was found dead at the foot of the building.

observed on his lake, near Wymondham;* and from the Rev. H. H. Lubbock, of Hanworth, I learn that they occasionally, in hard weather, visit the lake in Lord Suffield's park, at Gunton, with goosanders and other sea-fowl. I have known immature birds shot as far up the Yare as Rockland Broad; and in the severe winter of 1838 Mr. J. H. Gurney remembers a pair having been shot at on the ice, in the river at Earlham, some two miles above Norwich. The sheld drake also appeared during the flood of 1852-53 in the Southery district, when Mr. A. Newton informs me that he saw a pair, one of which was shot by Mr. Newcome on May 5th, 1853, in Hockwold Fen.

Mr. Selby, from his own observation of the habits of this species upon the Northumbrian coast, states that the males do not pair until their plumage is perfected in the second year, but, once paired, remain constant to the same mate. In the male, also, at the commencement of the breeding season, the fleshy knob at the base of the upper mandible, scarcely perceptible in autumn and winter, "begins to swell and acquires a beautiful *crimson* hue, and when at its full development, is nearly as large as a marble." The nests are formed of "bent grass and other dry vegetable materials," lined with soft down from the old birds' breasts, and the eggs, from twelve to sixteen in number, "of a pure white or slightly tinged with green," are incubated in thirty days, and are sometimes ten or twelve feet from the entrance of the burrow. The male sits on the eggs when the female is off feeding, and both birds, like the partridge and wild duck, will feign lameness, and adopt other stratagems to decoy intruders from the vicinity of

---

* See Lord Kimberley's paper on "Birds observed on the Kimberley Estate since 1847."—Trans. Norf and Nor Nat. Society," vol. ii., part 1, p. 46.

their young when able to quit their nesting holes, hence probably the name of "sly goose" applied to this species in some localities. The young are sometimes carried in the bills of their parents down to the sea. St. John, in his "Natural History and Sport in Moray" (p. 293), refers to the strange instinct which enables the female, sitting on her eggs many feet under ground, and more or less distant from the sea, to know to a moment when the tide begins to ebb, and then and then only to betake herself to the freshly exposed feeding grounds. The males of this species vary much in size as may also the females, but the latter are always smaller than the males as well as less brilliant in colour; but, unlike the true ducks, both sexes in the genus *Tadorna* are alike in plumage, and retain it when once fully acquired. The flesh of the sheld drake is coarse and unpalatable, and its food consists, according to Selby, of "marine vegetables, molluscous shell-fish, insects, &c.," but so minute are some of the forms of mollusca which afford them a meal, and so great their consumption, that Thompson, in his "Birds of Ireland" (vol. iii., p. 69), describes the crop and stomach of one of these birds as containing by a careful computation not less than twenty thousand minute mollusca, *Montacuta purpurea*, *Skenea depressa*, and *Paludina muriatica*. *Skenea* being about the size of a clover seed; *Montacuta* one-twelfth of an inch broad *when large*.

The RUDDY SHELD DRAKE, *Tadorna rutila* (Pallas), though more than once recorded as killed in Norfolk, has not occurred to my knowledge in an undoubtedly wild state.* In one or two instances birds supposed to

---

* Yarrell states that a specimen was shot at Iken, near Orford, Suffolk, in January, 1834, "which passed into the possession of Mr. Manning, of Woodbridge," but this, as Mr. J. H. Gurney, jun., has pointed out to me, is evidently the bird thus recorded by the

be of this species in private collections, have proved only large and strangely coloured varieties of the common marsh duck. A fine adult male, however, shot at Snettisham, near Lynn, on the 26th of March 1869, as recorded in the "Zoologist," p. 1910, and I think in the "Field" at the same time, was subsequently ascertained to have been an escaped bird, belonging to Mr. J. H. Coldham, of Anmer. Mr. Coldham had purchased a pair of these birds in London, which had come from Russia, but the female died shortly after, it was supposed from the operation of pinioning. The male, therefore, had only the wing feathers cut, and when these were renewed after the moult, the bird, restless from the loss of its mate, took to flight, and had been missed nearly a week before it met with its death on the sea shore at Snettisham. The above satisfactorily accounted for the perfect state of its plumage when sent me from Lynn for identification.

ANAS CLYPEATA, Linnæus.

THE SHOVELER.

If the broads and meres of Norfolk had not been such a *terra incognita* to the earlier writers on British

---

Rev. W. B. Clarke, in the "Mag. Nat. Hist.," vol. vii., p. 151. "A ferruginous duck or ruddy goose (*Anas rutila*, Fauna Suecica) was shot a few days since at Iken, near Orford. It is in the possession of Mr. Manning, chemist, Woodbridge ('Ipswich Journal, Jan. 11th, 1834), W. B. C. [See vi., 141.]" On turning to vol. vi., p. 141, of the "Mag. Nat. Hist.," I find a note on a pair of ferruginous ducks killed near Orford, plainly showing that this species of duck, and not the ruddy sheld drake, was referred to in Mr. Clarke's communication.

One is stated by Mr. Fenwick Hele, in his "Notes about Aldeburgh," to have been seen in 1864, "in company with several common sheldrakes, near Blackstakes."

ornithology, they would scarcely have doubted whether the Shoveler nested in England, and the fact of its being still a resident in, as well as a migratory visitant to, this county, may fairly lead to the conclusion that in the palmy days of the decoy system and long antecedent to their introduction, this species bred plentifully in our then un-reclaimed marshes.

From the commencement of the present century at least, our local records* supply ample proof of the shoveler

---

* If the term "popeler" was really applied, in olden times, to the shoveler duck, as stated by Mr. Albert Way in the Camden Society's edition of the "Promptorium parvulorum" (Tom. ii., 1851), the earliest record of its occurrence in Norfolk is contained in the following brief entries in the "Household and Privy Purse Accounts" of the Lestranges, of Hunstanton, for the year 1533:— "Itm iij popelers of store," "Itm v herns and a popeler of store"; articles of "store," as I have elsewhere stated, being supposed to represent the produce of the home farm or provisions previously purchased, whilst articles of "gyst" were undoubtedly given in lieu of rent. It is noticeable, however, that in no case is any price put to articles of "store" or "gyst," whether poultry and pigs from the farm, or cranes and plovers from the surrounding marshes; and thus in two separate entries we find the following strange mixture of edibles:—"Itm a Goos, a Pygge, a Crane, iiij Conyes, and a loyn of veile of Gyst." "Itm a Goos, iij Malards, ij Telys, and iij Conyes ot store." I see no reason, therefore, to suppose that the popelers (whatever they may have been) were semi-domesticated birds, but as wild as the mallards, the bustard, the crane, and the hernsewe, all specially mentioned as "kylled wt ye crosbowe," and like them, no doubt, brought in by the fowler on the estate who, with the falconer, is frequently mentioned in the "Accounts." As these two "Itms," also, occur in the fifth and sixth weeks *after* the 29th day of March in that year (according to the accountant's method of reckoning), we may fix the date as early in May, the very month when the two "spowys" (whimbrel) received of "gyste," at the same time, would be passing northwards to breed, as the "May bird" of the Blakeney and Breydon gunners does to this day.

As to the identity between the word popeler and shoveler, Mr. Alfred Newton has kindly supplied me with the following note:—

being a resident species in Norfolk, and of its breeding in some numbers in the neighbourhood of Yarmouth. Hunt in his "British Ornithology" (1815) referring to Montagu's statement that all his enquiries had failed to establish the fact of the shoveler still nesting in Lincolnshire,

---

"It seems to me that popeler is only a corruption (by common metathesis) of "lopeler"—*i.e.*, Lepelar Dutch, Lepler German, which are both equivalent to Löffler—from Löffel, a spoon or shovel. I find among German names for *Anas clypeata*, Löffelente (*i.e.*, spoon-duck), and Leppelschnute (*i.e.*, shovel-snout), while, according to Bechstein, " *Anas glaucion*" is also called Spatelente, Löffelente, and Leppelschente (the last looking like a misprint for Lepelschnute). "Lepeleend," *i.e.*, shovel-duck, is the Dutch name, or one of them, for *Anas clypeata*." The same authority, however, points out that, on the other hand, the words shoveler and spoonbill have been so completely " mixed up" by early authors that it is next to impossible to determine whether the species referred to by any writer is the duck or the wader—*Anas Clypeata* or *Platalea leucorodia*. In the above edition of the " *Prompt. parv.*" I find the following entries:—"POPELERE byrd (or schovelerd infra), *Populus*"; and in a foot note Mr. Way remarks, "It appears subsequently that the POPELERE was considered by the compiler of the Promptorium to be the same as the shoveler duck, *Anas clypeata*, Linn." The only other passage, however, in the work, so far as I can ascertain, in which the word popeler occurs and to which, I presume, Mr. Way alludes by the expression " subsequently," appears under the head of schovelerd, thus, " *schovelerd*, or popler, byrd (schoveler, or poplerd, s. schoues bee, or popler byrd, p.) *Populus*." And if on this is founded Mr. Way's belief that the compiler—[a Dominican Friar, of Lynn Episcopi, Norfolk, A.D. 1440, who might be presumed, therefore, to know something of the provincial names of birds on the Lynn and Hunstanton coast]—associated the term popeler with *Anas clypeata*, I imagine most ornithologists would decide in favour of the spoonbill *(Platalea)*, supported, as such impression would be by the association of popelers with herons in the L'Estrange Accounts ; whilst Sir Thomas Browne (*circa*, 1666) plainly distinguishes " the *Platea* or shovelarde, which build upon the tops of high trees," from " *Anas platyrhinchos*, a remarkably broad billed duck," the only allusion to this species in his list of Norfolk birds. To whichever species then we may assign

remarks, "we know for a certainty that they continue in the county of Norfolk the whole year, and breed in the marshes at Winterton, Caister, &c. We have had the young ones repeatedly from the age of a few days to that of their being able to fly." The same account is confirmed by Messrs. Sheppard and Whitear in 1825, with the important statement that "in the spring of 1818 the warrener at Winterton found several nests belonging to this species, containing in the whole fifty-six eggs." From that time till 1834, probably but little change took place in their nesting habits in that neighbourhood, although then briefly noticed by Messrs. Paget as "not uncommon." Again, in 1845, Mr. Lubbock, who speaks of the shoveler as occasionally breeding in the Norfolk marshes, specially notes the fact that "a brood or two are hatched every year in the marshes round Winterton decoy," and that he had known young fowl of this species during August, at Horsey, and had seen broods of them upon Hickling broad.

Besides the "Broad" district, however, we have records of the shoveler nesting formerly, as it does now, on the meres of the south-western part of the

---

the term popeler, it is curious, as pointed out by Mr. Way, that "in medieval decorations, such birds were not unfrequently represented." Thus, in the Caistor inventory of Sir John Fastolfe's effects, taken 1459, appears "Clothis of Arras and of Tapstre warke. Item, ij clothis portrayed full of popelers;" and, again, in one of the bed chambers, "Item, j hangyng clothe of Popelers" Archæol., xxi., pp. 258, 264. Sir Richard de Scrop, also in 1400, bequeathed "*Aulam de* poplers *tentam, et lectum integrum cum costeris de rubeo, cum* poplers *et armis meis broudatum.*" And that at least one species of duck was similarly pourtrayed in former times, Mr. Way cites "the vestments discovered at Durham, attributed to St. Cuthbert, and the entry in the Bursar's accounts, given by Mr. Raine, respecting an altar there, on '*le rerdos*' of which were depicted the eider ducks, termed the birds of St. Cuthbert."

T

county, and at least one authentic instance is known of its breeding on Scoulton mere, near Hingham, although both the shoveler and pochard have long ceased to frequent that chief stronghold of the black-headed gull, in Norfolk. Mr. Alfred Newton was informed by the late Miss Hamond, of Swaffham, that the nest of eight eggs of the shoveler in the collection of her brother, the Rev. Robert Hamond (in the possession of Mr. Robert Elwes, of Congham House), were taken many years back at Scoulton.

The late Mr. J. D. Salmon in his MS. diary of natural history occurrences in the neighbourhood of Thetford (now in the Norwich Museum) has the following entries respecting shovelers seen, during the summer months, from 1834 to 1837, on what he then terms a small piece of water on Stanford warren:—

1834. June 6th. A male shoveler and young ones seen in the pond, but the female not visible.

1835. May 10th. Eight shovelers' eggs found on the warren (Stanford), placed on the ground with scarcely any nest, found within a few days of hatching.

1836. April 20th. Saw a pair of shovelers at Stanford, and again on May 7th. May 18th, a nest found with one egg. May 25th, saw the male bird. June 1st, the young all hatched.

1837. May 3rd. A male shoveler seen at Stanford pit.

These notes refer of course to the same birds, which Mr. Salmon, in a communication to Loudon's "Mag. Nat. Hist." for 1826, describes as having "hitherto annually bred amongst some green rushes, on the warren at Stanford;" and to which Mr. Hewitson ("Brit. Bds.' Eggs," vol. ii.), alludes in acknowledging the receipt of a shoveler's egg from Norfolk, from Mr. Salmon; nor is there any reason to doubt that the descendants of those birds have ever since frequented the same locality.

STANFORD AND THOMPSON MERES.

Strictly preserved, and artificially extended beyond its former limits, Stanford mere, presents a fine sheet of water of about five and thirty acres in extent, lying in the midst of a rich meadow land, with the warren adjoining, on which at some little distance are two or three belts of plantation, where wild ducks, teal, and shovelers, resort for nesting purposes, returning with their young broods to the mere almost as soon as they are hatched. The broom coverts on the heath, though at a considerable distance from the water, have also, I understand, similar attractions for these three species of fowl. Through the kindness of Lord Walsingham, on whose estate both this and the adjoining mere at Thompson are situated, I had the opportunity, on the 8th of June, 1875, of visiting Stanford and making notes on the different species of ducks that frequent its waters, and nothing certainly could be more attractive to fowl, generally, than the shelving margins of the lake, fringed here and there with flags and sedges; having at one end a thicket of dwarf trees and tangled undergrowth, with a bordering swamp of reeds and rushes; at the other a small island, rich in aquatic herbage and dotted with sallow and other bushes. The keeper informed me that he knew of five shovelers' nests that summer, but the birds had all hatched,* and with the glass I distinguished three pairs on the water, amongst the other fowl, disturbed by the presence of strangers. One hen bird kept persistently near to a large clump of rushes on the further side of the mere, and had good cause, no doubt, in the presence

---

* Lord Walsingham, writing to Mr. Gould ("Birds of Great Britain") from Merton, in 1869, says, "not less than eight or ten pairs of shovelers are in the habit of breeding here every year"; referring, no doubt, both to Stanford and Thompson meres.

of her brood close at hand.* On the following day I visited Thompson mere, which lies within easy flight of Stanford water, and, including some small pools or "pulk"-holes connected with it, extends over fifty acres. Here, also, the various species of fowl have ample protection and shelter for their nests. On the side skirting the roadway the margin is somewhat bare, but to the left, at the upper end, a furzy common slopes down to the water's edge where wild ducks and teal nest under the bushes; and the full length of the banks opposite is broadly fringed with beds of reeds and rushes, with rough grassy mounds, extending to the arable and heath land beyond, from whence one looks down upon the fish ponds or pools that indent the shore, each circled with a rich growth of sedge, flag, and reed, whilst the water is scarcely visible through the profusion of lilies and other surface loving plants. There is no island here, but a small patch of land at the end next the heath is thickly planted with young firs and shrubs, attractive to pheasants as well as fowl, and divided only from the mere by a wide belt of reeds. Much, of course, has been done of late years to adapt both these localities to the requirements of a preserve for fish as well as fowl; and these natural meres which, like those of Wretham heath, were affected by drought, have their waters now artificially kept up, with the means at hand for drawing off either one or the other in a few hours' time.

---

* Mr. Salmon speaks of one nest of eggs being "within a few days of hatching" as early as the 10th of May, but in two other instances he did not find the young hatched till the 1st and 6th of June; a difference to be accounted for perhaps in some cases by a first clutch of eggs being taken. St. John, in Morayshire, found nests of this species, with eggs, on the 19th and 28th of May; and in Denmark, Mr. Dresser ("Birds of Europe") states that the eggs are laid from the 2nd to the 26th of May; and a nest of eight eggs had been found as late as the 24th of July.

The pike, which run to a large size in both meres, are the chief enemies the fowl have to contend with, and neither ducklings nor young coots whilst in their downy stage have much chance of escape from these fresh water sharks. Several mallards and certainly one cock shoveler, which left the mere as soon as a boat was launched, flew off in the direction of Stanford, and did not return, but two hen shovelers, flushed from the reeds, had evidently young concealed amongst the herbage, and the keeper showed me afterwards two nests from which the young had been hatched within a few days. The first was in a wheat field about fifty yards from the grassy borders of the mere—a slight hollow in the ground with no other lining than a sprinkling of fine grass and down from the bird's breast, mottled grey and white, and which I have found to agree with that on the breast of a hen shoveler in my collection. The bird had been flushed within a week from her eggs, the fragments of which remained. The second nest, some sixty or seventy yards from the first, was hidden amongst the long grass on the raised embankment of the mere, and, according to the keeper, contained eleven eggs* when he last saw it, a few days before. That these had been hatched very recently was evident from a dead nestling, partly out of the shell, being still fresh and not fly-blown; another youngster had died in the egg, and plenty of broken shells filled the downy lining of the nest, which

---

* Authors differ strangely as to the number of eggs laid by this species. Lubbock says from seven to nine; Selby ten to twelve; and St. John ("Nat. Hist. and Sport in Moray") found a nest with eleven eggs. Messrs. Sheppard and Whitear give the usual number laid as eight or nine, but state that in one instance they knew of thirteen in one nest, whilst Dresser, naming ten or twelve as the ordinary amount, quotes Naumann as an authority for its sometimes laying fourteen.

in character exactly resembled the first, and in the scantiness of the materials used, exposing the bare soil, agreed exactly with Mr. Salmon's description. There could be little doubt that the owner of this nest was the female shoveler that rose within a few yards of it from a clump of sedges, and flew round and over us in a great state of excitement, settling at no great distance on the water as we pulled out into the mere, and, after two or three more hesitating flights, dropping suddenly into the very shelter from which we had disturbed her. Messrs. Sheppard and Whitear, also, describe two nests found in the Winterton marshes as each "placed in a tuft of grass where the ground was quite dry, and made of fine grass," and add, "when the female begins to sit she covers her eggs with down plucked from her own body," but the amount used, judging by the two nests I found at Thompson, is but little in comparison with the snug warm nest of the garganey.

From Thompson, the same day, I drove to Wretham, distant about four miles, where, from the extreme drought, I found the three meres* on the open heath—Ringmere, Langmere, and Foulmere—completely dried up, but here, as well as on the more private waters adjoining Mr. Birch's residence, no doubt the shoveler nested formerly, as at the present time, in favourable seasons. Mr. Edward Newton, when staying at Wretham in the autumn of 1853, was told by Mr. Birch that a hen shoveler had hatched and brought off her young near the hall some years before, and others were known to be on East Wretham heath that year. A pair had been seen late in the spring, and orders were given to the keeper not to shoot them. Mr. A. W. Partridge observed this species at Wretham

---

*For a further description of these meres see "Trans. Norf. and Nor. Nat. Society" vol. i.—1869-70.

in the summer of 1868, and on the 8th of August, 1869, I visited the meres on the heath for the first time. Langmere then presented a fair sheet of water, off which from ten to fifteen couple of fowl rose, startled by the noisy exit of turtle doves from some fir-trees, under cover of which and a raised grassy mound, I was attempting an undisturbed view. Duck and mallard were alike in plumage in the autumnal change, and cock shovelers would thus easily pass unnoticed, but whilst these winged their way in the direction of Mr. Birch's decoy, I got my glass to bear upon several half-grown flappers and smaller nestlings, being escorted by anxious mothers amongst the sheltering rushes. In one of these I recognised a hen shoveler, whose nervous actions, evinced maternal solicitude in the most unmistakeable manner. On Foulmere and Ringmere, the only birds seen were, a pair of teal and some coots on the former, and several little grebes on the latter. Mr. Southwell has, however, furnished me with the following graphic description of a pair of shovelers, which he discovered on Foulmere when visiting all the Wretham meres, on the 24th of May, 1874:—" As I showed myself," he writes, " the fowl left the water and the coots retired into the sedges, but I noticed one drake shoveler still hanging about the mere. As I walked round he changed his quarters so as to be at a safe distance, but on approaching a particular spot he became more anxious and flew round and round. At a patch of sedge and rushes a sallow inclined over the water, and as soon as I put my foot on the stem of the prostrate tree an old duck shoveler bustled out of the green shelter in a great state of consternation followed into the open water by several young ones. They swam out and crossed the water to another patch of vegetation, and as soon as they were safely under cover the drake, which had been flying round all the time, occasionally so near that I could see his eye, disappeared altogether." From this, and the fact of my finding cock shovelers, both on Stanford and Thompson, when the nestlings were hatched, it seems that though the males, like the common mallard, may keep apart whilst the hens are sitting, still the care of the young brood is shared between them. Mr. Southwell, also, identified shovelers amongst other fowl on Langmere,* which was almost dried up, and believes that he also saw this species on the meres near the hall.

At no great distance from the above named haunts, Mr. Beverley Leeds informs me that shovelers breed annually at West Tofts, on Mrs. Lyne-Stephens' estate, upon a small patch of common land of not more than an acre in extent, where the cows graze quite up to their nests. The chief attraction for them in such a spot is, he supposes, the abundance of water snails, of which they are very fond. He has shot them in September, and found scarcely any other food in their stomachs.† The male birds, he believes, whilst the females are sitting, resort during the day to the mere at Stanford; and it is probable that birds of this species, which Mr. H. M. Upcher tells me may be seen, during the summer, on the river near Didlington, are either stragglers from Stanford and Tofts, or have found a home for themselves in that

---

* On the 15th of April, 1876, Mr. Southwell flushed eight drake shovelers on Langmere, the waters of which were unusually extended. The females were, no doubt, not far off, but he saw no shovelers on Foulmere.

† Hunt states that those he examined had their stomachs generally full of periwinkles which shows that marine and fresh water mollusca, of some size, form a chief part of their diet, which is usually stated by authors to consist of worms, insects, and the seeds of grasses. Audubon, describes them, in the United States, as consuming, besides vegetable matter, leeches, small fish, large ground worms and snails.

neighbourhood. Further to the south and west they are certainly not resident now, and, from enquiries made on the spot by Mr. Alfred Newton some twenty years ago, it seems doubtful whether this or any other kind of "half fowl"—as any species of duck, not the common wild duck, is termed in Norfolk—bred in the "fens" about Hockwold and Feltwell, even when their physical aspect justified the name they still retain. One other haunt, however, of the shoveler on this side of the county, but at a considerable distance to the north and west, is Roydon fen, near Lynn, where, as Mr. J. H. Gurney, jun., was informed by Mr. Burlingham, it bred as recently as 1872, and I have no reason to doubt its continuing to do so. From thence, passing eastward along the entire coast line to Winterton and Yarmouth, the Holkham marshes are perhaps the only locality where the shoveler may still be met with as a resident species, now that almost every decoy is closed or entirely done away with. Here, as Lord Leicester informs me, a good many fowl are reared every year, and shovelers, he believes, amongst others,* but he does not allow these grounds to be disturbed during the breeding season. The lake at Holkham affords no shelter for nesting purposes, with the exception of three small islands, which are always taken possession of by the Canada geese to the exclusion of everything else.

Throughout the "Broad" district, as I know from my own observations and those of correspondents, the shoveler is pretty generally distributed during the summer, yet, whilst the garganey, so far as I can ascertain, but rarely breeds on the meres of this county, the shoveler, at the same season, is far out-

---

* This seems confirmed by a note in the "Field," of August 29th 1874, in which Mr. J. A. Howell states that he shot a young drake shoveler, on the 14th of August of that year, in the Burnham marshes, together with an equally juvenile common mallard.

numbered on the broads by the garganey. On the Yare, between Surlingham and Rockland, a few pairs may be found annually, in spring and summer, at a retired and well preserved spot, known as "Rudd's waters," situated in the marshes between the two broads; and on the opposite side of the river, as recently as the autumn of 1860, young shovelers were killed with other flappers in the marshes at Strumpshaw, no doubt bred on what was once a broad, though now almost entirely grown up. Rockland, as an "open" broad, affords no protection, and, although of late years I have seen shovelers on Surlingham Broad in May and June, and the young birds have been shot there in autumn, the marshmen have always doubted their nesting nearer than the locality I have mentioned.* Fifteen or twenty years ago, I think it probable that they bred occasionally on the drier marshes round the broad itself, as I can remember, on more than one occasion, as I pulled on to the broad in the early morning, watching the low sweeping flight of a cock shoveler, repeated again and again, over some rough tussocky ground lying between the broad and the river; agreeing well with the following description of its habits in the breeding season, contained in a recent letter from Mr. Beverley Leeds: "the male shovelers get to the hens at night, and if put up early in the morning, almost always, in flying round, pass low over the spot where the female is sitting." At that time, however, I was under the impression that their nests were placed in the almost inaccessible swamp. On the Bure and its tributaries I have evidence of its still nesting in the vicinity of many of the broads in that neighbourhood, and under the protection of a "close time" Act sufficiently stringent in its

---

* Mr. Robert Pratt informs me that in the summer of 1877, shovelers were bred on this Broad, which he considers a very unusual circumstance.

penalties, there is no reason why this and other species of fowl should not breed with us as numerously as at any period within the present century. On Hoveton Broad I saw a cock shoveler, on the 27th of April, 1874; and a pair were seen there by Mr. Silcock, on the 12th of May, 1877. Mr. Blofeld, also, informs me that young birds are almost always killed in July with other flappers, but neither old nor young are ever seen in winter. About Stalham and Sutton, though the broads there exist rather in name than reality, being nearly grown up, Mr. B. C. Silcock believes that the shoveler still breeds, at least occasionally, being a regular visitant in spring. He saw a pair at Sutton as late as the 18th of April, in 1872, and at Stalham on the 16th March, 1876, but has never found the nest. At Irstead, Mr. F. Norgate knew of one nest in 1874, but doubts if this species has bred since either at Barton or Irstead; but at Neatishead he heard of a "shovel-billed duck's" nest as recently as June, 1876. Ludham broad, in the same neighbourhood, is now "a thing of the past," no longer, Mr. Silcock informs me, a haunt for fowl. In the summer of 1872, when visiting the the two broads at South Walsham I made particular enquiries after this species and the garganey, of Mr. Jary's keeper who knew of no recent instance of either nesting there; but at Ranworth shovelers are still met with in summer, and protected, as I believe they are on their arrival in spring, will have cause to rejoice over the dismantled pipes of the decoy in which, on the 25th of March, 1863, I saw three or four couples with other fowl, several having been netted and sent to Norwich market only the week before.

Mr. E. T. Booth when at Potter Heigham early in July, 1875, found a "coil" of seven shovelers which, with young garganey, teal, and wild ducks, seen at the same time, were bred, no doubt, in the marshes adjoining

Heigham "Sounds," or the broad at Hickling, where an adult male was killed late in March of the same year. Neither Horsey mere, however, nor that chain of broads at Rollesby, Ormesby, and Filby, near Yarmouth, can now be included, I fear, amongst the regular breeding haunts of the shoveler. As regards the latter, although Mr. Frere, of Yarmouth, has ascertained that young shovelers have been shot at Rollesby, of late years, with other flappers, the Rev. C. T. Lucas, of Burgh, is unable to furnish me with any similar evidence respecting other portions of the same waters; whilst Mr. Rising informed me in 1862 (the same year, I believe, that Winterton decoy was closed), that since he had let the shooting on his estate, both this species and the garganey had ceased to nest at Horsey. Mr. Lubbock, in a communication to Yarrell, speaks of having seen two nests there at different times, "placed in a very dry part of the marsh, at a considerable distance from the broad."

Mr. Fielding Harmer, in some notes kindly supplied me in 1871 on fowl and other birds, observed by himself on Breydon, remarks the scarcity of shovelers both in spring and autumn on those waters of late years. The absence of the young birds at the latter season would be accounted for by the species having ceased to breed in the immediate neighbourhood of Yarmouth.

On the lakes at Gunton, Westwick, Kimberley, and Blickling, all in the eastern division of the county, it appears to be only an occasional visitant with other fowl in spring and autumn, or more rarely still in winter. I have also known it occur in December on the Hempstead ponds, near Holt, where a small decoy, now closed, was formerly worked with great success.

Yarrell remarks that "the shoveler is to be considered generally as a winter visitor to this country," but my own experience, as regards Norfolk, agrees rather with that of Mr. Lubbock, who describes it as "more frequent with us in spring and summer than in the depth of winter"; and Mr. Fenwick Hele, of Aldeburgh, speaks of it as occurring in Suffolk "most frequently in the early spring." On the other hand Mr. Cordeaux, in his "Birds of the Humber district," shows its abundance, in winter, in North Lincolnshire, by the numbers taken in the Ashby decoy,* and adds, "these however, represent only a portion of the shovelers visiting the Ashby pond during that period, as many of the flocks would leave again without entering the nets."

Having carefully tabulated all the specimens entered in my note books from 1850 to 1875 inclusive, I find that I have seen this duck, in a wild state, either alive or freshly killed in every month of the year, but out of a total of sixty examples of both sexes, twenty occurred in March and about half that number in April, mostly in pairs. Of those killed in November, December, January, and February, the numbers are four, seven, two, and one respectively, of which nearly all were males, a circumstance which seems to indicate that in this, as in several other kinds of fowl, there is, to a certain extent, a separation of sexes during the winter months, the ducks and young birds passing south in advance of the drakes; as exemplified, more particularly, in the case of goosanders, merganzers, smews, and golden eyes, of which the old males rarely appear on our coast till after a long and severe frost. Such statistics are, of course, only valuable as indicating the relative proportions in which these birds have visited us, at different seasons, during a period of twenty-five years, and though lately the bulk of the fowl killed in this county has been sent

---

* In thirty-five years, from 1833-34 to 1867-68, two hundred and eighty-five shovelers were netted in this small decoy, the largest number in any one winter (1860-61) being 34.

direct from Yarmouth and other parts of the coast to London, the beauty of the cock shoveler, when in perfect plumage, brings it more generally into our bird-stuffers' hands or the local markets, as a "specimen" than as an article of food.

Until the winter of 1862-3 I had always considered this species as a rarity in winter, having met with it only two or three times in December, and then not in "hard weather," but since then I have known adult males occur during intense frost, both in December and January, more likely, I think, to be birds which have come down to us from more northern localities* than our summer residents still lingering on the coast. Our "home breds," as a rule, leave their inland haunts in autumn; and at Blakeney, Mr. Dowell has met with them in the salt marshes in October, the southward migration commencing probably about the close of that month;† their return, in spring, commencing usually with the second week in March. The only bird (an old male) I ever saw in February, was killed about the middle of the month on Ludham Broad; and the weather being extremely mild at the time had probably something to do with its early appearance.

---

* Mr. Cordeaux gives instances of this species breeding in Yorkshire, Mr. Hancock ("Birds of Northumberland and Durham," pp. 150, 151), of its having done so in Northumberland, and Mr. R. Gray ("Birds of the West of Scotland"), of its occurrence, both in summer and winter, in the north of Scotland. St. John, also, described it as not only breeding but remaining all winter at the loch of Spynie on the Moray Firth, and though apparently not known in either the Orkney or Shetland Isles, stragglers from some of the above localities may visit us during a prolonged frost.

† Colonel Irby ("Ornithology of the Straits of Gibraltar") speaks of the shoveler as mostly arriving on the Spanish side of the straits in October; and Mr. Dresser ("Birds of Europe") gives various authorities for the southward migration of this species occurring on the continent in November.

As long since as the spring of 1818, the late Mr. Youell, of Yarmouth, tried the experiment of rearing young shovelers, by placing eggs, taken in the marshes at Winterton, under domestic fowls, and according to the Rev. W. Whitear* he procured that year upwards of thirty eggs, most of which were hatched out, but succeeded in rearing only two young ones. He also satisfied himself that the bills of young shovelers, when first hatched, are not, as was asserted by some authors of that date, disproportionately large for their size; the beaks of his nestlings, when a few days old, not being "longer than those of the domestic duck," but "at the age of three weeks they had obviously increased in length more than those of the common duckling." Mr. Beverley Leeds, who has recently succeeded in rearing the young of this species, hatched out in like manner, informs me that they are extremely difficult to get on to food, at first refusing corn or meal of any sort, swimming about the pond as if looking for something on the water. He has only succeeded in keeping them alive, the first two or three days, by throwing coarse dry barley meal in the water, which floats on the surface. If they can get insect food or worms they will eat nothing else, but soon die. They will thrive fast on barley and oatmeal, but are always shy birds, "and never seem happy at sundown."

Mr. Cordeaux is the only author, I believe, who refers to the singular habit in this species of swimming round and round each other in circles, but as he speaks only from the observation of others, I have much pleasure in giving Mr. Alfred Newton's version of it from his own experience. It is no amatory action, but the real and only object, he considers, is that of procuring food, as a pair, when feeding, "get opposite to

---

* "Transactions" of the Linnean Society, vol. xiii., p. 615.

one another, and swim round in a circle, holding their heads towards its centre and their bills plunged into the water perpendicularly and up to the base, while their mandibles are employed in 'bibbling,' to use a Norfolk term. They will swim in this way for ten minutes together, always preserving their relative position on the circumference of the circle they are describing, then after a pause, and perhaps a slight removal of a yard or two, they will resume their occupation."

Lubbock gives the word "beck" as a provincial name of this species in Norfolk, and his friend, the late C. S. Girdlestone, of Yarmouth, in a letter to Selby, in 1824, (before referred to) on the provincial names of fowl in this county, speaks of shovelers as "always called becks." The same term also occurs in a communication to Bewick ("Brit. Bds.," vol. ii., p. 328, 1826, and later editions), by Mr. Bonfellow, of Stockton, Norfolk, who, in naming the various kinds of fowl and the months in which each are netted in decoys, gives "becks" amongst others, as taken in March and April. At the present time, however, and indeed within the last quarter of a century, I have never known our marshmen or shore gunners allude to this species otherwise than as shovel-spoon-or broad-bills.*

The cock shoveler, in common with all the males of the fresh water, and most of the more strictly sea

---

* This name for the shoveler and those of "arps" and "cricks" for tufted ducks and teal, also mentioned by Girdlestone and Bonfellow, were names used, I imagine, mainly by decoymen, and probably were introduced into the neighbourhood of Yarmouth, years ago, by old Skelton, a noted Lincolnshire decoyman, who made and worked the Winterton and other decoys in Norfolk early in the present century. These provincialisms do not occur in Ray's "collection of English words not generally used," either in his "List of North," or "South and East country words," published in the same volume as his "Catalogue of English Birds."

ducks, acquires that strange assimilation of plumage to the sombre garb of the duck, at the close of the breeding season, so happily termed by Waterton in the case of the mallard, the state of "eclipse." The cause of this remarkable change of plumage is still unaccounted for, and the time of its commencement, completion, and abandonment, varies considerably in different individuals. One male, that I examined carefully through a telescope on Stanford mere, showed scarcely any trace of this change coming on the 8th of June, but another, seen at Thompson,* the following day, had a very perceptible tinge of brown pervading the glossy green feathers of the head. Several common mallards on the same waters were equally variable, one or two of them being apparently in perfect plumage. The following table,

Species under examination.	Time of beginning to lose the full male plumage.	Time of having completed the process of losing full male plumage.	Time of beginning to reassume the full male plumage.	Time of having completed the re-assumption of full male plumage.
Hybrid between a Mallard and Pintail	May 16	July 2	August 8	October 22
Gadwall	June 2	,, 7	,, 4	,, 5
Mallard	,, 6	August 4	Not observed	,, 5
Garganey	,, 16	,, 4	December 28	February 18
Widgeon	,, 17	,, 20	September 19	November 10
Carolina Wood Duck	,, 17	July 23	August 8	October 1
Pintail	,, 20	,, 18	November 8	January 9
Shoveller	,, 30	August 10	,, 10	February 18
Teal	July 2	July 27	October 1	November 3

the result of the daily observation of certain pinioned birds on a pond at Easton, near Norwich, was published in the "Zoologist" for 1851 (p. 3116) by Mr. J. H. Gurney, from notes supplied him by his gardener, Mr. Annes (a very reliable authority), and though relating to fowl in

---

* Though commonly so called the proper name of this locality, I am informed, is Tomston.

a semi-domesticated state is still of considerable interest as showing the relative dates of commencement and abandonment of this remarkable change of plumage in the various species named, the opportunity of making such observations, also, rarely occurring. In this instance it will be seen that whilst the common mallard commenced its change by the 6th of June, and was himself again by the 5th of October, the drake shoveler did not change till the 30th of June, and, like its companion the garganey, had not re-assumed its proper dress till the 18th of February; the latest of all the species under examination. Mr. Alfred Newton also tells me he had an old cock shoveler whose custom it was not to be in full breeding dress till after Christmas. Judging, however, from my own notes respecting male shovelers seen in our markets, or in the bird-stuffers' hands, in autumn and winter, I come to the conclusion that in a wild state the full nuptial dress is re-assumed towards the close of November or early in the following month. Thus, in 1867, I find an adult male entered as "still in change," on the 12th of October; two adult males as "perfect" on the 23rd and 25th of November, 1865; one as "still showing some light feathers on the head and neck, and a few crescent-shaped feathers on the upper part of the breast on each side," on December 5th, 1867; and another as "perfect" in January, 1868. These, I believe, so far as I was able to judge at the time, were all old males, but it is sometimes difficult and requires a little consideration to discriminate between adult males recovering from their "eclipse," and young males, of the year, assuming their breeding plumage for the first time.

One of the two survivors of Mr. Youell's brood, hatched out in confinement in the summer of 1848, proved to be a male, and having "lived till it was ten months old, had then attained, in a considerable degree, the adult plumage of the shoveler." Now supposing that this bird was hatched by the end of May (Mr. Salmon found a nest of eight eggs on the 10th) it would not have acquired its mature dress before the end of the following March or beginning of April; yet even this would not have been so abnormally late, in this respect, as the old male shoveler on the Easton pond was in the re-assumption of its breeding plumage, since I have seen a young male in about half change on the 24th of January, and one with still slight traces of immaturity in the breast feathers in the last week in March. It would seem, therefore, that whilst the old males in November and December are rapidly perfecting their distinctive garb, the young males of the year at the same period still combine the characteristics of both sexes. Two important points of difference between the adult male shoveler in its post nuptial dress and the female are thus given in Dresser's "Birds of Europe": in the male "the bright green speculum and blue on the wings which it never casts"; in the female the iris being hazel brown instead of yellow as in the male.

Mr. J. H. Gurney, jun., has kindly supplied me with notes on the peculiar ruddy tinge observed by himself on the breast of hen shovelers in the London markets, between November and April. Under date of March 22nd, 1871, he writes, "I saw at least fifty! the breasts of all the females were more or less rufous. I do not think it was the rusty stain which is often seen in pintails, teal, &c., in the spring especially, but the real colour of the plumage." In this I quite concur, having seen one or two of these rufous females in Mr. Gurney's collection, but, whilst the majority of the shovelers seen by him in Leadenhall and Newgate markets were, no doubt, Dutch fowl, it is somewhat singular that I can recall no instance of a rufous-breasted hen shoveler coming under my notice in this county. In 1870, as early as the 28th of August, Mr. Gurney saw one with a red breast, but has met with others in October, November, and March, with the ordinary light brown colour on the under-parts, and, as the rufous tinged birds in November have also bluish wings, they might, as he remarks, be easily mistaken, without dissection, for young males. Yarrell does not allude to this peculiarity of plumage in the female, but Montagu describes a specimen sent to him on the 24th of November, 1809, which, from its weight and size and rufous colour on the breast, he concluded was a male, till satisfied to the contrary by dissection. This proved, by the worn appearance of some feathers still unmoulted, to be a fully adult bird; yet a female in my possession, killed on the 28th of March, 1861, has the same indications of age, but with the breast of the ordinary tint. If not, then, a matter of age, is it a seasonal change, only, in individuals, or a climatal variation in specimens from certain parts of the continent?

In size and weight both male and female shovelers vary considerably. Of a pair purchased at Lynn, in April, 1869, by Mr. Wilson, a bird preserver, the male was but eighteen inches in length, and the female less; the former weighing one pound and a quarter and the latter one pound. Montagu also describes a small pair, taken in a Lincolnshire decoy, of which "the male was fat, and yet weighed only seventeen ounces, the female was rather poor, and weighed no more than ten ounces and a half, which is less than that of a teal." His rufous-breasted female weighed twenty ounces and a half, and measured twenty inches in length. The ordinary length of the male is from twenty to twenty-one inches, of the female about eighteen inches, and the weight, according to Hunt, from seventeen to twenty-two ounces.

---

* Supplement to the "Ornithological Dictionary"—1813.

As to the qualities of the shoveler as an article of diet, I can endorse the high encomiums of most authors, and can only imagine them "very inferior eating," as described in Shelley's "Birds of Egypt" (an opinion concurred in by Mr. J. H. Gurney, jun., from recent experience in the same country), when killed out of season or in brackish marshes.

ANAS STREPERA, Linnæus.

GADWALL.

The Gadwall, though nowhere abundant in comparison with many other species, is apparently a more regular visitant to the coast of Norfolk, during the months of winter and spring, than to any other of our maritime counties, and the fact of its being not uncommon on the opposite shores of Holland* in winter, may account for the majority of specimens, obtained in this county, occurring in the neighbourhood of Yarmouth and in the Broad district generally. Mr. Lubbock says of this species (1845), "it is generally to be seen in the Norwich Market once or twice in the winter," and such would probably be the case now were it not that the greater

---

* M. Julian Deby, in his interesting notes on the "Birds of Belgium" ("Zool.," 1845, p. 1189), describes this duck as "common on our marshes, rivers, and lakes," arriving with the first frost, and leaving in March and April ; which agrees also with the experience of Baron de Selys Longchamps.

The extensive range of this species, in Europe, Asia, Africa, and throughout a large portion of Northern and Central America, is exhaustively shown in Dresser's "Birds of Europe," and he quotes Naumann to the effect that in Germany "small flocks of eight or ten individuals are more often seen than single pairs; but, on the other hand, as large flocks as from thirty to fifty individuals are more seldom observed."

portion of the fowl, taken on our coast, is sent by rail to London, where, as well as at Brighton, the demand is greater; and Norfolk rarities, when the dealers' statements can be relied on, are quite as likely to be met with in Leadenhall Market as at Norwich or Lynn. Nearly all that I have seen of late years have been in our birdstuffers' hands, and occasionally a season or two passes over without a single specimen being thus noticed. The Messrs. Paget, writing in 1834, remark that "two or three are generally shot every year on Breydon;" and still earlier records prove their partiality then, as at the present time, for the fresh waters of the more inland broads. The specimens from which Hunt's drawings ("Brit. Ornithology," vol. ii.), were made, were killed at Ormesby in April, 1818; and a pair which I purchased at the sale of the late Mr. Spalding's collection at Westleton, in 1872, were bought by him, in the flesh, at Yarmouth, about fifty years ago. Mr. Lubbock, who received an adult male from Sutton, in January, 1832, also records three as taken with other fowl in Ranworth decoy in the winter of 1841.

Of some five and twenty examples of this duck that have come under my notice since 1851,* two-thirds, at least, have occurred on the Yarmouth side of the county, Hickling and Ludham Broads (the latter now grown up), next to the tidal waters of Breydon, having afforded apparently the chief attractions; whilst occasional stragglers have been met with very far inland. In January, 1852, an adult male was shot on the river at Heigham, near Norwich; another in February of the the same year on Lord Kimberley's lake, near Wymond-

---

* Mr. Cordeaux describes it as one of the rarest ducks in the Humber district, and states that only *twenty-two* are recorded in the Ashby decoy book as taken "between the winters of 1833-34 and 1867-68 inclusive, three being the largest number in any one year."

ham; and in January, 1865, a single bird was shot on the Yare, near the entrance to Surlingham Broad, and one at Reedham, on the same stream, in March, 1870. An old male, in my own collection, was killed at Ludham, in April, 1858, but adult males, as in several other species of wildfowl, are less frequently met with than females and young birds; and, judging from the records in my note books of such only as may fairly be considered migrants, I should say that a few make their appearance on our coast in November and December, more, as a rule, in January and February, and a few pairs occasionally in March and April, as late even, in some seasons, as the 19th and 23rd of the latter month.

Mr. Dowell informs me that in the month of May, 1846, he had two male gadwalls brought to him in the flesh, and in his notes describes this species as neither common nor yet very rare at that time, in the Blakeney marshes, in winter and spring, where it is known, he says, as the "grey duck;"* and Overton, a then noted gunner, used to describe it as "the yellow-legged mallard, with shell markings on the breast." One was seen in the Holkham marshes, in April, 1854, but from this point of the coast, passing westward towards Lynn, my records of specimens obtained are few and far between. A single bird was shot at Brancaster in

---

* In Ray's "Catalogue of English Birds" (1674) this species is entered as "the gadwall or gray *Boschas minor torquata*." Hunt, also, gives the term "rodge" as a provincialism in Norfolk, but it is apparently obsolete at the present time.

In Spain, in the Coto de Donana, Lord Lilford ascertained that the local name of this species was "Frisa," a quaint term (referring, no doubt, to the appearance, not the texture, of certain portions of its plumage) signifying coarse cloth or frieze. See Irby's "Ornithology of the Straits of Gibraltar."

January, 1870; and an adult male at Hunstanton, during the severe frost that prevailed in February, 1871;* but from the fact that for more than twenty years a naturalized race of gadwalls—the descendants of a pair of wild birds taken in a decoy and turned off, when pinioned, on the lake at Narford—have bred regularly and numerously in the neighbourhood, it is impossible, at least on that side of the county, to determine true migrants from those reared in that part of Norfolk.

For the history of these Norfolk gadwalls, I am indebted to the Rev. John Fountaine, of Southacre, who informs me that, so far as his memory serves him, the original pair were taken, about five-and-twenty years ago, in Dersingham decoy (near Lynn), and were given to him by George Skelton, who made and resided at that decoy, which is now destroyed. He says, "I cut a very small portion off the pinions of these two birds, so that they were able to fly for a considerable time, but no doubt dared not trust themselves to the regular spring flight of migration when the other fowl left; the result was that they bred upon the lake at Narford, and their progeny have continued to do so ever since. The greatest number I ever counted in any one year was seventy, but of course their numbers were much reduced by shooting in the neighbourhood, or there would have been many more. This, in a thickly populated country like England, renders the introduction of wild species almost impossible."

That many of these ducks, as Mr. Fountaine states,

---

* Mr. Baker, of Cambridge, in some notes on Norfolk birds sent to him for preservation, mentions a mature male and two female gadwalls, shot at Littleport, March 5th, 1865, but this locality is in the Isle of Ely beyond the western limits of our "Fen" district.

THE BIRDS OF NORFOLK;

BEGUN BY THE LATE

HENRY STEVENSON, F.L.S.,

AND AFTER HIS DEATH CONTINUED BY

THOMAS SOUTHWELL.

[*All efforts to find the remaining portion of the article printed in 1877, which breaks off at the end of the last page, having failed, though it is known that the article had been completed, I have done my best from other sources to supply the deficiency.*]

The letter from the Rev. John Fountaine, quoted by Mr. Stevenson on the preceding page, is dated May 8th, 1875, which fixes the year 1850 as the date of the successful introduction of the gadwall at Narford.

Since that time the number of these birds breeding in Norfolk has greatly increased, as also the area over which they are spread. Whether or not the large number of gadwalls which now nest yearly in the south-western portion of the county are all descendants of Mr. Fountaine's birds, or, as seems probable, their numbers have been increased by wild birds attracted by those which had become naturalised, it is certain that a large number of these fowl breed every season in particular localities, and that the number of summer residents far exceeds that of the autumn and winter visitants.

On the 6th June, 1881, one of Lord Walsingham's keepers showed me two nests of the gadwall at Stanford, and told me he had found nine nests that season. Those which I saw were placed on sedge tussocks, one, in a very boggy situation, containing ten eggs; the other, which had four eggs, was also on the top of a tussock of sedge, growing out of the shallow water some yards from the shore. Both nests were deep and compact, constructed of dry grasses and sedges, but contained very little down or feathers. In 1882 I paid two visits to the meres on Wretham Heath; the second time, on the 27th June, I was accompanied by Mr. Stevenson and Mr. J. H. Gurney, jun., and for the first time in that locality we had the pleasure of seeing the gadwall. It is not improbable, as not one of us had seen these birds on Wretham Heath, that it was the first year of their nesting there.

On the 22nd April, 1884, Professor Newton was shown two nests at Tottington, and on the 29th May of the same year I found a nest of newly-hatched young on Wretham Heath (some miles from the mere visited by Professor Newton), from which I put off the old bird. The young scattered in all directions, but I caught one, and while I held it in my hand the old duck showed her anxiety by fluttering, apparently broken-winged, almost at my feet, at the same time calling loudly.

In 1886 others were seen by Mr. Frank Norgate, who had exceptionally good opportunities of observing them, and has very kindly supplied me with some valuable notes on this and other species of ducks, which I shall frequently have to refer to. Mr. Norgate first identified a pair on May 8th. On May 18th he saw a nest containing eleven fresh cream-coloured eggs, from which he put off a duck gadwall, it was constructed chiefly of dead leaves with very little down, and situated on the shore of an island. On May 22nd Mr. Norgate again saw the pair of gadwalls to which this nest belonged, flying and swimming together, and remarks that "each in swimming appeared to have one leg resting over the tail. When flying they soared to a great height, playing with each other in the air, and then descended almost perpendicularly to the water, on which they float high as if more buoyant than most other ducks." Professor Newton ("Encyclop. Brit.," ed. 9, Article Gadwall) remarks upon the peculiar seat upon the water of this species, as follows: "Its appearance on the water is very different [from the wild duck], its small head, flat back, elongated form, and elevated stern rendering it recognisable by the fowler, even at such a distance as hinders him from seeing its very distinct plumage."

I have never seen the nest of a gadwall far from the the water; it is generally placed either in a very boggy spot, or in a tussock of sedge, by which it is raised above the shallow water itself. In such situations it is constructed of dead grass or sedges, and very sparingly lined with down. The usual complement of eggs seems to be from ten to thirteen.

I am not aware that the gadwall has been actually detected nesting in the Broads, but Mr. A. H. Evans was told that two of these birds were seen on Hickling Broad for the week or two preceding May 22nd, 1880. On the 31st January, 1889, Lord Walsingham killed at Stanford forty-one ducks, six of which were gadwalls.

Locally this species is known as the "Grey-duck" or "Heart-duck."

ANAS ACUTA, Linnaeus.

PINTAIL.

Although sparingly met with in most years from October to March, this elegant duck is at no time numerous in Norfolk. Occasionally it is taken in decoys, but I have records of only eighteen such occurrences. Seven of these were as follows: one in October, two in November, one in December, two in January, and one in March, so that they seem to be pretty equally distributed over the season. Its well-known preference for fresh water would render the more frequent capture of this species in decoys probable.

The pintail breeds readily in confinement, and many hybrids with the common duck have been known.* Lord Walsingham informs me that for the last two years (1887 and 1888) a pinioned pair have bred on his waters at Stanford, and it is much to be hoped that he may be as successful in naturalising this pretty species as Mr. Fountaine was with the gadwall.

The Rev. E. W. Dowell thirty years ago used to find the pintail not uncommon on the marshes at Blakeney in winter, and has seen them going north, in the month of March, already paired, and at such times in very poor condition.

In the cold spring of 1855 Mr. Stevenson says pintails were seen on Ranworth Broad as late as the 7th April, and Mr. J. H. Gurney, jun., has a male in full plumage, showing some of the rusty colour underneath, which was killed at Horsey as late as the 14th May, 1842. Although there is no positive knowledge of its having bred in this county in a state of nature, Hunt ("British Ornithology" ii., p. 293) believed that such was occasionally the case, and says, "we have seen several specimens exposed for sale in Norwich Market, during the months of June and July, at which times they appeared to be in an intermediate state of plumage between the two sexes." Mr. Booth writes ("Rough Notes," part xiii.) under date of the 28th May, 1878, "a pair were often observed on and around the 'Hills' on Hickling Broad, and that latterly the drake (an exceedingly brightly marked bird) was usually seen alone. I also ascertained that, a short time previously, a duck's nest with eight eggs, supposed to be that of a shoveler, had been taken by a marshman on the same hill these birds frequented. As it is improbable that the natives were well acquainted with the eggs of either species, this clutch may possibly have belonged to the pintail."

It was probably this species which Sir Thomas Browne calls the "sea phaysant, holding some resemblance unto that bird in some fethers in the tayle,"

---

*Cf. Newton, Proc. Zool. Soc., 1860, p. 338, pl. 1·8, for a remarkable instance of the hybrid offspring of a common duck and a pintail proving fertile *inter se*.

a name by which it is even now known, although there is nothing to prove that it was not the long-tailed duck to which he applied the name. From a gastronomic point of view both Folkhard and Payne-Gallwey give the pintail a foremost place, and the latter says they are always looked upon by wild-fowl shooters as a great prize.

ANAS BOSCAS, Linnæus.

THE WILD DUCK.

This species, although never found in the large flocks so characteristic of the wigeon, is very numerous as a winter visitant, and forms the staple of the decoyman's return. In the early autumn the home-bred birds resort to the larger waters of the Broads or inland lakes, soon, however, to move on south, their places being taken by "foreign" fowl, which arrive from the north, and in their turn continue their southern course, should the weather prove severe. On the lake at Holkham, Mr. Alexander Napier tells me, the home-bred mallards put in an appearance early in August, being the first fowl which arrive; the main body of the foreign-bred mallards do not arrive till the beginning of December, sometimes even later, but their movements, like those of the wigeon, are greatly governed by the weather.

It is seldom the wild duck is distressed for food, and even when frozen out, the supply of acorns, which are almost always to be had, even in snowy weather, proves a great attraction.

A large number of wild ducks nest every year in Norfolk, generally dispersed over the county, and, although the greater number are produced in the Broad district, there is scarcely a stream or piece of water of any extent which does not form a nursery for a brood or two at some time. Our sluggish rivers, meandering through a flat country, and in many places flanked by damp woods and cars, are a source of great attraction to these birds, but many nest on the dry, open heaths, at a distance of a couple of miles from any water, under the shelter of a whin bush or a clump of brakes, whence the old birds lead their young ones to the nearest water.

The nest is generally constructed of dry grass, sedge, or dead leaves, sometimes of moss. To this the bird, as soon as she commences to set, adds down from her own breast till at last the nest becomes thickly lined with feathers and down. Mr. Lubbock, writing fifty years ago, states that in some years from 1,200 to 1,300 fowl were bred at Ranworth, which is exceptionally situated, the rising ground, covered with wood and thicket round the Broad, forming a great attraction for ducks; but taking the Broad district as a whole, I do not think there are at present so many ducks bred there as one would be led to expect; they seem to prefer less extensive waters and drier ground than is generally to be found in the marshes contiguous to the Broads. One curious circumstance in connection with the nesting of the wild duck is the frequency of the occurrence of pheasants' or partridges' eggs in their nests; many such instances have come under my observation, and I have frequently heard of others. The partridge occasionally makes use of the comfortable nest of the duck as a receptacle for its eggs, but not, I believe, so frequently as the pheasant. When the proud mother marshals her young ones, to conduct them to the water, great must be her surprise at the ugly ducklings which form part of her brood.

It not unfrequently happens that ducks depart from their usual habit of nesting on the ground, and resort to trees for that purpose. Hunt ("British Ornithology," ii., p. 322) mentions that a huntsman to the then Mr. T. W. Coke, in the year 1782, killed a duck from her nest in a lofty Scotch fir in Holkham Park, much to his surprise, fully expecting to have killed a hawk. Mr. Stevenson, in his journal, mentions a similar case: "Near Fakenham, in 1850, three ducks were found in a plantation by the gamekeeper breeding in tall trees about twenty feet from the ground, and using the deserted nests of woodpigeons for that purpose; one had laid twelve, one seven, and one five eggs." This, if my memory serves me correctly, occurred at Raynham; and, at the same place, I was shown, five and twenty years ago, the nest of a wild duck in some ivy on the top of a wall, near the moat at Raynham Hall. Mr. J. H. Gurney, jun., tells me, a wild duck hatched out in a pollard willow tree, near Keswick Mill, in 1888 ; and Mr. Norgate, on 20th April, 1884, " saw a wild duck fly off its downy nest of ten eggs on the top of the trunk of a pollard willow ten feet from the ground."

The wild duck is a very early breeder, but the date of nesting varies considerably. Lord Lilford records a duck's nest with eggs, I think at Lilford, on the 26th February, 1885. I have known them on the 16th March with a full complement of eggs, and others on 17th March (fourteen eggs), 6th April (eighteen eggs).

Mr. F. Norgate has also kindly favoured me with the following dates of wild ducks' nests found by himself in various parts of the county :—April 2nd, twelve eggs ; April 9th, five eggs ; April 9th or 10th, eleven eggs, sat on ; April 15th, nine eggs ; April 20th, ten eggs ; April 21st, already hatched off ; same date, one with eleven eggs ; and on the same day (but not the same year), another with eleven ; April 30th, thirteen eggs ; May 14th, eleven eggs ; May 28th (young) ; June 14th, twelve eggs. Mr. Norgate has also seen wild ducks which had paired, and probably chosen their nesting-place, on the 8th, 21st, and 28th February. Mr. J. H. Gurney, jun., has given me notes of young ducks at Hempstead, hatched on 7th April, others seen on 10th, and young birds seen in the water on the 28th of the same month. It is evident from these dates that many ducks must have paired and chosen their nesting-places in February, and, probably, had eggs in the middle of March.

This is a matter of considerable importance, as it should of course be the chief guide in fixing the date on which the close time ought to commence. The 1st of March, as at present arranged, does not appear at all too early for this species, for it must be borne in mind that many migrating wild ducks are passing northward after our home birds have settled their domestic arrangements and commenced housekeeping, and as it is impossible to distinguish between home birds and migrants, we must either be content to allow these migratory birds to pass unmolested, or, in bringing them to bag, we shall drive away the resident birds, with the result that there will be no flapper shooting in the autumn.

A great enemy to ducks in the nesting time is the rat; those arrant poachers, more worthy of the gamekeeper's attention than the birds of prey, which are so useful in thinning their numbers, are exceedingly destructive to the eggs of this species; it is not surprising, therefore, that their nests should be so frequently found to contain nothing but empty egg shells. Nor are the young birds out of danger when they take to the water; for, doubtless, many in their early days fall a prey to the monster pike which are so common in our waters; it is quite possible also that the eels claim their tribute. These causes are a serious check upon the increase of the various water fowl, and, in some cases, I believe, have proved destructive to whole broods.

The remarkable change which takes place at certain periods of the year in the plumage of the males of many species of duck, by which they assume for a time the appearance of the females of their species, or, as Charles Waterton so aptly expresses it, "the drake goes, as it were, into an eclipse," is very observable in this species. Some interesting remaks upon the subject will be found *ante*, p. 153.

As might be expected with a species which is so readily domesticated, hybrids are not at all uncommon; it has been recorded as breeding with more than a dozen others, particulars of which have been collected by Baron de Selys Longchamps ("Bull. de l'Acad. Roy. de Belgique," 1845, 1856), and Mons. Suchetet ("Note sur les hybrides des Anatidés." Rouen, 1888).

Varieties of this species are not unfrequently met with; albinos are said to be very rare indeed, and I am not aware of such an occurrence in this county, but in some loose notes by the late Mr. Robert Rising, of Horsey, now, I believe, in the possession of Mr. Colman, he mentions two cream-coloured mallards killed at Horsey on 2nd September, 1853. Mr. Lubbock, in an interleaved copy of Bewick, mentions that a black variety, "a genuine wild bird," was taken in 1857 by his friend, Mr. Kerrison, in the decoy at Ranworth, "coal-black, with bottle-green reflections on the head and neck." In February, 1838, Mr. J. H. Gurney procured a variety of the wild duck from Palling-by-the-Sea, quite black except the head, which was green, as usual, with a white neck ring; and Colonel Feilden has a curious variety which was shot at "flighting" near Herringfleet, in February, 1886. It is a female, rather undersized, and the whole of the under surface, from the base of the lower mandible to the vent, he describes as being of a buff colour, approaching that of the goosander, paler towards the vent.

"In the winter of 1854," says an entry in Mr. Stevenson's note-book, "a singular looking drake was netted with other fowl in the Ranworth decoy, and, being preserved and pinioned at the time, from its unusual appearance, has been since kept alive upon a pond in Mr. Kerrison's garden. Through the kindness of that gentleman I had an opportunity of examining this bird in the following winter, when it had once more resumed its distinctive markings, after assuming with other drakes the female plumage during the nesting season. At that time the top of the head and back of the neck were green, mixed with buff colour on the cheeks, but no white ring. Two yellow streaks extended backwards from the beak over each eye; the throat, and under parts to the vent, reddish brown; sides speckled grey, back varied with light and dark shades of grey, legs orange red, and no curl feathers in the tail. The bill resembled the wild duck's, but the voice was different from the common mallard's. This bird paired in 1856 with a domestic duck, which hatched off nine young ones from her eggs, all the young ones showing more or less in their first plumage the light streaks over the eyes, the most marked feature in the old male. All the brood but one duck were, unfortunately, killed by rats. She escaped from the pond and paired with a wild mallard. In 1857 one or two of her brood were taken, and were found to exhibit traces of the original stock. In the same year the old drake paired with a wild duck, and ten young ones were hatched, still showing the eyebrows distinct, but these all died from cold."

Mr. Stevenson adds, "were this bird merely a variety of the common wild duck, it is scarcely probable that its marked peculiarities, more particularly the clearly defined streaks over the eye before alluded to, would not only be reproduced in its offspring even to the second generation, but also reappear year after year in its own plumage after every moult. Supposing it, however, to be really a hybrid, the question as to its parentage is one of no little difficulty, but if the teal and the wild duck have consorted together (see Bimaculated duck), why not also the latter with the garganey?" This singular bird was stolen from Mr. Kerrison in December, 1865, after having lived in captivity eleven years. Two of the young ones in the down, hatched in 1857, were long in Mr. Stevenson's collection, and were sold at his second auction on 21st March, 1889. They showed the light eye-stripe very distinctly.

On the 7th February, 1883, a female wild duck died at Northrepps, aged twenty-nine years; it was hatched there in 1854, and was blind for several months before its death. For about eight years it had been in nearly complete drake's plumage, the exception being a few brown feathers mingled with the green on the side of the head and neck, and a few normal feathers on the flanks.

Mr. Norgate tells me that the parasites which he found infesting the plumage of the wild duck, in Norfolk, were identified by the Rev. C. R. N. Burrows as *Docophorus icterodes*.

DECOYS IN NORFOLK.

Sir Thomas Browne, writing about the year 1663, refers to "the condition of the county [of Norfolk] and the very many decoys, especially between Norwich and the sea, making this place very much to abound in wild fowl." It may seem strange to speak of the decoy, perhaps the most deadly engine ever invented for the purpose of luring wild-fowl to their destruction, as being at the same time favourable to their abundance, but it is strictly in accordance with fact. The great attraction of the decoy-pond is its absolute seclusion; here the fowl which return in the early morning from their nocturnal feeding-grounds find perfect rest, and pass their time peacefully in happy unconsciousness of the slaughter which may be going on within a few yards. It must be remembered the number killed in any decoy is seldom more than a tithe of that frequenting the pond, and doubtless the quiet which all sorts of fowl enjoyed in the precincts of the decoy-pond at a time when the gun was used indiscriminately at all seasons of the year, and when the shooting of breeding ducks was thought no sin, has resulted in saving yet a remnant of home-bred birds to continue their race in the happier times which have succeeded. Since the passing of the Wild Fowl Act, and the imposition of the not less salutary licence on guns, the quantity of ducks, both of species and individuals, breeding in Norfolk, has increased surprisingly, and some species of marsh-breeding birds have responded to the protection afforded them in a manner most encouraging to those who have been wise enough to shelter them.

Norfolk is still the land of decoys. It has at present time seven active decoys, including those in the district known as Lothingland, which, for ornithological purposes, must be taken as belonging to this county, although politically it is assigned to Suffolk, viz., one at Southacre, on Mr. Fountaine's estate; one at Westwick, belonging to Mrs. Petre; a third at Wretham; another at Didlington, the seat of Mr. W. A. Tyssen Amherst, M.P.; a recently constructed one at Merton, on Lord Walsingham's estate; and two others on the lake at Fritton.

There are also twenty-four decoys which have become disused in more or less recent times. These were situated at Winterton, Waxham, Ranworth, Mautby, Acle, Woodbastwick, Hemsby, Gunton, Cawston Woodrow, Wolverton, Hillington, Hempstead, Langham (worked by the celebrated nautical novelist, Captain Marryatt), Holkham, Dersingham, Wormegay, Narford, Stow Bardolph, Hilgay, Northwold, Hockwold, Lakenheath (on the Suffolk side of the boundary river), Besthorpe, and Flixton, near Lowestoft, surely a goodly array for one county and its immediate vicinity.

Some of these decoys have been in times past very productive, and even now in favourable seasons the number of fowl taken in the comparatively few still in use is very considerable.

Many interesting particulars with regard to the formation and history of these local decoys will be found in Mr. Lubbock's "Fauna of Norfolk," 2nd ed., pp. 134 and 220; also in a paper printed in the "Transactions of the Norfolk and Norwich Naturalists' Society," ii., p. 538,* to which I must refer the reader for further information on this branch of the subject, as the accounts there given are much more complete than space will here permit of, but I should like to describe very briefly the mode in which a decoy is worked, and for this purpose will give the actual experience of a recent visit to Fritton.

Through the right-hand front reed-screen of the decoy is thrust a flat slip of wood about an inch wide; by turning this slip in a horizontal position a slight opening is made in the screen through which the decoy-man has an excellent view of the pond and all that is taking place in front of, or on the "banks" leading to the mouth of the "pipe." Here he makes his inspection before deciding whether or not to commence operations. I know of no prettier sight, or one more instructive to a naturalist, than a peep through this "spy-hole," whether in summer or winter. Here, as Mr. Lubbock remarks, "you see nature as she really is," and such birds or beasts as are on the move may be watched probably within a few feet of the spectator, engaged in their various occupations, perfectly unconscious of the close proximity of their most dreaded enemy—man. One lovely summer's evening I thus made the near acquaintance of a pair of coots—the most watchful of birds—which were busily foraging in the shallow water; on the right-hand "bank," close to the water's edge, was a pied wagtail intent on securing a supper of gnats; whilst from the opposite bank, as I looked, came leisurely swimming across, picking its way through a bed of *Polygonum*

---

* For a very complete description of the mode of constructing and working a decoy, as well as many interesting accounts of decoys ancient and modern, see "The Book of Duck Decoys," by Sir Ralph Payne-Galwey, Bart. London, Van Voorst, 1886.

amphibium, the lovely rosy-red blossoms of which, lit up by the oblique rays of the sun, were indescribably beautiful, a large water-rat, looking quite a monster as its long coat floated on the surface of the water; on reaching the shore it mounted on the broken-down leaf of a great water-dock, and deliberately commenced making its toilet in the prettiest manner conceivable, its bright eyes glancing round with an air of perfect confidence and security, and this, as I have said, almost within reach of my hand. Very different was the scene presented at the same peep-hole on the occasion of another visit. About three o'clock on a December afternoon, in company with Page, the decoyman, we came down to inspect the water, with a view, if possible, to making a capture. Out on the lake were many fowl, but they were too far off to render it probable that they would enter the pipe. Much closer were several wild ducks, floating quietly on the surface, apparently sleeping peacefully with their heads comfortably tucked back under their feathers. We put the dog round the front screen, but it was useless, the birds would not "work." In the distance, however, not far from the entrance to the next pipe, we saw a bunch of teal which looked more promising, so we went round to try them. On looking through the screen, there were the lazy fowl out in the distance, whilst upon the bank a duck and mallard sat intent on making what proved to be their last toilet, and the teal were well placed in the foreground, wide-awake and busy. Through went the dog, running round the wing screen and coming to hand beautifully. In a moment all was excitement, the teal had their heads up, and the ducks on the bank slipped into the water, with "what was it?" plainly expressed in their whole bearing. They were not left long in suspense; round went the dog at the next screen, and, as if by magic, the two ducks and all the teal headed up towards the pipe; on went the dog, passing round screen after screen obedient to the motions of the decoyman, the birds following him up now in hot haste, till the fatal spot in the curving pipe, which hid them from the view of the fowl outside, was passed, when the decoyman ran rapidly back under cover of the outer screen, till he was behind the fowl, and showing himself, handkerchief waving in hand, in the rear of the birds, up they all rose in haste, still heading up the pipe, rapidly followed by the man, who showed himself at every screen in succession, till they had rushed headlong into the tunnel net, which, by a rapid motion of his hand, the decoyman disconnected from the tail of the pipe, at the same time giving it a twist, and, barring their retreat, the poor birds were thus hopelessly entrapped. There they lay, nineteen teal and two ducks, with, I fancied, a curiously dazed expression in their beautiful bright eyes, as though the whole thing were too sudden and confusing for their comprehension. An amusing anecdote is told of the late Rev. John Fountaine at the opening of Didlington decoy. The Duke of Leeds being present, Mr. Fountaine was anxious to make a successful catch; but, alas! by some unfortunate mistake, he forgot to attach the tunnel-net, and all the fowl rushed out at the tail of the pipe! The duke is said to have remarked encouragingly, "Well done, Mr. Fountaine! now, can you put them through again, may I ask?" I must plead guilty to a wish, when I saw the poor birds in durance vile, that Page had made the same mistake; but no, there they were, and from the last scene of all I turned away with the consolation that the operation would be performed so skilfully as to inflict the minimum amount of suffering.

Of course the mallard, the wigeon, and the teal form the chief product of the decoys, although various other species are occasionally taken. Thus, in 1864, at the Ranworth decoy, now disused, there were secured eight hundred and seventy-seven ducks, seventy teal, eight wigeon, three shovelers, one pintail, one tufted duck, and one goosander, an unusually great variety. Occasionally garganey were taken at Ranworth in the month of March, and various other strange fowl sometimes entered the net, as, for instance, woodcocks and pheasants. The winter of 1879-80 was a very favourable one for decoys, and at Fritton that season 2,218 ducks, 123 teal, and 70 wigeon were netted, 1,613 of these in December, 1879. As a rule a few home-bred ducks are taken in October and November, as, should food be plentiful and the place quiet, the "flappers" which have collected in the decoy pond, are reluctant to leave for the south, but in November and December the foreign birds arrive, and it is during these months most of the teal are taken. The wigeon here pass on early, and do not occur in any numbers on their return journey until early in March; consequently, owing to the operation of the Wild Birds Protection Act, very few, as a rule, are now captured.

Some idea of the number of fowl killed in this county in former times may be formed from Messrs. Paget's statement that in the winter of 1829 one game dealer, in Yarmouth, had in a single day "no less than 400 wild-fowl of different kinds, 500 snipe, and 150 golden plover," brought in for sale; and that from the same district in ordinary years about fifty per week, on an average, were sent to London throughout the season, which generally lasts five months, viz., from October to the beginning of March or April! Even in the present day a sharp winter brings a great influx of fowl. Mr. Stevenson was indebted to Mr. Bellin, of Yarmouth, for the following list of wild-fowl received by one only of the game dealers in that town, between the 14th and 28th of December, 1878, when we were visited by an exceptionally severe spell of weather :—

Wildfowl, Waders, &c., received from December 14th to 21st, 1878.

Full Snipe	447	Herons	4
Jack Snipe	21	Kingfishers	3
Green and Golden Plover	206	Teal	55
Grey Plover	3	Golden-eyes and other fowl	147
Woodcocks	14		
Water-hens	41	Duck and Mallard (220 from Decoy)	421
Rails	2		
Water Rails	17	Great Plover	1
Coots	43	Eared Grebe	1
Stints	133	Rough-legged Buzzards	2
Owls	13	Smews (male and female)	2
Hawks (various)	4	Sundries	29
Grebes	9		
Curlews	2		1,600

From 21st December to 28th December.

Full Snipe	207	Owls	7
Jack Snipe	6	Grebes	13
Woodcocks	11	Herons	7
Green and Golden Plover	100	Teal	37
Water-hens	57	Golden-eyes, &c.	73
Coots	17	Duck and Mallard (87	
Water Rails	14	from Decoy)	197
Land Rails	1		
Stints	360		1,107

```
From December 14th to December 21st  ...  1,600
     „        21st      „        28th  ...  1,107
                                            -----
                                            2,707
```

Of the twenty-two grebes, three were dabchicks, one "eared" grebe (probably Sclavonian), the remaining eighteen were great crested; the "golden-eyes" were almost all tufted ducks, the former being a common name for that species amongst the Norfolk gunners. In the month of December, 1884, during a very severe frost, a game dealer, at Yarmouth, also sent away to London, late in the month, 700 fowl, including a large number of pochards.

In former days the quantity of ducks resorting to the south-western corner of the county in winter must have been very large. A man, named Williams, who hired the Lakenheath decoy, which ceased to be worked after the railway from Brandon to Ely had been opened, is said to have cleared £1000 in a single year, and to have sent a ton and a-half of fowl to London four times a week. Sir Edward Newton was told by the late Mr. Birch, in 1853, that when Lord Paget (afterwards the celebrated Lord Anglesey) lived at Wretham, which was prior to the year 1815, George Turner (of bustard-killing notoriety) is said to have killed the extraordinary number of 130 ducks, on the great mere, at one discharge of his big gun. But, for number and variety, perhaps the following mixed bag, made by Lord Walsingham to his own gun (the particulars of which

I have his lordship's permission to give), has probably never been exceeded:—

"*A Skirmish to finish the Season. Stanford, 31st January, 1889.*

The bag to my gun consisted of:—

1	Pheasants	39	13 Teals	7
2	Partridges	5	14 Swans	3
3	Red-legged Partridge	1	15 Wood Pigeon	1
4	Hares	9	16 Herons	2
5	Rabbits	16	17 Coots	63
6	Woodcock	1	18 Moorhens	2
7	Snipe	1	19 Otter	1
8	Jack Snipes	2	20 Pike	1
9	Wild Ducks	23	21 Various (a very large Rat)	1
10	Gadwalls	6		
11	Pochards	4		
12	Golden Eye	1	Total	189

Amongst other Birds noticed during the day were:— Pintails, Tufted Duck, Wigeons, Shovellers, Sandgrouse (*Syrrhaptes paradoxus*), Water Rails, Gulls, Kingfishers, &c.

WALSINGHAM."

ANAS CIRCIA, Linnæus.

GARGANEY.

In reading the very interesting notes left us by the naturalists and sportsmen of the early part of the present century, there are two things which cannot fail to strike us. The first is the cold-blooded way in which they shot every member of the *Anatidæ* they met with in summer, even from their nests. The second is, the very little real information they possessed as to the breeding of the birds of this family, which probably nested in their midst. Possibly the latter of these two circumstances arose out of the former; but, whether or not such was the case, we have now no means of deciding. It is certain, however, that in the present day, when we should blush to use a gun among breeding birds, we either have a much larger number and variety of ducks nesting with us, or, the telescope having taken the place of the gun, we know much more about them.

This charming little duck is a case in point. Hunt remarks (ii., p. 311), "We had a specimen of the male of this species [garganey] sent us to be preserved in the month of May, 1819. And a pair of these birds were shot on the 6th May, 1817, at Hockwold, in the county of Norfolk; the female had a perfect egg in her*; from which circumstance they would doubtless have bred in that neighbourhood." He then continues: "This fact proves what has not been before ascertained, that some of this species continue in England the *whole year*," a conclusion now known to be erroneous, and probably arrived at on the assumption that its winter habits coincided with those of its near relative *Anas crecca*. Mr. Lubbock says that "a friend [Mr. Girdlestone] received a pair alive in March, 1822, from the Winterton decoy, the female of which deposited an egg in the basket during the journey;" and then, still quoting his friend Mr. Girdlestone, adds: "Garganey breed often in Norfolk; but, as they deposit their eggs in the most inaccessible reed bushes, their nests are never discovered;" the former part of which statement is perfectly true, but the latter, although it has been frequently quoted, is certainly not in accordance with their habits in the present day. Mr. Lubbock himself does not seem to have been possessed of more information in 1845, as he only knew of the fact of its breeding here from finding the young in autumn, and remarks that it "is rarely seen during severity of weather;" indeed, he adds, "I cannot recollect a single instance," seemingly, like his predecessors, not being fully aware that with us this species is strictly a summer migrant. Messrs. Sheppard and Whitear, in their "Catalogue of Norfolk and Suffolk Birds," 1825, quote the Hockwold incident mentioned above to support their statement, that "it seems probable the garganey sometimes breeds in Norfolk." The brothers C. and J. Paget, in their "Natural History of Yarmouth and its Neighbourhood," 1834, quote Mr. Girdle-

---

* Professor Newton informs me that among the eggs in the late Mr. Scales's collection, which were not burnt, and came into his possession, is one marked "Garganey, cut out of the bird." This is quite likely to be the egg mentioned above.

stone as their authority for its "occasionally breeding in Norfolk;" and even Messrs. Gurney and Fisher, in 1846, do not add anything from their personal knowledge. General authors about the same period either quote the above authorities, or are even less informed on the subject.

Very late in February, or early in March, just as the fowl which have wintered with us are taking their departure, the "summer teal," as it is generally called in the Broad district, puts in an appearance, and soon settles to nest. At this time it is tolerably abundant where found, but is very local. It is singular that where the garganey most abounds, a locality which for obvious reasons it is not desirable to indicate too precisely, the teal is a rare bird; but at Ranworth the latter species is common enough. The summer teal is also met with in smaller numbers in several other localities in the Broad district, and I have frequently seen it not many miles from Norwich. On the other side of the county Mr. J. H. Gurney, jun., and myself believed we saw this species at Wretham, in 1882; but, in 1884, it was fully identified by Lord Walsingham, at Tomston mere, near Merton, where it is perfectly safe from molestation. I have already said that a pair were shot at Hockwold in 1817. On the coast it is not often met with, but it has been shot, according to Mr. Dowell, on the Salthouse and Blakeney salt marshes. In the latter locality on the 27th April, 1857, Mr. Dowell killed a pair of these birds, the female of which contained an egg. Mr. Norgate has given me a note of a garganey killed at Beeston Regis, on April 7th, 1863; and of a female and male shot at West Lexham, on the 21st and 22nd July, 1865, respectively. It thus appears that this species is widely, though sparingly, distributed over the county.

A note in the late Rev. T. J. Blofeld's interleaved "Yarrell" (kindly lent me by Mr. T. C. Blofeld), states that the garganey breeds frequently on his estate, and describes the nest as deeply cup-shaped, composed of grass, and lined with a quantity of down from the parent bird, remarkably warm and snug, and adds, "in every instance within my memory it has been placed in a tuft of coarse grass in the drained marsh." The female sits very close, and it is difficult to disturb her. On the 25th April, 1862, Mr. Blofeld examined a nest containing fourteen eggs, and, on the 30th April, another with ten eggs. On 29th May, 1871, I found a nest of this species exceptionally situated on the edge of the " rond " near a reed-bed ; I had a good view of the bird as she left her nest. On the 10th May, 1881, I saw a garganey's nest in the dry grazing marshes, in a tuft of rough grass on the bank of a Broad drain. Mr. A. H. Evans describes a nest which he saw in 1882, containing seven eggs, as situated in a " grass field."

Mr. F. Frere, in a letter to Mr. Stevenson, says that " Platten, who was Captain Ensor's keeper, says garganeys bred on his marshes all his time, and that in June, 1874 (the last year he was with Captain Ensor), he knew of a nest about one hundred yards from the Broad, in a rough marsh, from which twelve young ones were hatched, and were about till the month of August, when several were shot with flappers."

On the 29th May, 1884, Sir E. Newton saw a nest of this species near one of the Broads, which had contained five eggs, two of which had been taken ; it was situated on the " wall," " a bank raised some three feet above the level of the marsh, and, perhaps, twenty feet wide. The nest was in some long grass, well concealed, but within a yard of a path, which was daily used by mowers." On the 30th April, 1886, he was shown another nest in the same neighbourhood, containing eight eggs. This nest was situated " on a fairly well-drained grass field, with a few tufts of rushes here and there, in one of which was the nest."

Mr. J. H. Gurney, jun., on May 8th, 1885, saw a nest containing only one egg, which was quite covered by the downy lining; the inside diameter of the nest was only six inches. Close by was a second nest, which had probably belonged to the same pair of birds, but the eggs had been destroyed ; it exactly resembled the previous nest, and in each case he noticed a little run or pathway leading up to the nest. At the same time he was shown another nest from which nine eggs had been taken ; it was well lined with black down, each filament of the down having a white centre. On the 1st June, 1886, Mr. Gurney saw the skins of some nestlings, which had been hatched in a nest situated in the stump of a willow tree by the side of a road, and across which they were being marshalled by the old bird towards a ditch when they were discovered. Subsequent experience has proved to me that I was wrong in assuming (Lubbock's "Fauna of Norfolk," 2nd ed., note 139, p. 155) that the nest was never far from a reed-bed; this was my belief at the time, but such a situation is certainly exceptional. Mr. Booth is quite correct in stating that "now and then the birds were known to have bred among the long coarse grass and tufts of rushes on the dryer portion of the hills* surrounding the broads; but, as a rule, they go further from their usual haunts." ("Rough Notes.")

The garganey was occasionally captured in the Ranworth decoy; on the 30th March, 1864, four of these birds are entered in the decoy book, the return for the month being twenty-nine ducks, two teal, fifteen wigeon, one pintail, four garganey, and one pheasant; the 1st and 2nd April produced ten wigeon, which ended the season. In April, 1865, three shovelers and four mallard experienced an abrupt termination to their honeymoon in this decoy.

I have no certain information as to the date of departure of the garganey, but the "flappers" are strong on the wing in July, and they, probably, leave us in August. The young of this species are very liable to be mistaken for those of the teal, but may be readily distinguished by the absence of a speculum on the wing. The neck and bill are also rather larger than in the commoner species.

Mr. Gurney possesses a pair of these ducks, killed at one shot at Strumpshaw, in July, 1858, one of which is a pure white; when fresh killed its bill was pink, and the legs pinkish brown.

---

* A marshman's "hill" is of the very slightest, generally a spot rather less wet than its surroundings.

ANAS CRECCA, Linnæus.
TEAL.

As a resident this species is fairly numerous and generally distributed over the county, although not found in such large numbers as formerly, when a greater extent of country suitable to its habits existed. Sir Thomas Browne speaks of the great abundance of teal in East Norfolk in his time, and in more recent days, as will be seen, they were also numerous. In October and November large numbers of teal arrive from abroad, and in these months they are most abundant; after that time, should the weather prove severe, they continue their journey southward, and comparatively few are met with till the return of open weather.

On the Holkham lake Mr. Napier tells me the teal begin to arrive towards the end of September, but he has never observed them there in large numbers.

On their first arrival they are very easily taken in the decoys, no bird working better to a dog than the teal, and very large numbers used formerly to be secured. In 1834 the following note occurs in Miss Anna Gurney's journal:—"October 14th. The decoyman at Hempstead reports that nearly 1,000 teal have come over;" and in January, 1835, "Above a thousand teal taken this winter in Hempstead decoy." I have already mentioned that Skelton, at the Winterton decoy, took in six consecutive days 1,010 of these birds, and at the same place he caught more than 200 at once in a single pipe. In Mr. Dowell's notes on Langham decoy, he states that up to December 8th, in the year 1851, Capt. Marryatt had already killed 300 of these birds. The numbers caught in the decoys at the present day are very much smaller than formerly, although, perhaps, not less in proportion than those of other fowl, but the greatest takes are still made in the month of November.

Hunt does not seem to have had any certain knowledge of the teal breeding in this county, but he states that he has not the least doubt that they do so in several places. Sheppard and Whitear say that a few breed in Norfolk, and give an instance, near Rudham, in 1817;

also mentioning Ranworth Broad and Scoulton Mere as breeding places. It is well known now that a considerable number of teal breed yearly in the neighbourhood of the inland waters of Scoulton, Wretham, Stanford, and other places, as well as in the Broad district proper. Mr. Lubbock speaks of young teal being found on Sculthorpe Moor, near Fakenham, and of their breeding in 1838, near the river between Harling and Larlingford; also [in 1840] at Old Buckenham Fen, near Attleborough, whilst the latter part of the country remained unchanged by drainage and the enclosure of waste lands. A pair used to breed annually at Hempstead, near Holt, where the decoy mentioned by Miss Gurney was situated; they also breed regularly at Winterton. The nest of the teal is a very snug structure, plentifully lined with the dark-coloured down from the old bird's breast; it is generally placed in rather a dry situation, and sometimes in a tussock of grass or heather, or at the bottom of a whin bush, and the first eggs are generally laid about the middle of April. On the 22nd of April, 1867, the late Rev. T. J. Blofeld found a teal's nest placed in an alder stump, profusely lined with down, and filling up the centre of the stool; it contained eleven eggs. Mr. Blofeld remarks that, although he knew this species must breed at Hoveton, from having frequently killed the young in June and July, this was the first nest he had ever seen there. On the 13th May I saw a teal's nest at Ranworth containing ten eggs, from which the keeper had taken two pheasant's eggs; another nest, near the same place, which was destroyed by rats, had also contained two pheasant's eggs. Mr. Norgate saw a teal's nest, on the 19th April, on Santon Warren, which contained eight teal's, one duck's, and several pheasant's eggs. The old bird is very much attached to the nest, especially when near hatching. Mr. J. H. Gurney, jun., once found a nest at Hempstead, on June 13th, containing ten eggs, on which the old bird "sat like a stone" till he almost trod on her; and a good many years ago, in Invernesshire, I actually removed a teal from her nest with my hand, so close did she sit. The stratagems resorted to by this pretty little duck to draw off the attention of the intruder from its brood exhibit a charming display of maternal affection; the little ones, too, have a marvellous power of concealment. On one occasion I disturbed an old teal which was brooding over a large family: off went the old bird, fluttering away as if in the last agonies of death, and the young scattered in all directions, but keeping my eye fixed on one particular baby teal, I saw it squat down a few yards off, its neck stretched out and its little body close to the ground where some dead oak leaves were lying, the concealment so perfect that had I not actually seen it assume the position I should most certainly never have detected it, nor did it stir from the spot till I stooped and took it up in my hand.

Mr. Lubbock mentions that in 1842 a teal was seen at Ranworth with twelve young ones, which, he adds, is far the most numerous brood he ever heard of; but Salmon, in his diary, states that in May, 1835, he found a nest at Stanford with sixteen eggs. I think, however, the number of eggs rarely exceeds nine or ten. As soon as the young birds are able to fly they desert the marshes and congregate on the Broads and open waters, where they remain till they are forced to leave by stress of weather, and their places are taken by migrants from the north.

Mr. J. H. Gurney, jun., tells me that some years ago Mr. Dack, a bird stuffer, of Holt, had a teal with a distinct white ring round its neck like a mallard, and that he has seen two others, not Norfolk-killed, with considerable indications of a similar ring.

The note of the male teal is a clear musical whistle; the voice of the female, however, although, perhaps, not inharmonious, is decidedly unmusical.

The so-called "Bimaculated duck," now recognised as a hybrid between this species and the wigeon, is believed to have occurred in Norfolk, a specimen, said to have been sent from Yarmouth, having been purchased by Mr. Jones in Leadenhall Market, on the 9th December, 1846. It is described by that gentleman in the "Zoologist" for 1847, p. 1698; and subsequently noticed by Mr. W. R. Fisher, with an illustration, in the same magazine for 1848, p. 2026.

ANAS PENELOPE, Linnæus.

WIGEON.

The Wigeon, or (as it is known to our shore gunners) the "Smee," is one of the most regular, as also one of our most abundant visitors, arriving towards the end of September, and finally departing towards the end of March; the gunners in the Wash consider November the best month for wigeon passing south, and look for the commencement of their northern migration about the time of Lynn Mart, the 14th February, but their movements are greatly influenced by the prevailing weather; and Mr. Monement tells me he has known flocks to arrive from the north as late as February, notably February 12th, 1889. Large numbers of wigeons frequent the decoy ponds and other quiet places of resort upon their first arrival, but very few are at that time taken. Formerly the best season at the decoys for this species was early in March, when the birds congregate previous to their departure, and passing flocks often remain to rest or are detained by bad weather; now, however, owing to the close-time, the birds passing northward almost entirely escape, the number taken depending very much upon the season; thus in 1865-6, out of 631 fowl taken at Sir S. Crossley's decoy at Fritton, 274 were wigeons, whereas in 1867-8 there were only fifty wigeons out of 2,278 fowl; and in 1884-5 only three wigeons were taken in a total of 2,048.

On February 12th, 1880, I visited the Fritton decoys. Although December had been a very good month, the season was then too open, and, notwithstanding the large number of ducks, wigeons, and teals on the lake, food being abundant the fowl would not "pipe." The view of the birds through the decoy screen was charming in the extreme; several pochards were close to the pipe, and their bustling, active movements, as they dived and chased one another on the water, were very pretty. There were only two wigeons, a male and a female, within working distance, and these we took; most lovely birds they were, the chaste beauty of their plumage only equalled by the elegance of their forms.

At the Langham Decoy, which was closed about the year 1854, Mr. Dowell tells me very large numbers of wigeons were taken; he estimated the proportions as three wigeon to every two mallard and one teal, but since the closing of this decoy the wigeons have resorted to Holkham Lake as their day quarters, and may be seen there in thousands all through the winter. Prior to 1854, Mr. Dowell says that only mallards and a few stray teals and tufted ducks resorted to the Holkham Lake, but now at times the surface of the water is almost covered by fowl, chiefly wigeons. Mr. Alexander Napier has been kind enough to furnish me with the following particulars with regard to the wigeon which frequent the lake at Holkham. The lake itself, which is about $34\frac{1}{2}$ acres in extent, filling the bottom of a narrow valley for about a thousand yards, and with varying width, is fed by springs at the south end, and is less than a mile distant from the sea in a straight line; the banks are sloping and in parts wooded, and there are four small wooded islands, otherwise there is not much shelter for the fowl, which are quite visible to those walking or driving through that part of the park. The wigeon begin to arrive, says Mr. Napier, " early in November in very small companies, the main body not putting in an appearance until well on in December, and I should say there are always more to be seen on the lake from the middle of January to the end of February than at any other time of the year, but their movements are largely governed by the weather. If the weather be fine and open they do not show up so early, but sit out at sea. They begin to leave the lake at the commencement of March, but not in great numbers, the main body staying until quite the end of March, and there are always a few which stop well into May. At the present moment [21st April] there are from two to three hundred wigeon on the lake. After leaving the lake they sit out at sea for a day or so before taking their final departure to their breeding haunts. They do not leave the lake in severe weather until night time, unless there is a very severe frost, and the whole of the water is frozen over. As long as there is an open piece of water they remain on the lake, and I can assure you that in severe weather Holkham Lake is a sight to gladden the eyes of a naturalist. Hundreds, I may say thousands, of wigeon, mallard, teal, pochards, tufted ducks, with a few goosanders, smews, and golden-eyes, make the picture perfect. Lord Leicester has given strict orders that they are never to be disturbed, especially in severe weather; thus they have at all times a secure retreat."

On the 12th March, 1889, Mr. J. H. Gurney, jun., and myself, under the guidance of Mr. Napier, had an opportunity of seeing the fowl on the lake. There were a large number of mallards, some teals, tufted ducks, golden-eyes, and shovelers; but by far the greater number were wigeons, which appeared to keep very much to themselves. Hundreds of these birds were on the shore of the lake, where the grass was grazed as closely as on a goose green, but a far larger number were seated on the water, where flock after flock joined them as we disturbed them from their resting-place on the bank. On the water all was activity and incessant motion, the birds diving and changing places constantly, while the melodious, oft-repeated call of the wigeon, and the whistle of the teal, accompanied by the occasional deep note of the wild duck, added immensely to the charm. The note of the female wigeon is a low guttural sound difficult to describe, and appears not to be uttered nearly so often as that of the male.

On the shore, Mr. Monement tells me that in foggy weather and rain wigeon are restless and silent at night, but when the weather is bright and frosty they are usually noisy and more or less unsuspecting. As with wild duck the female of the wigeon is a more expert diver when wounded than the male, although the superiority is not so marked. He finds the wigeon's sense of smell to be less acute than that of the wild duck or teal, but it is nevertheless unsafe for the gunner to go directly to windward of them unless at a considerable distance.

Occasionally the wigeon has been known to stay the whole summer in this county, but I do not think there is any well authenticated instance of its breeding here. Mr. Booth ("Rough Notes") says that those birds met with by him at Hickling, in the summer, "always exhibited immature plumage, and were, doubtless, either weakly, or had suffered from wounds, and thus were unable to follow their stronger relations." Mr. A. Hamond informed Mr. Stevenson that in 1877 several wigeons remained all the summer on the lake at Narford. On the 25th April, 1875, Mr. Frank Norgate saw with a telescope several wigeons and one pochard on the lake at Fritton decoy. William Hewitt, Mr. Blofeld's marshman, told him that he had seen a wigeon on Hoveton Broad on the 3rd May, 1860, and on the next day Mr. Blofeld himself saw a bird in company with three mallards which he had no doubt was a male wigeon. Mr. Frank Norgate saw a male wigeon on the "Big water," at Heydon, which flew up within a few yards of him, on the 24th June, 1875; and, on the 4th July, the same gentleman saw "a pair of wigeons within a few yards" of him, on a pond at Cawston; finally, on the same water, on September 8th, 1880, he saw "four ducks, apparently wigeons, at a distance of about sixty yards." In 1868, the decoyman, Boyce, at one of the Fritton decoys, told Mr. Norgate that "wigeons sometimes drop a few white eggs on the decoy banks before they leave for their summer quarters." A similar statement was made to Mr. Stevenson from East Norfolk, by Mr. F. Frere, who says that at Potter Heigham wigeons have been known to drop their eggs before going north in spring, a habit which Montagu attributes to them also in confinement. In the "Zoologist" for 1847, p. 1785, is a note by Messrs. Gurney and Fisher, to the effect that "about the 17th May a nest containing four eggs, which, from their appearance and the description which was given of the old birds, are probably those of the wigeon, was taken on the edge of the river Bure." Lord Walsingham informs me that a wild male wigeon paired with a pinioned female, and remained to breed on Stanford water.*

---

* It is worthy of remark that the wigeon is, of all fresh-water ducks commonly kept in confinement, the one which most seldom breeds, and yet it very readily becomes tame. This fact is of importance, as showing how little we know of the causes which affect the domestication of species. At Elveden, Professor Newton and his brother had wigeons for many years, together

The wigeon is strictly vegetarian in diet, and where food is abundant is not so regular in leaving its daily resting-place as some other species; nor is it so entirely a nocturnal feeder. Prior to the year 1868, Mr. Blofeld states, in his notes before quoted, that this species was not found in considerable numbers on Hoveton broad, but in that year the "American weed" *(Anacharis alsinastrum)* made its appearance at Hoveton, and proved very attractive to the wigeons, which came in large numbers to feed upon it, and were in excellent condition. The Ranworth decoy was at that time worked, and the decoyman has since told me that he found it impossible to entice the wigeons, they were so well fed and lazy. The same influx of these birds occurred again at Hoveton in 1869, through the entire autumn and winter, when the weed so filled the broad that even a small boat could hardly be "quanted" round it. In the following summer the water became so stagnant that the fish died; after this the weed rotted and vanished entirely, leaving the broad perfectly clear of weeds of all kinds. This weed, so attractive to wigeons and swans, completely banished the diving birds, which could not penetrate the dense mass of vegetation.

Mr. J. H. Gurney, in a letter to Mr. Stevenson, expresses an opinion that the wigeon, on an average, does not regain its full dress until later in the year than any of our common ducks, and states that on the 21st November, 1867, he saw, in Leadenhall market, several of these birds, all males, which still retained in great part (many almost entirely) the duck's dress which they assume in summer. This late assumption of the "full

---

with wild ducks, pintails, and shovelers; and, while the last three bred pretty freely, he informs me the first hardly showed any desire of propagation. The pond in which they were kept was apparently better suited to wigeons than to the other species, as there was plenty of grass about it, whereon they were continually feeding, but they never showed any inclination to nest. Mr. Cecil Smith, in a paper on "The breeding of certain waterfowl in confinement" ("Zoologist," 1881, p. 448), states that both his father and himself have had wigeons on a pond ever since he could remember, but they never bred till 1872; since which time they have done so regularly.

dress" quite agrees with Professor Newton's experience with pinioned examples kept by himself.

At Mr. Stevenson's second sale, on 21st March, 1889, Mr. Gunn re-purchased what Mr. Stevenson speaks of in his note book as "a wonderful red wigeon," killed at Rockland, on the 5th May, 1871, which he himself had purchased of Mr. Gunn. The under parts are a rich yellowish buff, extending from the vent to the chin; the forehead is also strongly tinged with the same hue. Mr. Stevenson submitted some of the rusty feathers of this bird to Mr. F. Kitton for the purpose of testing them chemically; and I find with them a note from the latter stating that the ferruginous colour is due to the presence of an oxide of iron, as shown by the deep blue colour imparted to them when tested by ferrocyanuret of potassium. In the same envelope there are also some feathers of the pintail indicating the presence of iron stain in the same way.*

Mr. Stevenson mentions a very singular looking specimen, at first believed to have been the American wigeon *(A. americana)* in immature plumage, which was killed at South Walsham in 1852, and proved to be a male of the common species in a stage of plumage intermediate between the male and female. In Mr. Gurney's collection is a very beautiful variety of the wigeon, taken at Lynn, which has the ground colour of the plumage white, marked and freckled with light brownish buff.

A pinioned wigeon, belonging to Mr. Burroughes, lived on a private water, at Hoveton, for eighteen years.

SOMATERIA MOLLISSIMA (Linnæus).

EIDER DUCK.

This species, although it occurs in small numbers almost every winter, must be considered a rare bird on the Norfolk coast, more especially in the adult plumage.

---

* See *ante*, p. 78, with regard to the plumage of the mute swan.

It is usually met with in small flocks, consisting, perhaps, of members of the same family, which are almost invariably females or young birds. Among the few adults recorded as having been met with in this county may be mentioned a male in full plumage, shot off Wells, in January, 1820, in company with which were two others, as mentioned by Messrs. Sheppard and Whitear;* also an adult male, killed in the estuary, near Lynn, on the 13th November, 1868; a female, at Hunstanton, in January, 1871, and a nearly adult one at the same place a few days later; also a female mentioned below, now in the Norwich Museum.

In November and December, 1883, as recorded by Mr. Stevenson ("Trans. Norfolk and Norwich Nat. Soc.," iv., p. 138), a party of these birds visited Breydon, where probably they were shot at and dispersed inland, as on the 22nd of November a female was seen on Flegg Burgh Broad. On the 11th December, 1883, an adult female, now in the Norwich Museum, and said to have been seen with others in the same locality, was killed with a stone on a small stream, near Hellesdon mills, a few miles above Norwich; it was described as strangely tame. On the 12th December another, believed to be an immature female, seen in company with six others, was shot on Breydon. This bird also was remarkably tame, so much so that it came to a boat-yard, where some boys threw stones at it. On the 14th of the same month, another, said to have been a young male, was killed on Breydon. The plumage of these last two birds is described as very dark. Many other immature individuals have from time to time been recorded.

In July, 1885, an immature eider frequented the coast off Blakeney point, fishing along about fifty or sixty yards outside the breakers, from the entrance to Blakeney Harbour southward to Cley, with great regu-

---

* The late Mr. Whitear thus refers to this bird in his diary ("Trans. Norfolk and Norwich Nat. Soc.," iii., p. 252):—"1820, February 9th. I saw at Hunt's an old male eider duck in full plumage. Hunt informed me that this bird was killed last month (on the 27th) at Wells; there were three birds in company at the time. In the stomach of this bird there was a considerable quantity of *Echini* and crab's claws."

larity at certain states of the tide. Many attempts were made to shoot it from the beach, but it constantly kept just beyond gun shot. It stayed, I think, till the middle of October, when it took its departure. On its first arrival, before the close-time was ended, the bird is said to have been very tame, but after it had been pelted with stones by the boys from the beach, it kept at a safe distance from the shingle.*

SOMATERIA STELLERI (Pallas).†

STELLER'S DUCK.

The Norfolk and Norwich Museum possesses what was for fifteen years the only known British-killed specimen‡ of this beautiful arctic species—a male in almost perfectly adult plumage, which was shot at Caister, near Yarmouth, in February, 1830. The "Norfolk Chronicle" of the 20th of the same month, contains a paragraph stating that "One of the greatest

---

* Messrs. Paget record a female *Somateria spectabilis*, on the authority of Mr. Lilly Wigg, as having been shot on Breydon, July 25th, 1813. With regard to this supposed occurrence, Mr. Stevenson, some years ago, favoured me with the following note:— "It is singular that the common eider is not named in Paget's list, and 'king eider' may have been written by mistake. In the days before Yarrell, I question if Wigg, or any one at Yarmouth, would have recognised the female of the king eider as distinct from the more common species." The Suffolk examples rest on an equally shadowy foundation, and are now universally rejected.

† Pallas being (in 1769) the first describer of this bird, the specific name he gave it has to be adopted instead of *dispar*, subsequently (in 1786) bestowed by Sparrman. It is curious that by an accident, as noticed by Pallas himself ("Zoogr. Rosso-Asiat.," ii., p. 239), the plate accompanying his original description did not represent this species, though it has been commonly cited as doing so. The name "Western Duck," by which, under the mistaken impression that the bird was peculiar to Bering's Sea, it has also been known, has, of late years, been generally and advantageously dropped.

‡ A second example, and the only other known to have been obtained in the British Islands, was shot 15th August, 1845 ("Zoologist," 1846, p. 1249), at Filey, in Yorkshire.

treats for those interested in Natural History is to be seen at that able and zealous ornithologist's, Mr. J. Harvey. This northern straggler is *Anas stelleri* of Pallas, western duck of Pennant, described in his 'Arctic Zoology.' It was shot near this place [Yarmouth] on the 9th inst., and is one of the handsomest of the genus, except *A. spectabilis*, Linn. It has not been noticed by any author to have before visited this island." The occurrence was briefly announced by Mr. Yarrell to the Zoological Society on the 25th of January, 1831 ("Proc. Zool. Soc.," 1831, p. 35), and in the "Magazine of Natural History" for March of the same year (iv., p. 117) he recorded it, with the additional fact that the specimen was "in the possession of a gentleman at Acle;" who, there is every reason to believe, was the Rev. G. W. Steward, rector of Caister, the place where the bird was killed. At all events, in the Presentation Book of the Norwich Museum occurs the following entry, under the date of 5th July, 1831:—"Western duck (unique), by the Rev. G. W. Steward;" while in the Museum report for the same year the receipt is thus acknowledged:— "The Rev. G. W. Steward, rector of Caister, has presented a valuable collection of 152 specimens of British birds and quadrupeds, comprising among them the western duck of Pennant, *Anas stelleri*, the purple heron, and Caspian tern." The subsequent history of this specimen has been very uneventful, as it has only once left the charge of the curator, and that merely for the purpose of being re-mounted.

In the "Trans. of the Norfolk and Norwich Nat. Society" (ii., pp. 409-413) will be found three letters from Mr. Dawson Turner to Selby, who wished to borrow a drawing of this specimen, that he might issue it in his "Illustrations of British Ornithology," and eventually procured one from Mr. Turner "addressed to the care of Mr. Hewitson," and "consigned to one of the masters of our coalships"—a mode of conveyance which shows the difficulty of communication suffered by naturalists in those days..* These letters corroborate

---

* Nothing is now known of this drawing, which was no doubt returned, as requested by Mr. Turner. Mr. Harting, however,

the statements already quoted, and add the information that a French naturalist, believed to have been Delamotte, offered Mr. Steward one hundred French birds in exchange for his specimen. From the same source Mr. Turner also learned that, as mentioned by Selby, almost simultaneously another example had been taken in Denmark—a fact confirmed by Kjærbölling. The occurrence was also recorded in 1834 by the Messrs. Paget, but they give the 10th instead of the 9th of February as the date—a discrepancy of no material significance, and one which it would be hardly possible now to remove. Harvey, into whose possession, as will have been seen, the bird immediately passed, was a well-known dealer in wild-fowl, at Yarmouth, before mentioned in this work, and has long been dead; but Mr. Stevenson, with his usual care, enquired into the matter of his son, and left the result in writing. "The following," he says, "is young Harvey's account of Steller's duck from Caister. Harvey was quite a lad when he saw a gunner, named George Barrow, returning from shooting with the bird in his hand. He followed him to the alley (the name of which he told me) with Bessey [another gunner], who got the pratincoles [cf. vol. ii., p. 65] and another man, and then went home and fetched his father. When he [the father] arrived with him, Barrow was going into Bessey's house with the duck, and Harvey, senior, bought it, but did not know what it was."*

The Rev. E. W. Dowell informed Mr. Stevenson that, in September, 1835 or 1836, Harry Overton, a

---

possesses a copy of Sheppard and Whitear's "Catalogue," formerly belonging to Mr. Brightwell, of Norwich, containing three water-colour drawings, one of which represents this specimen, and was probably made by Miss Brightwell about the same time as the above. Yarrell also mentions another drawing of this specimen, sent to him for the use of his work by Mr. Charles Buckler.

* The details above given were, of course, unknown to Mr. Seebohm when referring ("Brit. Birds," iii., p. 613) to this specimen. He states that "there are several discrepancies in the details of its subsequent history, which throw some doubt on the authenticity of the alleged occurrence." It may be safely asserted that the history of few captures made so long ago, and of equal interest, rests upon better evidence.

gunner living at Blakeney, shot there four "very gay" ducks, which were taken to Mr. Charles Sparham, of that place, and left, it was supposed, to be preserved for his collection; but, he being from home, on their arrival the cook dressed them for dinner the same day. Overton afterwards, on looking over Gould's "Birds," at Holkham Hall, immediately recognised in the figure of Steller's duck the birds he had killed. Mr. Dowell, in a recent letter, thus speaks of Harry Overton: "He was about the best field naturalist I ever knew; he was noted for his very keen eyesight and accurate observation, among many instances of which I may mention that he ascertained for himself the difference between the Sclavonian and eared grebes in winter plumage (both rare birds on the Blakeney coast), and also that the kittiwake gull had but three toes, although he seldom saw them except on the oyster grounds. When I took him to Holkham and showed him Gould's birds, I did so on purpose to ascertain, if possible, what the 'gay' ducks were, but I gave him no lead whatever; I showed him the drawings of the shore birds as they came, and listened to his more or less interesting observations on on each. Directly he saw the western duck, he said, 'them's the gay ducks I have often told you about,' &c., and I, for one, have not the slightest doubt that 'them's them.'" What value is to be attached to this statement I must leave for others to decide, but I may remark that these men, who pass their whole life on the shore, are very observant, and quick to detect an unaccustomed bird which is likely to fetch a good price; and of both the intelligence and honesty of Overton Mr. Dowell has often spoken to me in the highest terms.

ŒDEMIA FUSCA (Linnæus).

VELVET SCOTER.

Scarcely a winter passes without the occurrence of a few of these birds on the coast, and they are met with not unfrequently in considerable numbers. There may be many of this and the next species out at sea, but it is only during and after heavy weather that they are driven in and fall to the shore gunners; at such times they not uncommonly resort to inland waters, and have been met with far from the sea. The Rev. H. T. Frere says that on two occasions the velvet scoter has been killed on Diss mere, a small piece of water of about six acres in extent, some thirty-five miles from the sea, and nearly surrounded by houses and gardens. Many other inland examples have occurred, as an adult male, shot on the river at Cossey, near Norwich; a specimen killed at Larlingford, &c. This species is frequently associated with the common scoter and long-tailed duck; and Mr. Cordeaux, as quoted by Mr. Saunders in "Yarrell" (ed. 4, iv., p. 476), observes that "in the Lynn and Boston 'Deeps' almost every flock of the common scoter have a pair or two of the velvet scoter swimming with them."

Mr. Lubbock speaks of upwards of twenty specimens coming into the hands of a bird preserver, in Norwich, in the winter of 1829-30; the next year the same man received only one. In Sir W. Hooker's MS., under the signature of T. Penrice, occurs the following entry:— "March 1st, 1832, J. Harvey brought me nine of these birds, eight drakes and one duck, all just shot on the coast; and a few days afterwards he told me he had just had ten more!" This remarkable abundance of velvet scoters is mentioned in the "Norwich Mercury" of 10th March, 1832. In February and March, 1855, one of the most severe seasons known for many years, a number of females or immature males were shot on different parts of the coast; 1859-60, also a very severe season, produced several, which were killed on the larger broads, and even further inland; but no adult males are recorded. A female was exposed for sale in our Fishmarket on 7th April, 1866. Mr. Stevenson mentions an unusual number off the coast in November, 1870 ("Zoologist," 1871, p. 2408), and others in 1871, on the 4th January of which year a fine old male was shot at Hunstanton, and a second adult male was exposed for sale in Norwich Market, on the 11th February. About the same time, the weather being very severe, females and immature males were frequent off the coast in flocks varying from ten to twelve ("Zoologist," 1871, pp. 2600-1). In 1874 several were met with. Two recorded by the Rev. Julian Tuck ("Zoologist," 1884, p. 488) were shot at Hunstanton as early as October 16th; early in the month of November, 1887, several were killed, and others at intervals during the winter on inland waters, and at different localities on the coast.

The very large majority of those met with are immature birds; adults of either sex, especially males, very rarely occur.*

ŒDEMIA NIGRA.

COMMON SCOTER.

The following is *verbatim* from Mr. Stevenson's note book:—"Although specimens of this duck have been killed at times in this county in almost every month of the year, it is strictly speaking a winter resident only, appearing in very large numbers on the coast during very sharp weather. However intense the frost they rarely quit the sea for inland waters, but I know no species that exhibits a more weather-beaten appearance in really hard winters than the black scoter.

"Mr. Yarrell, speaking of the appearance of this bird off the Isle of Wight and in Christchurch Bay, in June and July, says, 'it is not improbable that these were birds only twelve or fourteen months old, that would remain unable to breed till the following summer.' I am inclined to believe that the small flocks I have seen occasionally at Cromer, as late as the 5th July, may be

---

* Mr. J. H. Gurney, jun., has favoured me with the following note:—"Captain J. A. Vipan, of Stibbington Hall, Wansford, thinks he saw a surf scoter (*Œdemia perspicillata*), in Lynn Deeps, on December 11th, 1886, and it really seems likely that he was right. It was with a dozen velvet scoters, and had a white patch on its head; in size it seemed to be half-way between the velvet and common scoters. The surf scoter is not the rarity it was once supposed to be: in the Orkneys Saunders says it is of frequent, almost annual, occurrence ("Brit. Birds," iv., p. 482). Capt. Vipan, who, I believe, is well up in ducks, at once informed Lord Lilford of it."

accounted for in the same way, as I have also observed them myself on the south coast, in April and May, off Beachy Head, and one or two that I shot proved to be young birds. The almost annual appearance of stragglers on our inland waters in spring and summer is, however, more remarkable, from their strictly oceanic habits at other times, but I believe these occasional freshwater visitants to be merely migratory birds resting on their passage towards the far north, where, as in Scandinavia, they are known to breed very late in the season. The following is a list of some of the latest examples that have come under my notice during the last few years:—

1853.	April 6.	One, Barton Broad.
	April 22.	Pair, Scottow.
	May 24.	One, Burlingham.
	June 20.	One, Hickling.
1854.	July 1.	One, Surlingham.
1856.	April 2.	One, Thorpe.
	June 11.	One, Scottow.
1860.	April 5.	Pair at Blakeney.
	June 15.	Twenty or thirty seen off Cromer.
	July 5.	Twenty or thirty seen off Cromer.

These were chiefly adult females and young birds, but one or two old males appeared amongst them."

To these I could add several subsequent like occurrences, the most remarkable of which is recorded by Mr. Stevenson ("Trans. of the Norfolk and Norwich Nat. Soc.," ii., p. 210), "I may here mention that I have two notices of wigeon flushed in different localities in the county, in the middle of June, but still more remarkable, if true, yet on authority I can scarcely doubt, that a brood of young common scoters was seen on Hickling Broad throughout the summer [1875]. Mr. Booth, who was in that neighbourhood in July, tells me that he saw some fourteen or fifteen scoters flying over that broad towards the end of the month." This species has at least on one occasion (as well as the velvet scoter) been shot on Diss mere.

Mr. Youell, of Yarmouth, kept one of these birds alive for several months, which he fed upon barley ("Trans. Linnean Soc.," xiii., p. 616).

In November, 1887, and again in March, 1888, numbers of these birds found their way into the Norwich Market, probably from Hunstanton. The Wash, between the Lincolnshire and Norfolk coasts, is a favourite place of resort for the scoter, and Mr. Cresswell has usually found them in great numbers in the Lynn Roads between the Thief and the Whiting sands. Lord Lilford, in a letter to Mr. Gurney, dated 28th February, 1888, says that some years ago Mr. George Hunt, in a few days' shooting on the Wash, brought home with him some 300 " black ducks," much to the satisfaction of the cottage people, among whom he distributed them, as "they combined a fine vehicle for onions with a flavour of fish and fowl." The great extent of mussel scaup, off Hunstanton and Holme, possesses an irresistible attraction for the ocean ducks; here, in the shallow water, they find a rich feeding-ground, and sometimes congregate in astonishing numbers. Mr. Monement tells me that the fishermen notice where these birds "work," as they "till" where plaice are, and both plaice and black ducks frequent mussel beds. Off Weybourne the ducks have to dive about five fathoms to get at their food. In November, 1887, Dr. Whitty killed sixty-two common scoters on this part of the coast with a shoulder gun; and a correspondent from Hunstanton informed the Rev. Julian Tuck that up to February 15th, 1888, he knew of 284 of these birds having been brought in, and that he estimated three out of every ten " knocked down " were lost, and yet at that date they did not appear to have decreased in number, there being still " thousands left, although they do not decoy quite so well." These birds did not take their departure till quite the end of March.

FULIGULA RUFINA (Pallas).

RED-CRESTED POCHARD.

This beautiful duck has occurred in Norfolk some eight or nine times. The first recorded British speci-

men is thus referred to by Hunt in his British Birds:—
"The specimen from which our drawing was made was killed on Breydon in the month of July, 1818, and is now in the possession of Mr. Youell, of Yarmouth." This bird was a female, and he adds, "We are informed that a specimen of the *male* was killed in Norfolk a few years since, and was preserved in the London Museum." This formed lot 96 of the sale catalogue of Bullock's [London] Museum, 6th May, 1819, as I am informed by Professor Newton, but the entry (p. 33) is there printed, "Round-crested Duck, killed in Norfolk, extremely rare." His priced copy shows that it was bought for three guineas by Lord Temple. In the winter of 1826 two were shot, also on Breydon, and an immature male was killed at Surlingham in December, 1827, which Mr. Lombe says passed into the possession of Mr. Deen, of Bradistone, and subsequently to the late Mr. Thurtell, of Eaton, from whom it was purchased by Mr. Gurney, in whose collection it now is; it is a male, but not in full plumage. On January 12th, 1844, another of these birds was killed at Horsey by Mr. Rising, at the sale of whose collection it was purchased for the Norwich Museum; and in December, 1867, a female was shot at Hickling by Nudd out of a lot of pochards, fifteen of which he killed at the same time. This specimen, late in the Rev. S. N. Micklethwait's possession, was bought at the dispersal of that gentleman's collection, on the 21st of June, 1889, by Mr. Connop, of Caistor. There is also little doubt that a drake in perfect plumage, purchased by Mr. Gurney at Stephen Miller's sale in 1853 for £1 18s., was killed on Breydon, probably, Mr. Stevenson thinks, one of the 1826 birds. When Mr. Doubleday's collection was sold on 23rd August, 1871, Mr. Borrer, of Cowfield, became the possessor of one of these birds, a male in change, marked in the catalogue, "Yarmouth, very rare," respecting which Mr. J. H. Gurney, jun., informs me he has the following note in Mr. Doubleday's writing:—"This I also had in the flesh from Mr. Stevens [of Leadenhall Market], who assured me that it came from Yarmouth, and I have no doubt of the correctness of this statement, as I saw the package in which it came with other sea birds."

Mr. J. H. Gurney thus described the plumage of Mr. Rising's specimen, and, as the tints appear to be very evanescent, it may be desirable to reproduce his remarks:—" When newly killed it was as beautiful a bird as I have often seen; the beak was of a most splendid vermilion-red colour, the nail of the beak being also red, but paler than the rest. The colouring of the beak began to fade soon after the bird was mounted, as also did another beauty which was apparent when the bird was first killed, and which consisted of a wonderfully elegant tinge of rose colour, which pervaded the whole of the white part of the plumage, especially the two large patches on the back above the shoulders. The colouring of the other parts of the bird (which appears to be of a permanent character) agrees very well with the usual descriptions of this species in the adult male plumage." ("Zoologist," vol. ii., p. 576.)

It is unnecessary to enter in detail into the distribution of this species; and of its habits here, nothing, of course, is known except that the 1867 bird was shot in company with pochards, as previously mentioned, whereas Mr. Rising's specimen, which was killed early in the morning, was quite alone and very tame. It must be regarded as a rare winter visitor, as all those killed, with the singular exception of Hunt's 1818 bird, which was killed in July, have occurred at that season; and of five instances in which the sex has been recorded three were males and two females. The specimen now in the Norwich Museum was purchased for that institution at Mr. Rising's sale.

FULIGULA FERINA (Linnæus).

POCHARD.*

The pochard, or "dun bird," visits Norfolk regularly towards the end of autumn in considerable numbers,

---

* By the gunners on the Norfolk coast the scaup duck is known as the pochard, and the pochard as the "red-headed pochard," "porker," or "pocker," as it is variously pronounced.

which are augmented in severe weather. On all the broads, and at times on other inland waters, they congregate during the day, and are objects of considerable aversion to the decoymen, for in addition to it being almost impossible to take them in the decoys, their restlessness on the water is very disturbing to the other fowl. It is very pretty to see them chasing each other and diving for the sunken corn at the mouth of the pipe, but the decoyman well knows that they will neither enter themselves nor will they allow the other fowl to do so, and that should they by chance be found up the pipe, no matter how shallow the water is, they will dive back and escape. Various methods have been tried to secure them, such as sunken nets, or snares under the water; but should they be thus entrapped, the struggles of the drowning pochards beneath the water alarm the fowl outside, and fatally disturb the quiet of the decoy. It thus happens that although large numbers of these birds may be on the water they are very rarely captured, and those which come to market are killed by the punt gunners, who frequently get "lumping" shots at the first dawn of day, when they begin to "head up" together preparatory to their departure for their day quarters, or are killed at "flight time" in the morning and evening as they arrive at or depart from their daily resting-places.

Mr. W. B. Monement tells me that the pochard is not so plentiful at Blakeney as it used to be.* Twenty years ago they were generally to be found in small numbers both at Blakeney, and at Salthouse when the marshes were flooded. Now they are rarely seen, even during the most severe weather, but are much more plentiful in the Wash. The same applies to the scaup duck.

Mr. Lubbock (ed. 2., p. 137) states that immense flocks of pochards were wont to rest all day in the middle of Rollesby decoy, whence they poured forth to feed every evening. Hickling Broad, he states, was one great resort

---

* Mr. Monement states that the habit of partially submerging the body when an enemy is in sight is very marked with the pochard.

at their flight, the water being very shallow, and abounding with "a peculiar weed, *pochard grass* as it is called." What this pochard grass may be I have never been able satisfactorily to determine, no such name, so far as I can learn, being applied to any aquatic weed by the marshmen in the present day; in reply to my suggestion that it might possibly be a *Potamogeton*, Mr. A. Bennett, of Croydon, was kind enough to point out that *Chara aspera* covers the bottom of Hickling Broad for acres, to the exclusion of everything else, that patches of *Potamogeton pectinatus* appear only here and there, and that he thinks the *Chara* may be the attraction. Speaking of the fecundity of the semi-domesticated wigeon in Sir Ralph Payne-Gallwey's decoy at Thirkleby, in Yorkshire, a writer in the Yorkshire *Naturalist* (August, 1887, p. 243), says: "This fecundity is, we are informed, probably attributable to the weed *(Chara foetida* var *atrovirens)* which grows so profusely in the decoy, and which the wigeons are never tired of pulling and eating." It would at first sight seem improbable that so apparently unattractive a weed as *Chara* (owing to the calcareous nature of its skeleton, in addition to its habit of evolving sulphuretted hydrogen) should form a suitable food for diving ducks, but there seems reason to believe that Mr. Bennett's suggestion is correct.

The bulk of the pochards leave us, as a rule, about the middle of March; but, in late seasons, flocks are found till nearly the end of April, and every year some remain to breed in various parts of the county. This has long been known to be the case, and is thus referred to by Hunt ("Brit. Ornithology," ii., p. 308): "It has been doubted whether any of this species remain with us after the vernal migration to the north. We are assured by Mr. Smith, of Diss, that he has seen both sexes of the pochard during the breeding season on Scoulton mere, and that he found their nests and took their eggs, some of which are at this time in his collection. A female bird was shot in Norfolk, on the 14th July, 1818, so that it appears certain that a few, at least, remain with us the whole year."

Mr. Lubbock also speaks of having heard from "an accurate observer that he has shot young pochards at

Scoulton." The person referred to as the "accurate observer," is not unlikely to have been the late Mr. Blofeld, of Hoveton, whose old marshman, William Hewitt, told me that he had shot many "red-head" flappers at Scoulton with his master. Sheppard and Whitear also speak of this bird nesting at Scoulton; it has, however, long ceased to do so. A note in Salmon's diary, March, 1836, is to the effect that he was told by Mr. Carter, who gathered the gulls' eggs at Stanford, that the pochard always bred there. Lubbock quotes from his friend Girdlestone's memoranda as follows: "Upon Hickling Broad, August 16th, 1827, I found four pochards, three of which I shot. They turned out to be all young fowl, no doubt bred somewhere in the vicinity. On June 3rd, 1847, Mr. W. F. Bird records (Zool., 1847, p. 1782) a pochard shot at Tunstall, which he thinks had a nest. In Mr. Blofeld's copy of "Yarrell" are entries of a male pochard seen by him at Wroxham on 10th May, 1849, and a pair by W. Hewitt on Hoveton Broad on 20th July, 1868. Mr. Stevenson (Zool., 1885, p. 327) records three killed on Breydon in July, 1883.

But it is only in recent years that this species seems to have made its appearance in not inconsiderable numbers at the breeding season, and although doubtless a few pairs breed annually in various localities in the Broad district, it is in the neighbourhood of Stanford, where Salmon speaks of their breeding in 1836, that we must seek their present headquarters.

On May 30th, June 1st, and June 10th, 1850, Professor Newton and his brother saw a male pochard on a mere on Wretham Heath, and the former remarks in a letter to Mr. Stevenson:—"He was always on the same mere, and seemed perfectly at his ease, just indeed as a drake invariably does when he knows his wife to be comfortably sitting on her nest, as I strongly suspect was the case in this instance." Lord Walsingham, also, in a letter to Mr. Stevenson, dated 24th March, 1875, observes, "Pochards certainly bred [on his estate] three years ago, for my keeper there knows the birds well, and assured me he saw the nest." On the 25th May, 1874, I saw a number of ducks sunning themselves on the sandy margin of a piece of water in the south-western portion of the county, which were too far off to be identified with certainty, but I took them to be pochards. On the 15th April, 1876, I again, in a neighbouring piece of water, saw about a dozen ducks which, with the aid of a good glass, I had no doubt were pochards, but in 1875 Mr. Stevenson and Mr. H. M. Upcher had found both eggs and young of this species on Lord Walsingham's waters, whilst in 1876 Lord Walsingham and Professor Newton had ample evidence of the pochard nesting freely in the same locality, where it has continued to do so to the present time, and for some years I have met with it regularly in other localities in the same district.

Mr. Frank Norgate, who had exceptionally good opportunities of observing the nesting of the fowl in this portion of the county from 1885 to 1887, has kindly given me some valuable notes on this species, from which I make the following extracts, omitting precise localities for obvious reasons:—

1885. April 24. Saw several pairs of pochards on a pool, and again on May 26th.

June 22. I identified, and watched with a telescope, two broods of young pochards, one brood about three-fourths grown; these last, and the old pochard, always jumped nearly out of the water before diving, as if to give impetus to the dive.

1886. May 8. I watched five pairs of pochards, fifteen pairs of tufted ducks, one pair of gadwall, a male teal, &c., and found a wild duck's nest.

June 2. I found a pochard's nest of six incubated eggs in the crown of an isolated tussock of *Carex*; no feathers, only grey and brownish down, and dead sedge for nest. Saw the old female, dark-eyed, rusty-brown head, blotchy-grey and brown back. I had many good views of her close to this nest, both swimming and flying.

June 5. I found a pochard's nest of ten eggs in a bed of *Scirpus lacustris*. This nest was made of bits of dead *Scirpus*, and profusely lined with pale grey and brownish-grey down. I saw the female swim out of the rushes from the nest, and had many good views of her colours with and without a telescope. Close to this nest was one of four little grebe's eggs.

June 7. I found a third pochard's nest of nine eggs, near hatching, in the crown of an isolated tussock of *Carex*, and clearly identified the female as she flew up from the nest. The other clutch of ten eggs is now hatching, three or four ducklings are now out, and others half out of their shells. I saw another female pochard swimming with a brood of young.

1887. May 5. I saw five pochards, six mallards, and ten tufted, all drakes. After walking round the pool, and thus disturbing the fowl, I saw one mallard, and most of the others in pairs, showing that at my first view the ducks were at their nests.

May 24. I identified two female pochards nesting One nest was of dead *Carex*, the other of dead *Scirpus*, and placed in a clump of the same plant. Each nest contained eight eggs, varying in one nest from greenish brown to brown; the other eight eggs were all brown.

May 26. I had a good view of a female pochard as she flew from her nest of seven hard-sat-on eggs in a tussock of *Carex*. When I put them into the pool, these eggs progressed through the water with a regular and considerable oscillation; in fact, they not only floated with the big end above the water, but they swam well. About a foot from this nest I picked up two pochard's eggs from the bottom of the water, so that the clutch probably consisted of nine eggs. Near this nest I found another of six fresh eggs; but could not identify the female until the 30th, when I proved it to be a pochard. I also found, and clearly identified, three more female pochards as they left their nests of seven, seven, and six eggs, and saw a tufted duck's nest of eight eggs. These five pochard's and the tufted ducks' nests were all within a few yards of one another. I also found another little grebe's nest of four eggs in the same pool this day.

May 30. In the same pool I identified, as she flew from her nest of seven eggs, much sat-on, another pochard (as well as the one I failed to recognise on the 26th inst.) Also two more tufted ducks' nests of eight and seven eggs, all near together.

June 3. I found a nest of five sat-on eggs of pochard in *Carex*, and identified the female as she flew away—a thick, heavy bird, with grey, brown-mottled body, and dull rusty-brown head and neck.

The above extracts, although confined to one district, are not restricted to one "pool," and do not include Lord Walsingham's waters before referred to. It will be seen, therefore, that this species nests very freely in the part of the county referred to, where, I am happy to say, all the fowl are well protected.

The interesting particulars in Mr. Norgate's most careful and valuable notes, afford remarkable encouragement to those who have advocated recent legislation on behalf of our wild fowl—for it is needless to say that nothing of the kind was the case twenty years ago—and render it unnecessary to add any remarks upon either the structure or situation of the nest, or the number and description of eggs, and the date of their production.*

PAGET'S POCHARD. This handsome hybrid between the common and white-eyed pochards was first recorded in this country from a male showing very slight signs of immaturity, killed on Rollesby broad, on February 27th, 1845. Mr. W. R. Fisher, writing in the "Zoologist" of that year (p. 1137), describes "a variety of duck intermediate in size between the common pochard and the nyroca, or white-eyed pochard," rightly surmising it to be "a hybrid between these two species." It was observed to be much tamer than some wigeons with which it was in company, and swam very low in the water. In the same magazine for 1847 (p. 1778), Mr. Fisher gives a fuller description of that specimen, with a woodcut. In that year, also, Mr. A. D. Bartlett purchased an example of this bird, in Leadenhall Market, which is now in the Derby Museum at Liverpool; and, on these examples, he ("Proc. Zool. Soc.," 1847, p. 48) established a new species which he named *Fuligula ferinoides*; and, in compliment to the late Mr. C. J. Paget, one of the authors of the "Sketch of the Natural History of Great Yarmouth," called it "Paget's pochard." It is needless to say that Mr. Fisher's first surmise was the correct one, and that " Paget's pochard " is not now recognised as a distinct species.

On the 24th February, 1859, a second Norfolk specimen of this bird, an adult male, was killed at Little Waxham. Both these are now in Mr. Gurney's collec-

---

* Mr. J. H. Gurney has favoured me with the following note on the change of colour in the irides in a living pochard:—" I am not sure that I ever recorded the following fact. I was removing a pinioned adult male pochard from one pond to another, and, soon after taking the bird in my hand, I observed the bright cherry red of the irides gradually disappearing till it was replaced by yellow, hardly, if at all, tinged with red; this was probably the effect of fright, as, when the bird was released and placed in another small pond (near enough to be well within sight), the irides gradually, but speedily, resumed their normal cherry-red tint."

tion, and the latter is the original of the accompanying illustration drawn by Mr. Wolf, probably in 1862. The plate seems to have been finished early in 1865, and therefore, although only now issued twenty-five years after its completion, it was certainly the first English coloured representation of this bird ever made. A third example occurred on the 13th November, 1871; it was killed on Hickling Broad during a severe frost by Mr. Booth, and is now in that gentleman's collection. In the "Rough Notes" is a figure of this last bird side by side with the common pochard.*

FULIGULA NYROCA (Güldenstädt).

FERRUGINOUS OR WHITE-EYED DUCK.

Although the "white-eyed" duck has been killed in more than twenty instances in Norfolk, it must still be regarded as a rare and uncertain visitant to this county, occurring between the months of November and April, and generally in very severe weather.† Of fifteen indi-

---

* It may be remarked that this hybrid was noticed in Germany not many years after its first occurrence here, and, as in England, received recognition as a distinct species under the name of *Fuligula homeyeri*, Bädeker. Both sexes are described and figured in "Naumannia" for 1852 (Heft i., pp. 12-15), but, as in England, it is now there also regarded as a hybrid. The specimen of a duck obtained in a London market many years ago by Mr. Henry Doubleday, and referred by him and Mr. Yarrell to the American scaup duck, was figured by the latter in the first three editions of his "British Birds." It is now in Mr. Bond's collection, and is obviously a hybrid, but whether of the same cross as the so-called "Paget's pochard" has not been positively determined.

† As this was quite the reverse of what might have been expected, I furnished my friend Mr. A. W. Preston, F.R. Met. Soc., with a list of the dates on which this bird is known to have been met with on the Norfolk coast, asking him to be kind enough to inform me what had been the weather prevailing at or immediately previous to the occurrence; this he did with the result as above stated. Mr. Preston remarks, "On the other side I send you notes of the prevailing character of the weather in the east of England at the

SUPPOSED HYBRID DUCK.
"Paget's Pochard".

John Van Voorst, Paternoster Row 1865.

viduals of which I have dates, two were killed in November, five in December, one in January, one in February, two in March, and four in April.

Messrs. Sheppard and Whitear, writing in 1825, say that "Mr. Wigg had two specimens of the castaneous duck, both killed at different times in the neighbourhood of Yarmouth. One of these was presented to Mr. Youell [the other appears to have been eaten]. We have also been informed that the Rev. G. Glover had a bird of this species, which was shot in Norfolk, a few years since." According to Hunt, his figure was drawn from Mr. Youell's bird. The Messrs. Paget (1834) state that examples of this species had been killed "in a few instances on Breydon." Mr. Lubbock says that it was in one instance taken in the Ranworth decoy, and the late Mr. R. H. Gurney had a male in his possession, taken in Hempstead decoy, which died in February, 1851, having lived in captivity fifteen years. Mr. E. S. Preston, of Yarmouth, had one of these birds in his collection, which was killed on Breydon, in December, 1829; it was a male, not quite adult. This bird passed into the collection of the late Mr. Stevenson, and at his sale was purchased for the Norwich Museum.

Two are mentioned in Mr. Lombe's notes, one a female killed at Rockland, on November 25th, 1826; and the other a male, shot on 16th December, 1839, at Surlingham. Mr. Gurney, jun., tells me that a very fine adult male in the collection at Westwick House, was purchased in the flesh by the late Mr. Petre, at Salthouse, but I believe the date is not recorded. I know of no others till 1850, when one, in the late Mr. Rising's collection, was killed near Horsey Broad by Mr. George Frederick, on the 16th April; on the 6th of November, 1855, another was killed in the same locality. These two birds, both adult males, were cased together, and bought at Mr. Rising's sale by Mr. George Hunt. The spring of 1855, which was remarkable for its severity,

---

dates mentioned. It would seem that the white-eyed duck made its appearance either during great severity or after storms. All the dates with milder weather were preceded by heavy gales and storms, or the cold on the Continent was particularly severe."

produced four of these birds. The first, an adult male, was shot on the 12th February, and two others during the first week of April, and a fourth about the same time was exhibited for sale in the Norwich Fishmarket. Of the last three, two were males in very perfect plumage, of which one was purchased by the late Mr. Alfred Master, and the other by Mr. Stevenson. At the dispersal of Mr. Master's collection Mr. Stevenson became the possessor of the companion bird, and, when in turn Mr. Stevenson's birds were sold, both passed into the collection of Mr. Connop. About January 18th, 1867, an immature male was shot on Hickling Broad (Gunn, "Zoologist," 1867, p. 709); and on the 18th December of the same year, Mr. J. H. Gurney, jun., saw a ferruginous duck in Leadenhall Market, said to have been sent from Norfolk. A pair killed at Dersingham, near Lynn, on March the 20th and 21st, 1868, are now in Mr. H. M. Upcher's collection. One killed on Breydon, on the 23rd December, 1878, was sent to Mr. Gunn; and, finally, on the 30th December, 1886, a male, now in the possession of Mr. E. J. Boult, was shot at Potter Heigham.

FULIGULA MARILA (Linnæus).

SCAUP-DUCK.

Mr. Stevenson has summarised his notes on the scaup, apparently to the year 1861, as follows:—"Young birds of this species are met with at times as early as the middle of October, but their numbers depend much upon the season, and adult specimens are rarely killed until the winter has fairly set in. With all other fowl they were, of course, very plentiful in the springs of 1855 and 1861, and in the spring of 1856 a pair were killed on Hickling Broad as late as the 22nd April. A fine old female scaup, in somewhat unusual plumage, was purchased in our Fishmarket in March, 1858. This bird, in the possession of Mr. Alfred Master, beside having the white patch around the mandibles wider in extent and purer than usual, has the feathers of the head and neck brownish black, with a decided tinge of green on the sides of the head, the dark feathers terminating at the base of the throat in a perfect ring, as contrasted with the brown feathers of the back and breast, the stomach and vent pure white."

Except in very severe weather, the scaups spend the day far out at sea, coming in at evening to the shallows to feed on the crustaceans and mollusks which so abound on the "mussel scaups" (whence their name) on some parts of our coast; in hard weather they make their appearance in shore in considerable numbers, but even then adult birds are the exception. In Miss Gurney's notes (extracts from which will be found in the "Trans. Norfolk and Norwich Nat. Soc.," ii., p. 20) under date of February 4th, 1830, occurs the following :—"Three starved scaup ducks brought to us at Northrepps Cottage. Out of 80 ducks brought to Northrepps Hall about this date, 70 were scaups."

Mr. Dowell states that in hard weather these birds are not only numerous but tame, so that great numbers are sometimes killed by the gunners. He says "Overton once killed 120 in one day in a hard frost, at Salthouse, chasing them up and down in a wake in the ice." Except under such circumstances, this species, unlike the scoter, is not particularly abundant along our coast; and, when they do appear, it is generally in small parties. The fresh water of the broads seems to possess little attraction for scaups, and when met with inland, which occasionally happens, their stay is of short duration. Several adult scaups occurred in the winter of 1870.

FULIGULA CRISTATA (Leach).

TUFTED DUCK.

Mr. Stevenson writes of this species in his note book as follows :—"A very common species in autumn, winter, and spring, and, though belonging to the oceanic group, it regularly frequents our broads and rivers, where the young birds are amongst the earliest fowl obtained by the gunners in most seasons. Fine old males, with long crests and rich glossy plumage, are more generally killed during sharp weather. Mr. Lubbock remarks that at such times the tufted duck 'follows the course of the rivers in its evening flight, until it has occasionally come quite into the heart of Norwich, and been surprised by daylight.' An instance of this kind occurred in January, 1856, when a bird of this species was killed on the basin of Heigham Waterworks."

On the larger broads, in severe weather, old birds are not infrequent, but they are far outnumbered by those in immature plumage. Mr. Booth states that this species and golden-eyes are often found associated, and both alike are known as "black and white pokers" by the local gunners. They are very expert divers, and the late Mr. Thomas Edwards told Mr. Stevenson that, having occasion to remove some pinioned birds from a pond at Thickthorn, he had the greatest difficulty in capturing them by means of nets, with which he succeeded in surrounding them. One tufted duck he could see in the clear water dive and swim round and round to find an opening to avoid the net, "and attempt to go down into the mud at the bottom of the lake, and grub its way under the net like a rat." This was done some eight or ten times by the same bird, and the time it remained under water was quite extraordinary. It will readily be imagined that this species is equally as difficult to capture in the decoy as the pochard; and although, perhaps, less rarely than the latter, still it seldom figures in the decoyman's returns. Mr. Lubbock mentions that on one occasion, in the winter of 1848-9, his friend Mr. Kerrison took eleven of these ducks and a pochard at once in his decoy. These he pinioned and turned down, but they escaped into the broad.

The tufted duck takes its departure about the same time as the wigeon, and is frequently seen on the broads as late as the middle of March, on the 12th of which month, in 1889, Mr. J. H. Gurney, jun., and myself saw about ten, mostly magnificent males, with other fowl, on Holkham Lake; on the 18th March, 1881, Mr.

Gurney saw fifteen tufted ducks, not one of which was an adult male, near the old decoy pipes at Ranworth.*

Mr. Stevenson records a specimen killed at Rockland, which, like the bird described by Yarrell (ed. 4, iv., p. 434), had the feathers at the base of the upper mandible speckled with white, like the adult female of the scaup; and with some elongation of the occipital feathers. Yarrell seems to infer that this peculiarity is the result of age. Mr. J. H. Gurney, jun., has also observed the same admixture of white, and considers Yarrell was wrong in his conclusion, and that it was in reality a mark of youth. Mr. Blofeld also has a note to the effect that he has "killed females with a white spot at the base of the upper mandible," but gives no opinion as to their age. The late Mr. Girdlestone states that in some parts of Norfolk tufted ducks are, or were, known as "arps."

Hunt states that the tufted duck " is frequently seen in our fresh waters as late as the latter end of March or beginning of April," and Mr. Lubbock, after remarking that this species "has never been *proved* to breed in Norfolk," adds that " many years ago, as that practical ornithologist, the late Mr. Girdlestone, was passing up a narrow passage amidst a wilderness of reeds in one of the broads, an old duck of this species, and three young ones, passed close by the boat." Such was the impression of Mr. Girdlestone and of the fenman† who was with them in the boat; and, continues Mr. Lubbock, "I believe it is almost impossible that they could *both* have been mistaken." Messrs. Gurney and Fisher ("Zoologist," p. 2134) record the appearance of a tufted drake and three ducks, on the river Wensum, at Cossey, on the 25th April, 1858, and remark that " it is possible

---

* To show the excellent results of protection at this season of the year, I may mention that at Ranworth, on the same day, Mr. Gurney saw, in addition to the fifteen tufted ducks, a pair of adult golden-eyes, twelve wigeons, a shoveler, eight teal, twenty mallards, twelve great-crested grebes, one peregrine falcon, one marsh harrier, stock doves, golden plover, redshanks, and snipes.

† This occurred at Catfield, in 1825; the man who was with him was "Hewitt, our shooting factotum." Mr. Lubbock adds that he himself has seen the old ones on broads in the midst of summer.

they were resting on their northerly migration, upon which occasion they more frequently cross the land than in their southerly autumnal movements." On the 10th of May, 1849, Mr. Blofeld saw a male tufted duck, on Wroxham Broad, in company with a male pochard. Mr. Frank Norgate tells me he has several times seen small black and white ducks in April and May flying about in and out the reeds in the Norfolk Broads as if they were at home; and, although he has never had the opportunity and sufficient time for proving this, he feels certain they were tufted ducks, and that they nest there regularly.

It thus seems probable that the tufted duck has habitually nested in this county in small numbers, although it has generally escaped detection until recently, and that it only required the protection now accorded to this and other species, to induce it to make its home here in rather considerable numbers; but, like the pochard, and in a less degree the shoveler, it is on the inland waters, in South-west Norfolk, that we shall find it most numerous.

Lord Walsingham cannot speak with certainty as to the breeding of this species on his estate earlier than the year 1873, although he believes they bred there before 1871; in the former year, however, he saw the nest and flushed the bird from it. On June 8th, 1875, Mr. Stevenson and Mr. H. M. Upcher saw three pairs of tufted ducks on one of Lord Walsingham's meres, but, although no doubt existed as to their having nests, they were not successful in finding them. On the 26th May, 1876, Lord Walsingham, with Professor Newton, flushed a tufted duck from her nest in the same mere, in which were six eggs, and, at the same time and place, they saw two males and four females of this species. The breeding of the tufted duck on this estate has now become a regular occurrence, and both the numbers and the area over which they have spread appear to have increased annually; so that in the past season of 1889 Lord Walsingham tells me he believes not less than thirty or forty pairs have bred on Stanford mere, and a considerable number on Tomston, a few miles off. The proportion of tufted ducks to pochards in the breeding season, at the present time, Lord Walsingham estimates to be about double. Both Mr. Gurney, jun., and myself have had some excellent opportunities of observing the nesting habits of this species; but I cannot do better than,—as in the case of the pochard in the same waters,—quote the valuable notes Mr. Norgate has been so kind as to favour me with, which I reproduce in his own words, omitting, of course, precise localities :—

1883. June 1. In one pool I saw eight pairs of tufted ducks, and in another pool five pairs.
1885. April 24. In the same neighbourhood I saw a tufted duck or two, and in two other pools several other pairs.
May 26. Many pairs of tufted ducks.
June 22. I saw two or three broods of young tufted ducks with the old ones.
June 25. I saw six or seven tufted ducks with their young broods, and noticed the blue bills and yellow irides of the old ones.
1886. May 8. I identified fifteen pairs of tufted ducks. They make a noise somewhat like that of the black-headed gull when nesting, but not so shrill.
May 18. On the same pool I had a good view of a tufted duck as she slipped off her nest of eight brownish eggs in the top of an isolated tussock of *Carex*, and swam, showing me her blue bill and yellow irides. I had a good view of another tufted duck as she flew from her nest of ten eggs (one of which was nearly black) in a similar tuft of *Carex*. I also found six more ducks' nests, three of which appeared to be tufted ducks, but I could not identify the owners on that day.
May 26. At the same pool, on similar tussocks of *Carex*, I found six downy nests containing respectively seven, eight, eight, nine, ten, and ?, eggs of ducks; probably all were tufted ducks', for I identified the old tufted ducks leaving two of these nests, and saw four tufted ducks near the other four nests as if they had just left them.
June 1. I identified a third nest (the one of ten eggs) as the tufted duck, having a good view of the duck and her yellow irides as she flew up from the water a yard or two from the nest, immediately after she had slipped off the eggs, which she left uncovered in her haste. The eggs are usually covered with down when the duck leaves them without being suddenly frightened off.
June 2. In the same pool I found a tufted duck's nest of nearly black down and white feathers (as in all the other tufted ducks' nests). These white feathers must be from the breasts of the drakes. This nest contained fresh eggs not incubated. I saw the bird rise from the water close to the nest.

[On June 5th, 7th, and 17th, Mr. Norgate found other nests of the same species with nine, nine, nine, and four eggs respectively, lined with the characteristic dark down intermixed with white feathers.]

1887. May 5. I saw ten tufted drakes, after walking round the pool and disturbing the pool; most or all of these were joined by their ducks, which are probably all nesting here now.

May 26. On the same pool I saw thirteen or fourteen pairs of tufted ducks, and had a good view of a tufted duck as she flew off her nest of eight fresh eggs within a few yards of five nests of pochards.

[On the 30th May and 3rd June Mr. Norgate also saw other nests of this species containing eight, seven, and seven eggs; also several nests of the pochard.]

I have quoted Mr. Norgate's notes somewhat at length, for it would be impossible to condense them without detracting from their value, either as evidence of the increasing numbers of fowl now frequenting the rather extended district to which they and Lord Walsingham's notes refer, or as conveying information with regard to the nesting of birds which few British naturalists have had so excellent an opportunity of watching as Mr. Norgate, and which I am sure none can turn to better account.

FULIGULA GLACIALIS (Linnæus).

LONG-TAILED DUCK.

The month of November almost invariably brings the long-tailed duck to our coast, although in varying and uncertain numbers, but it is not so entirely a hard-weather bird as has been generally supposed. Indeed, this ocean loving species seems able to remain at sea in almost any weather, and may even be abundant in the offing when quite unsuspected and without coming under the notice of the shore gunners. Mr. Booth states that "immature birds in considerable numbers, as well as a few adults, annually work their way south on the approach of winter, and take up their quarters off our eastern and southern coasts, for the most part selecting such situations over mussel-banks and other feeding-grounds as are suitable to scoters and other diving ducks. On the 28th November, 1879, while steaming along the shore, and also through the gatways between Yarmouth, Lowestoft, and on towards the south as far as Benacre and Southwold, I met with several large flocks near each of the places mentioned. Common scoters were by far the most numerous of the fowl gathered off this attractive portion of the coast, though velvet scoters, long-tails, and golden-eyes were well represented. . . . Every winter when spending my time at sea off that part of the coast, I have observed long-tailed ducks."

By far the greater number of those obtained are immature birds; mature males are decidedly rare, and their occurrence seems almost accidental, winters of great severity sometimes passing without a single individual being obtained, whereas, in exceptionally mild winters, they may occur several times.

Messrs. Sheppard and Whitear speak of the long-tailed duck being unusually numerous in the winter of 1819-20, particularly at Yarmouth, where many of them were obtained; they also mention that this species has been taken in Herringfleet decoy, a statement which seems extremely improbable, as, in addition to its not being likely to occur on the lake, it is so expert a diver that some special contrivance would probably be required to effect its capture; possibly there was some confusion between the long-tailed duck and the pintail. Hunt ("British Ornithology," ii., p. 303, foot note) also says, "we have been gratified with the examination of several specimens which were exposed for sale in the Norwich Market in the early part of November, 1819. It is the only instance we remember of this species being shot in Norfolk." The "Norwich Mercury" for 10th March, 1832, after stating that "it might naturally have been supposed that the present unusually mild winter would have been little calculated to tempt the more rare species of wild-fowl to quit their northern hibernacula and migrate to our shores," goes on to state that no less than twenty velvet ducks had been sent to Mr. J. Harvey, of Yarmouth, in the course of the week, and that with them "four of a still more rare species, the

A. glacialis or long-tailed duck." On 6th January, 1866, an adult female is recorded by Mr. Stevenson, as having been taken at Salthouse ("Zoologist," 1866, p. 260) which, he adds, "confirms my opinion that these birds occur at times in mild winters when least expected, and are often altogether absent in the hardest weather." November, 1887, was unusually productive of this species, both in Norfolk and elsewhere, some occurring in this county as far inland as Rockland Broad, near Norwich ("Zoologist," 1888, p. 287); and the Rev. J. G. Tuck tells me that in the winter of 1887-8 one of his correspondents from Hunstanton, in answer to his enquiry, told him the number brought home up to 15th February was twenty-seven, and he intimated that he had lost quite ten or twelve others, owing to wind and tide. He added, "there is one fine old bird about the feeding-ground, but up to the present he has not given me a good shot at him." Shortly after, Mr. Tuck had a second letter from the same correspondent (received on the 27th March, 1888, and quoted in "Land and Water," March 31st, 1888), in which he gives the following additional information:—"I was very sorry I could not get the old male long-tail; he is still about here, and, strange to say, three others have joined him, quite as good specimens. There is a large flock of them; since we have left off shooting they keep all the time just opposite my house, about four or five hundred yards out at sea. They are there every day feeding . . . just in a line with the bathing post. I should think there are quite 300 of them. I can see these old males quite plainly with the glasses. There are still thousands of scoters here." Lastly, on the 10th December, 1888, Mr. Cole had a fine adult male sent him from Wells for preservation. These are only a few of the many examples which have been recorded, anything like a complete list of which it would be impossible to give.

Although I have spoken of the occurrence of adult long-tailed ducks in open weather as not by any means so unusual as was formerly believed, it must not be supposed that such occurrences are the rule, the large majority of the full plumaged birds having undoubtedly been met with in severe weather. In two instances, however, this species has been killed in Norfolk in summer; on the 2nd of June, 1856, an old male, in full nuptial dress, was shot on Hickling Broad; and a magnificent drake in full summer plumage, now in the collection of Mr. J. H. Gurney, was killed at Acle about June 14th, 1885. It was found by some boys in an exhausted state in a ditch, on Acle marshes, and killed by stones. These are, I believe, the only occurrences at this season of the year, in Norfolk, on record. The Rev. E. W. Dowell says that the immature long-tailed duck is known to the Blakeney gunners as the "little mealy duck," and the adult as the "sea pheasant."*

FULIGULA CLANGULA (Linnæus).
GOLDEN-EYE.

The golden-eye is known to the shore gunners as the "rattle-wing," and, according to Mr. Dowell, to the Blakeney fowlers as the "wigeon"—the tufted duck, as before stated, being there almost always called golden-eye. As a winter visitor, both to the shore and to the larger broads, it is not uncommon—although not so numerous as formerly—the majority being immature birds. It seldom appears far inland, but has been shot as high up the River Yare as Earlham. The old birds do not often make their appearance before the middle of December, and are rarely seen after March. They are, as a rule, the most wary of all fowl, but Mr. Booth says the young birds are easily approached, for the most part evincing little distrust till repeatedly alarmed. Mr. Monement saw five, two of them mature drakes, in a

---

*The Harlequin Duck *(Fuligula histrionica)* is reported by the Messrs. Paget ("Nat. Hist. Yarmouth," p. 12), and Messrs. Gurney and Fisher (("Zoologist," p. 1380), to have been killed twice near Yarmouth; but Mr. J. H. Gurney, jun. ("Rambles of a Naturalist," p. 266), has shown that in neither of these cases, nor in a supposed third instance, mentioned by himself, can the records be depended upon, and, accordingly, further notice of that species is omitted from the present work.

pond at Weybourne, in March, 1889. Mr. Booth met with this species on Heigham Sounds in March, and on Hickling Broad in April, 1873; on April 6th, 1888, Mr. J. H. Gurney, jun., saw an old male at Fritton, which was very tame, allowing him to row to within forty yards; William Hewitt reported to Mr. Blofeld that he saw two on Hoveton Broad in the middle of May, 1862; and Mr. J. H. Gurney, jun. ("Zoologist," 1880, p. 404), saw one in Norwich market in the month of July, they have however never been known to breed here. Mr. Lubbock strongly suspected a pair of nesting on Horsey mere, where he constantly saw them after all the other fowl had departed. "The last time I saw them together," says Mr. Lubbock, "was on May 12th; on the 26th I saw the drake just in the same part of the water, swimming alone. I took great pains to find the nest which I think must have existed, but could not." At that time Mr. Lubbock was not aware of the peculiar site chosen for its nest by this species, and subsequently attributed his want of success to confining his search to the marsh. Then follow some remarks, almost prophetic in their nature, which are so characteristic of Mr. Lubbock as a naturalist, and so descriptive of some parts of the broads, even at the present time, that they are quite worth quoting. "I have been particular on this point, as, in this district, no part of ornithology demands more attention than the nests of water birds. Many marshes, which are regularly traversed by means of boats, and the ditches cut, in the autumn and winter, are, in the summer, pathless wilds of water, sedge, and reed: 'We can't get about there till the marshes are mown,' is a phrase familiar to all who have known the Norfolk broads. Even the heron sometimes nests in these situations, and the eggs are unmolested, as no one can reach them.

"When we consider that new species are continually discovered amongst us—when we know what observation every year brings to light—there is just room for hope that the increased ardour with which natural history is now followed, will show *some* cases in which birds not now suspected of it, occasionally nest in the Norfolk fens."

It was in such "pathless wilds of water, sedge, and reed" that the bittern made its nest, its presence being only made known by its summer cry, or the fact of its having nested by the fledglings brought by the retriever in the early days of flapper shooting.

Four examples of the hybrid, as it is now generally regarded, between the golden-eye and the smew, have been recorded, and though none of them was observed in this country, yet it is obvious that such a bird might almost any year occur in the British Islands; and, as very little notice of this cross has been taken by writers on British ornithology, particulars of the several captures may be of interest here. Professor Newton has, therefore, at my request, kindly furnished me with the following note on the subject:—

"The first example, a male, in apparently full plumage, killed on the river Oker, near Brunswick, in the year 1825, was described and figured under the name of *Mergus anatarius*, as belonging to a new species by Eimbeck ('Isis,' 1831, pp. 299-301, pl. iii.). The specimen, which is in the Museum at Brunswick, was subsequently refigured by Naumann in his 'Vögel Deutschlands' (vol. xii., frontispiece), and again by Dr. Rudolf Blasius, in the 'Monatsschrift des deutschen Vereins zum Schutze der Vogelwelt' for 1887. The second, a young male, was shot in February, 1843, in the Isefjord, on the north coast of Zealand (Denmark), as described by Kjærbölling, under the name of *Anas (Clangula) mergoides* ('Naumannia,' 1853, pp. 321-331, 'Journ. für Ornithologie,' 1853, Extrah. pp. 29-32), and figured by him in his 'Skandinaviens Fugle' (Suppl. pl. 29), he having already given in the same work (pl. 55) a copy of Naumann's plate. This specimen is now in the Royal Museum at Copenhagen. The third example, also a male, to all appearance in full plumage, was shot out of a flock of ducks, near the island Pöl, on the north coast of Germany, at the end of February, 1865, though not recorded until ten years later by its then owner, Franz Schmidt, of Wismar, in the 'Archiv der Naturgeschichte in Mecklenburg' (1875, pp. 145, 146). It is now in the possession of Herr Oscar Wolschke, of Annaberg, who gave a figure of it in the 'VII. Jahresbericht des Annaberg-Buchholzer Vereins für Naturkunde,' which has since been reproduced by Dr. Blasius in the work already named. The fourth was shot in Kalmar Sound, between Sweden and Œland, 20th November, 1881, and sent to the Museum of the University of Upsala, as described and figured by Herr Gustav Kolthoff (Œfversigt af K. Vetenskaps-Akademiens Förhandlingar' 1884, pp. 185-190, pls. xxxi., xxxii.).

"The association of a male smew with a female golden-eye, on the water in St. James's Park, was long ago noticed by Yarrell; and, from the fact that both the golden-eye and the smew, as ascertained by Mr. Wolley ("Ibis," 1859, pp. 69-76), have the same habit of nesting in holes of trees, or in logs purposely hollowed out and set up by the people in countries where both species breed, it is, perhaps, to be wondered at that more examples of hybrids between them are not known. To judge by the figures which have been given of the specimens captured, they present an appearance almost exactly intermediate between the two species. The Swedish example has the bill and legs black as in the smew, while these parts in the others have the yellow colour of the golden-eye; but the plumage of all shows little difference."

FULIGULA ALBEOLA (Linnæus).

BUFFLE-HEADED DUCK.

Mr. Stevenson has left the following note on this species:—"But one specimen of this rare American duck is known to have been killed in Norfolk. This bird, an adult male, formerly in the collection of the late Mr. Stephen Miller, but now in the possession of Mr. Rising, of Horsey, was shot near Yarmouth, about the year 1830." The first notice of this example, which, at the sale of Mr. Rising's birds, in 1885, was secured for the Norfolk and Norwich Museum, seems to be that of the

Messrs. Paget in their "Sketch of the Natural History of Yarmouth" (p. 11., foot note) as follows:—"Mr. Miller has a specimen which he considers proves that the morillon *(Anas glaucion*, of Linn.) is different from the golden-eye. It was an old male bird, but is fully one-third less than the males of the golden-eye, and the bill is considerably shorter; besides which the plumage is rather different."

It is now well known that at that time (1834), and even later, the various plumages presented by the golden-eye, and the long period the male takes to assume his full dress, were but little understood by some of the best-informed ornithologists. The gunners of Yarmouth, and perhaps elsewhere, spoke of two kinds, by the names of "rattle-wing" and "little rattle-wing" respectively, the former being the adult golden-eye, and the latter the so-called "morillon" of Bewick and other writers, now universally recognised as the immature stage of the same bird. It is, therefore, not surprising that the Pagets and Lubbock were unable to clear up the confusion, and the last, as may be seen from his conversation (quoted by Yarrell in the account of the buffle-headed duck given in his earlier editions), as well as from information received through Mr. Girdlestone, and published in the "Fauna of Norfolk,"[*] was inclined to believe that the "little rattle-wing"—a common bird enough—was the present rare species, the complete verification of the occurrence of which, according to Lubbock, we owe to Yarrell, though the fact was not made known to the world until July, 1842, when part xxxi. of the first edition of his "British Birds" appeared.

It must be remarked that the British Museum contains a specimen of the buffle-head duck, with the assigned locality, "Norfolk, from Mr. Hubbard's collection;" but, as has been elsewhere remarked (Lubbock, "Fauna of Norfolk," ed. 2, p. 165, note 156), its authenticity is more than doubtful.

---

[*] As regards this matter, see also a letter from Girdlestone to Lubbock, printed in the "Transactions of the Norfolk and Norwich Naturalists' Society" (vol. ii., pp. 398, 399).

MERGUS ALBELLUS (Linnæus).

SMEW.

Immature specimens of this beautiful bird are obtained nearly every winter, and in some seasons are even tolerably numerous, but, in the fully adult plumage, it is of much less frequent occurrence. Old males in their lovely white plumage are somewhat inappropriately (so far as sex is concerned) known as "Nuns,"* and are very rarely met with till the months of January and February. A male killed on Breydon, on January 15th, 1881, Mr. J. H. Gurney, jun., tells me had two-thirds assumed the adult plumage. It would be impossible to give a list of the numerous instances in which adult smews have been killed in this county. I shall, therefore, only mention some occasions on which they were unusually abundant. Messrs. Sheppard and Whitear state they were plentiful in the winter of 1819-20 at Yarmouth. Several were killed in the cold spring of 1855; four adult males were shot at Burgh near Yarmouth, Rockland, and other parts of the county, also several females and young birds. One old male, in the collection of Mr. J. H. Gurney, jun., was shot at Burgh, in February, 1860; and the following spring of 1861, during the intense cold which prevailed at that time, a very unusual number of smews, both adult and immature, says Mr. Stevenson, appeared on our coast; a fine old male, in his own collection, was shot at Surlingham, on the 25th January of that year, in company with another male and a female; and Mr. Ellis, bird stuffer, of Swaffham, received three in perfect plumage, together with twelve or thirteen females and young birds, killed in January and February, in the neighbourhood of Lynn. In 1867-8 old females and young birds were

---

* This, however, seems to be a name at least as old as, if not older, than smew. Merrett ("Pinax," p. 183) has "Non est avis aquatica . . . . caput cristatum unde forsan ei nomen, sc. a monacha velata"; and Willughby's translator says of the bird ("Ornithology," p. 337), "We may call it, with the Germans, the White Nun." He adds that a female "was sent to us by Mr. Dent, from Cambridge, by the name of a Smew."

somewhat numerous, but not a single adult male was recorded. In the winter of 1869-70 several were met with, a full plumaged male being shot at Burgh, near Yarmouth; two or three young ones were also sent to the Norwich Market. Again, in 1870-71, smews were frequent, with an unusual proportion of adults; and Mr. Stevenson records ("Zoologist," 1871, pp. 2499, 2600, and 2828) many shot in various parts of the county (one female as far inland as Wretham, in March), the weather being at the time extremely severe, with prolonged frost and snow. Between 1871 and 1879 very few good birds appear to have been met with, but, in the latter year, two adult females were shot on Breydon, on the 8th December, and two more on the 28th. In January, 1881, I have notes of four old females and three young birds, one of the latter killed at Taverham, near Norwich. In March of the same year two old males in fine plumage were killed at Hickling, on the 6th; and, on the 25th of April, a remarkably late occurrence, two were shot at Ranworth. In September another was killed at Hickling. On February 14th, 1889, a female was shot at Cley and sent to Mr. Pashley of that place for preservation, where it was seen by Mr. J. H. Gurney, jun.

The smew shows considerable partiality for fresh water, and it will be noticed that in the foregoing list of occurrences some are said to have been killed at Ranworth, Rockland, Surlingham, and even at Taverham, on the River Wensum, some thirty-five miles from the sea, following the course of the river. They are occasionally seen on all the larger broads and inland pieces of water. Mr. Alexander Napier tells me that a small party frequent the lake at Holkham annually; he remembers their being there for fifteen years, arriving about the end of December and departing sometimes as late as the first week in April, but the time both of their arrival and departure is very much governed by the weather. In rough weather they spend their time on the lake, always giving preference to the deep end, but on calm days do not as a rule leave the sea; the flock consists of from five to seven, and Mr. Napier has never known them to exceed the latter number; one or more males are invariably present, and in the past season (1889) there have been three males and two females. Although Mr. Napier only remembers smews frequenting the lake at Holkham for the past fifteen years, he thinks it probable that they have done so much longer, as there is one in Lord Leicester's collection, killed there in January 1854.

Sir Thomas Browne probably refers to this species under the name of *Mustela variegata* :* "The variegated or partiecoloured wesell, so called from the resemblance it beareth vnto a wesell in the head."

MERGUS SERRATOR, Linnæus.

RED-BREASTED MERGANSER.

All the birds of the genus *Mergus* are known to our gunners as "Sawyers," and, says Sir Thomas Browne, are "distinguished from other diuers by a notable sawe bill, to retaine its slipperie pray as liuing much upon eeles, whereof wee haue seldome fayled to find some in their bellies." Probably those examined by the learned doctor were killed in fresh water, to which they frequently resort, occasionally following the courses of rivers far inland, although they certainly do not with us affect rivers and lakes so much as the goosander.

Like all hard-weather fowl the young birds are much more frequent than the adults, old males in their beautiful mature plumage being decidedly scarce, and, as a rule, only met with during severe weather, although, as will be seen, this is not invariably the case. A notable exception to this occurred in the third week in July, 1852, when an adult male in full plumage was observed by Mr. J. H. Gurney flying over the Denes bordering the beach to the north of the town of Lowestoft. The

---

* This and other quotations from Sir Thomas Browne's papers, have been corrected for me, according to the original MS., by Professor Newton.

young birds arrive on our coast late in autumn or in the early winter, and depart about the middle of March, but owing to the continuance of cold weather in the spring of 1845, many winter birds stayed with us longer than usual, and Mr. Gurney had the opportunity of examining an adult pair of red-breasted mergansers at a Norwich bird-stuffer's, on 12th April of that year; and an instance is recorded by Mr. Stevenson of an immature merganser having been killed here as late as the 28th April. In Sir William Hooker's MS. occurs a note initialled D[awson] T[urner], which states that many specimens of this bird were "shot near Yarmouth, December, 1829, and January and February, 1830. Before that time very seldom seen." Mr. Stevenson remarks that it is somewhat singular that in the extraordinary winter of 1860-1 not a single old male, as far as he could learn, was killed in this county, though females and young were plentiful, but, in the less severe spring of 1855, between the 28th February and the end of March, four magnificent males, in their rich adult plumage, were shot at Surlingham, Hickling, and Yarmouth; a fine pair were also shot at Salthouse, in January, 1854. Mr. Dowell mentions that they were unusually abundant at Blakeney in 1846-47; and again, in November, 1852, a greater number than usual made their appearance in that locality, but all immature birds. They usually make their appearance at Blakeney towards the end of October, and are known to the gunners of that coast as "steel-ducks," from the blue colour which is most observable when they are on the wing. Several old males were killed in this county about the middle of February, 1870, as well as a number of immature birds; and in January and February, 1871, other adults were obtained, and immature birds were rather numerous. No remarkable visitation has occurred since 1871, to my knowledge; in fact of late years this species in all stages has been somewhat of a rarity with us, the reverse, as will be seen, of the next species, and I believe this remark also applies to the Yorkshire coast.

A very remarkable instance of the assumption, to a certain extent, of the male plumage by a female red-breasted merganser is recorded by Mr. Gurney in the "Zoologist" (p. 4252). To all appearance it was a young male bird beginning to assume the adult dress, but upon dissection Mr. Gurney found it to be a female, and without any signs of either disease or exhaustion of the ovarium.

Mr. J. H. Gurney tells me that in the winter of 1845 he had an opportunity of tasting the flesh of an old male red-breasted merganser, and that he found it well flavoured, and not at all fishy.

This merganser breeds regularly in many parts of Scotland, where, however, it is very much disliked on account of its fish-eating habits, and is accordingly in many places persecuted. Its comparative rarity of late years in Norfolk may perhaps be thereby accounted for.

The HOODED MERGANSER *(Mergus cucullatus,* Linn.), a North American species, has until recently always been accorded a place in the list of Norfolk birds, and that without hesitation, on the strength of an undoubted specimen of *Mergus cucullatus,* sent by Mr. J. W. Elton to Selby, who informed him that it was killed at Yarmouth in the winter of 1829. The bird in question is still in the collection at Twizell with Elton's name upon it, but there are circumstances which render the origin of the specimen doubtful; and it has been expunged from all the county lists published since 1877. Full reasons for this step will be found in a foot-note appended to a copy of the original letter from Elton to Selby, accompanying the bird, and printed in the "Trans. of the Norfolk and Norwich Naturalists' Society" (ii., p. 408). It is only right, however, to mention that there is a second candidate for insertion in the list of Norfolk birds. In Neville Wood's "Naturalist" (vol. iii., p. 413) for 1838, is a paper by the late Edward Blyth, entitled "Notice of Rare Birds obtained during the Winter of 1837-8," in which he writes as follows:—"Mr. Hoy informs me that a beautiful male hooded merganser *(Mergus cucullatus)*, in thoroughly mature plumage, has been secured in the county of Norfolk, being the first known instance of this bird occurring in its adult garb in Britain." In this, as in the previous case, the species is a certainty, and the only room for doubt is as to its origin. It will be observed that Hoy was very explicit with regard to the sex and age of the bird, and, although he does not actually state that it was in his own collection, the inference is that it was so. That such a bird was in the collection of the late Mr. Hoy we have the evidence of Dr. Bree, who, when describing a visit to Stoke Nayland ("Field," 14th December, 1867, p. 504), speaks of a "Hooded Merganser, under the name of Crested Smew," which, he remarks, was "doubtless the specimen recorded as obtained by Mr. Hoy in the 'Naturalist.'" It thus seems, I think, more than probable that the bird in Mr. Hoy's collection was a genuine Norfolk killed specimen.

MERGUS MERGANSER, Linnæus.

GOOSANDER.

"This species," says Mr. Stevenson, "like the preceding, occurs in its immature plumage nearly every winter, and somewhat earlier in the season, appearing at times towards the end of October. Old birds are killed only in sharp weather or just preceding a change to frost and snow, but they are, on the whole, less rare than the [red-breasted] merganser in fully adult plumage." The above note was written probably about the year 1860, and subsequent observation has tended to render the difference between the two species more apparent; while the red-breasted merganser appears for some years past to have occurred decidedly less often than formerly, the goosander, on the contrary, has increased in frequency, particularly in its adult form, and is now a regular visitant to certain pieces of inland water as well as being occasionally met with on almost all the rivers and broads of the county.

Hunt, in Stacy's "History of Norfolk," says that a pair of these birds was killed in the winter of 1837, in the neighbourhood of Gunton, near Cromer, a locality to which I shall presently have occasion to refer, but he speaks of the male in full plumage as rare. The brothers Paget state that it is "occasionally here in severe winters," and Mr. Lubbock also mentions it as rare in perfect plumage. An entry in the Hooker MS. states that two specimens of the goosander were "taken alive by a fishing boat off the coast, December 29th, 1830, very savage, attacking spontaneously the men that had them, and lacerating extremely their fingers by their bites." In the winter of 1838 this species is said to have been plentiful, when adult males were killed at Heigham and Costessey, near Norwich. A fine old male was killed at Yarmouth on 27th January, 1854, and six others at different parts of the county, in the winter of 1855, together with several females and young birds. These adult birds were in perfect plumage, and when first shot had, as is usual with this species, the most delicate salmon colour on the under parts, which, however, soon disappeared after death. Mr. Stevenson states that he has observed this tint in some immature birds, young males, according to Macgillivray, in their second year, still resembling the female, "with the exception of having the breast of a beautiful pink buff, as in the adult." Some of the examples above mentioned were killed as late as the 13th March, when the weather had become quite mild and open, but it is not unusual with the duck tribe, after a severe winter, for the adult birds to be met with on their return northward late in the following spring. In February, 1856, an old male was shot on the river Yare, near Trowse, and some ten or twelve females and young birds, but in the extremely sharp winter of 1860-1 only two old males occurred. Mr. Gurney has a female which was shot at Hellesdon out of a flock of three, on the 9th January, 1864. Many were shot in the severe snowy weather in January and February, 1867, and one or two in the autumn of the same year. In 1870 an adult male was shot at Stalham in February, and females and young birds were not uncommon in the severe frost which then prevailed. A fine pair were shot on Gunton lake on the 7th March, and two females on the 20th of the same month. In 1871, between January and the 11th of March, many were killed; in the former month a flock of about twenty frequented the lake at Gunton, eleven were seen at Breydon, five of which were shot, others were killed at Yarmouth, Burgh, Hickling, Rockland, and two on the river at Feltwell. Most of the male birds were in perfect plumage, and the old females were quite as rich in the salmon tint of their breast feathers when first brought in as were the males. In the winter of 1880-1 both adult males and females were met with at various points on the coast and on some of the inland waters, as well as immature birds. The winter of 1883-4 also produced a considerable number, and two early arrivals were said to have been seen on Breydon, on the 25th September, 1883. Again, in 1886, several good birds were met with, an adult female on the Wensum at Tatterford, near Fakenham, and others in the winter of 1886 and '7; in fact, scarcely a winter passes without their being observed in greater or less numbers.

But it is not only as an accidental winter visitor to this county that the goosander is known, for it may be said to be a regular winter resident on some of the inland waters. Writing in February, 1869, to Mr. Stevenson, Lord Suffield says, "I have invariably in former years seen some seven or eight goosanders together [on the lake at Gunton] in the month of October; they always arrive about the same time, the middle of the month. We generally see them now and then throughout the winter, but never in large numbers. The upper water at Gunton is not more than four miles, as the crow flies, from the sea, and the feeding ground is the best I have seen for wild-fowl in Norfolk; in consequence we always have a great variety of fowl at all times." The Rev. H. H. Lubbock, writing about the same time, says they are never found in larger flocks than nine or ten, and that at that time (February, 1859) "seven had been killed on the Gunton lake, two more are wounded on the water now, and seven more are flying about." Since that time, and probably for many years before, small parties of goosanders have frequented the lake at Gunton almost every winter. Twenty were seen there in January, 1871, and Mr. J. H. Gurney, jun., saw thirteen females or young males, on the same water, on 22nd January, 1887. The goosanders have been observed occasionally to leave the Gunton lake and to resort to a smaller piece of water in the neighbouring parish of Antingham.

On the lake at Holkham they are regular visitors, and here Mr. J. H. Gurney, jun., saw a pair on the 13th March, 1889; also to a fine piece of water in the park at Kimberley. Lord Kimberley, in a communication to the Norfolk and Norwich Naturalists' Society (iv., p. 604) says, "The goosander has visited us in unusual numbers of late. In December, 1884, I saw 13 together, several of them old male birds. They stayed here a long time. Every year since there have been several here, and always some old male birds. Last spring [1888] several stayed as late as May. Of course I never molest them." Mr. T. E. Gunn ("Zoologist," 1885, p. 56), after stating that goosanders were unusually abundant in this county in the winter of 1883-4, says that when pike-fishing on the lake at Kimberley, on 4th February, 1884, he counted fourteen of these birds, one being an adult male. The keeper told him the goosanders had been there several weeks, and seemed quite able to take care of themselves, something like 200 yards being the nearest approach they allowed. Mr. Gunn also mentions having received an adult male goosander, on February 20th, 1883, which had been killed on Gunton lake, where it was found in company with six or seven grey birds of the same species.

The frequent mention of small parties of from seven to ten "females or young birds," and the presence of one adult drake amongst them, would seem to indicate that they were family parties, which would probably remain in consort till they returned northward in spring.

In the Ranworth decoy books is one entry of the capture of this bird, dated March 8th, 1865.

The arrival and departure of these birds is, of course, to a great extent governed by the weather, but they have certainly been met with from October to early in April, and were undisturbed, as stated by Lord Kimberley, as late as the month of May.

PODICIPES* CRISTATUS, Linnæus.

GREAT CRESTED GREBE.

The following extract from Mr. Stevenson's notes, written probably about the year 1861, with a paragraph added in the spring of 1863, therefore all prior to the passing in 1869 of the Sea Birds Preservation Act (in which protection was luckily assured to this beautiful bird), will be found interesting, not only for the observations it contains on the life-history of a species so eminently characteristic of the Norfolk waters, but as showing what must inevitably have been its fate, so far as this country was concerned, had salutary legislation been long delayed. What is added as to the later condition of the "loon," will be found, I trust, a pleasing contrast to the melancholy record and anticipations of my predecessor.

"Of the many admirable sketches of bird life which render Lubbock's 'Fauna' so charming to the naturalist, the biography of the grebe or loon is decidedly one of the most interesting, and the accuracy of its descriptive power becomes more and more valuable as the bird itself becomes scarcer through constant persecution. Writing in 1845, Mr. Lubbock says, 'It will not happen in our time, but perhaps the next generation may speak of this bird, as we now do of the bustard, in the past tense. It is sometimes shot for the sake of the feathers, sometimes as pernicious to fish. The eggs are always taken when found; I have known thirty or forty collected from one broad. Surely there are common fish enough in our extensive waters, and a

---

* Dr. Babington, in his "Birds of Suffolk" (p. 200, note) has pointed out that the ordinary spelling *(Podiceps)* of this word, as commonly adopted by modern authors, possibly out of regard to euphony, but in utter disregard of Latin, is a contraction of, or misprint for, *Podicipes*, used by Linnæus (Syst. Nat., Ed. 10, p. 136). He states that, as pointed out to him by Professor Newton, both Willughby and Catesby use the latter word, and that he has accordingly restored the older and orthographic form as above given, in which, on such excellent authority, I do not hesitate to follow him.

2 G

few might be spared for this bird, the greatest ornament of the Norfolk broads.' With so many and such relentless enemies, the wonder is that a single loon remains at the present time to grace our waters, than that their numbers for the last twenty years should have steadily but surely decreased.* Every spring, though in smaller and smaller numbers, their eggs are exhibited for sale in our markets, and the value of their skins as ornamental trimming for ladies' dresses, ensures the destruction of every unlucky bird that shows itself within reach of a marshman's gun. The following note by Mr. Strangways, of London, from the 'Zoologist' for 1851, p. 3209, will probably enlighten the proprietors of our broads as to the cause of the disappearance of so many of these beautiful creatures from our reedy waters. In the months of April and May last (he said) I collected twenty-nine of these birds in full summer plumage, *all shot in Norfolk*. Four of them I preserved, and they are now in the Great Exhibition in Hyde Park, where they are exhibited by Messrs. Robert Clarke and Sons, the furriers, in class 18, to which they very appropriately belong, as the breast of this bird has become a fashionable and very beautiful substitute for furs. The rest of the skins I have had manufactured into ladies boas and muffs, and may, perhaps, say they are the first British specimens used for this purpose. The market for grebe is chiefly supplied from southern Europe.' This fashion, unfortunately, has by no means died out, and the demand for skins being greater than the supply, as much as ten

---

* Sir Thomas Browne thus refers to this bird, "*Mergus acutirostris speciosus* or Loone, an handsome and specious fowle, cristated, and with diuided finne feet, placed very backward, and after the manner of all such which the Duch call *Arsvoote*, they have a peculiar formation in the legge bone, which hath a long and sharpe processe extending above the thigh bone. They come about April, and breed in the broad waters, so making their nest on the water, that theire egges are seldom drye, while they are sett on." Graves, in his "British Ornithology" (2nd Edit., 1821), vol. iii., says, "on the extensive broads (as they are termed) in Norfolk and Suffolk they [the Great Crested Grebe] are extremely common; we counted twenty-six, at one time, on Filby Broad, near Yarmouth."—T. S.

shillings each is generally given,* except, therefore, where strictly preserved, the unhappy loon is doomed to destruction.

"A few nests are still found annually on Hickling Broad and, owing probably to the extreme mildness of the previous winter, an unusual number of loons appeared on these waters in the spring of 1863, but, I am sorry to say, that even this chance of [recovering] their former numbers has been wholly frustrated, one individual alone having received, for the mere purpose of barter, six dozen eggs from that locality.† At Horsey, as well as on some of the smaller broads, and from the same causes, they have wholly ceased to breed, the fenmen themselves apparently wanting the sense to preserve the goose with the golden eggs. At Ranworth, Hoveton, and some other localities, however, where care is taken of both birds and eggs, they may still be found during the breeding season. On the former waters I have had many opportunities of there watching their habits during the spring months, when they occasionally rise on the wing and fly round and round with a strong steady flight before settling again on some distant part of the water;‡ and at Hoveton the great interest taken of late years by the proprietor in their careful preservation has been rewarded by the presence of several pairs, accompanied by their young ones, disporting themselves in perfect security, as well as other species of fowl which had previously ceased to breed in that neighbourhood.

"These birds usually dive off their nests at the approach of a boat, with the least possible motion of the reeds, leaving their eggs lightly covered with loose weed, which agrees so nearly with the surrounding herbage as to pass wholly undetected, unless carefully looked for. When unmolested they are by no means shy, but at a

---

* Bewick, edit. 1804, ii., p.144, says, "The skin of one of the species sells as high as fourteen shillings."—T. S.

† In the spring of 1867, Watson, of Yarmouth, sent to London seven dozen eggs of the great crested and little grebes.—T. S.

‡ When on the wing their appearance is very singular, the neck being carried quite straight, and the feet thrust out behind; the small wings work with great rapidity, and the flight is direct and powerful.—T. S.

respectful distance may be watched for any length of time diving after their food or preening their glossy feathers, the rich chestnut crest of the old birds glistening in the sun as they shake the moisture from their silky plumage. The eggs, which are usually laid in April, are of an extremely chalky texture externally, and, though naturally white in colour, are often much stained by contact with the moist materials of the nest. This structure is thus accurately described by Mr. Lubbock in a communication to Mr. Hewitson, 'British Birds' Eggs' (3rd edit., ii., p. 441):* 'Great portion of the nest is under water; that which is above is conical in some degree, and on the top, in a slight cavity, are deposited the eggs, of a whitish colour by nature, but often so stained by the damp of the locality as to present quite a different appearance. These eggs vary in number. I have seen nests with only three, all nearly hatched; four is a common number, and sometimes there are five, but one at least is generally addled, so that three young Loons are generally seen following the old one.'

"The same writer also refers in his 'Fauna' to the habit of this bird of diving with its young under its wing, instancing the case of a loon shot on Rockland Broad, by whose side, when floating dead on the water, 'lay a little one not more than a week old.'†

"In winter the loons quit the broads altogether, and betake themselves to the vicinity of the coast, where they are not unfrequently killed on Breydon and other waters kept open during the sharpest frost by the action of the tide. I have never seen them actually at sea on this coast, but in the south of England it is not unusual, as well as on the more tranquil waters of the various bays."

So far Mr. Stevenson's notes; and for some years after, the same indiscriminate destruction of these birds continued. He recorded in the "Zoologist" for 1866

---

* This communication was made through Salmon, and appeared in the 1st edition (1st November, 1835), vol. ii., text to pl. xcii.—T.S.

† Other instances are given in the 2nd edit. of the "Fauna of Norfolk," p. 125, note; and Mr. Frank Norgate tells me that a similar circumstance came under his personal observation on the 27th of July, 1865, on Hickling Broad.—T. S.

(p. 263), that eight freshly arrived birds had been killed and sent "for ladies' plumes" to a bird stuffer in Norwich; and to the same journal for 1869 (p. 1490), he also contributed a note to the effect that at a coot shooting party on Hickling Broad on the 10th April, 1868, "no less than nine splendid specimens of this grebe, only just returned from the coast to their summer quarters, were ruthlessly slaughtered, and of course subsequently made into "plumes."

The destruction immediately before or during the breeding season, to which this species was subjected, as shown by the instances just cited—and many others could be found—was stopped by the Act of Parliament already mentioned.* This measure seems first to have brought the perilous state of things to the notice of the indifferent or the ignorant; and, public attention being aroused, it was followed in 1872, 1876, and 1880 by other Acts, which, if not all that could be wished, have certainly saved to the British Fauna several species of birds that were more or less threatened with extermination, and have subsequently been the means of allowing some of them, as before stated with regard to the gadwall, pochard, and tufted duck (pp. 162, 204, and 214), to increase beyond expectation. But in attaining this end, so agreeable to all naturalists, the willingness to carry out the provisions of those Acts

---

* In this connection it may not be inappropriate to mention that it was at the Norwich Meeting of the British Association, in 1868, in a paper read by Professor Newton "On the Zoological Aspect of Game Laws," that scientific attention was originally drawn in a practical form to the destruction of our birds generally, and at that meeting what was afterwards known as the "Close Time Committee" was first appointed. The good work which this committee did in after years, and especially at the time the Sea Birds Protection Bill was going through Parliament, none can doubt. The committee took an immense deal of trouble about the matter, and was able by authority of the information it collected to modify the very extravagant demands of the ultra sentimentalists—from whom there was at all times great danger to be feared, for their action, if uncontrolled, would have been almost worse than that of the destructionists, as they were ready to perpetrate almost any absurdity, regardless of consequences, and such an Act as they wanted would, from its stringency, have been disregarded at once, or, if enforced, very speedily repealed.

evinced by many large proprietors of land, and nowhere, perhaps, more than in this county, must not be overlooked; while, in districts where the ownership is much divided, the imposition of a gun licence has proved of great utility in checking the indiscriminate shooting which was before so baneful to many of our rarer birds.

Owing to the operation of these causes this "greatest ornament of the Norfolk Broads," as it was aptly styled by Mr. Lubbock, has for some years experienced happier times, and is once more fairly numerous on the broads and inland waters throughout the county, having reappeared in some localities from which it had long been driven. The loon habitually breeds on Tomston and Stanford waters, but not, so far as I am aware, on Scoulton mere, though at Hingham Sea Mere, within a few miles of the latter, I have seen their nests; and on some of the Wretham meres they breed regularly. On Heigham Sounds and Hickling Broad, at one time favourite resorts of this species, but where at present it is much harried, it is still found, but in reduced numbers. The same may be said of Horsey and Wroxham Broads, in which latter fine piece of water I have known it chased by a steam launch, but where the British tourist is strictly excluded, as at Ranworth and Hoveton, their numbers have largely increased of late years. The keeper at Ranworth believes that eight pairs have hatched off there this year, and that there are at least twenty-five birds, old and young, at present on his water. Mr. J. H. Gurney, jun., was fishing there on the 31st July last (1889), and tells me that the young were then as big as their parents, and made a constant whistling sound like young ducks. At Hoveton, also, there have been a good number of nests this year.

The loons return to their homes early in March, but are generally accompanied by a number of others which pass on to distant breeding places,* while those that remain to nest appear, in most cases, to have already paired, and lose no time in commencing their

---

* On the 18th March, 1881, Mr. J. H. Gurney, jun., saw twelve of these birds on Ranworth Broad, some in breeding plumage, others still in winter dress.

domestic arrangements. Both sexes assume the ruff in the breeding season, and do not differ greatly from each other in appearance. Mr. J. H. Gurney, jun., tells me that a great crested grebe, killed at Ranworth, on the 30th December, 1884, had, immediately after death, yellowish buff eyes, with a narrow inner circle of white, but the next morning the iris had changed to a bright red.

The preliminary courtship is marked by the abandonment, at least by day, of the rapid flights from one part of the water to another in which the birds had hitherto indulged. The engaged couple choose some suitable corner or bay near a reed bed, and there pass their time in close company, diving and swimming side by side— or when not feeding, and the fit takes them—in wide circles with many gesticulations of the neck, which is at times extended for some moments upright in the air, at times laid flat on the water, then suddenly raised, and as suddenly lowered. Many of these movements are simultaneously performed by each partner, others by each alternately, the one seeming to imitate the other, and the jerks are occasionally exchanged for ludicrously solemn bows. Now and then a playful attack is made by one bird on its mate, and in the scuffle that ensues the water is thrown up so as to hide both from the onlooker, who sees in place of the grebes a splash or series of splashes, as though a cannon-ball had struck or glanced along the surface. Their actions, and especially the last, are accompanied by no little noise, for the cries of the loon are loud and far-reaching; harsh some people would doubtless call them, but they are in thorough harmony with

—— the water lapping on the *marge*,
And the long ripple washing in the reeds.

All this is very pretty to witness, but soon the labour of nest building—which, from the quantity of material required, must be very considerable—occupies their attention, and, if not disturbed, they will, as a rule, have eggs by the first or second week in April; some pairs, however, are considerably later than this date before they have laid; it may be that these are younger birds, or those that had not paired before their arrival. The nest is constructed of dead vegetable matter, and either built up from the bottom, or, if floating, it is securely moored to the surrounding growth of reeds, and is generally situated a short distance within the fringe of a reed-bed, but never where the reeds are so close as to offer any obstruction to the old birds leaving the nest by diving. From the constant sliding on and off of the old birds, after the nest has been used some time, it becomes smooth and closely compacted on the surface, and has the appearance of a mass of muddy weed with a slight depression in the centre, and very little raised above the surface of the water; here the eggs are placed, and a most wet and uncomfortable structure it looks. Mr. Lubbock says, "opinions have varied as to whether fermentation has aught to do with the warmth furnished to the eggs of this bird." With a view to test this I tried the temperature of three nests this year, and found that a nest with three perfectly fresh eggs, into the substance of which my thermometer was thrust three inches and a half indicated 67 degrees; the second nest contained three eggs which were considerably incubated, and the thermometer registered 72 degrees; and in a third, in which there were three eggs and a newly hatched young one, the thermometer rose to 73 degrees, showing that the nest, so far from being the cold and uncomfortable structure by some supposed, was a real hot-bed. On inserting the thermometer into a beautifully neat and dry coot's nest, which the bird had just left, I found the temperature to be 61 degrees. The day was wet and cheerless, and the maximum reading of the thermometer in the shade was 58 degrees.

In almost all cases one or more of what are known to the marshmen as "cocks' nests" will be found at no great distance from the true nest, especially after the birds have begun to set; these have been variously referred to as "play nests" or "look-out platforms" for the male bird; they are mere platforms of reed or other material, slightly constructed, and altogether different in structure from the true nest, and I am strongly of opinion, although I have had no opportunity of proving such to be the case, that they are constructed as resting-places for the young birds. No such resting-places could possibly be required by the old birds, whose natural element is the water, and under whose weight they would be completely submerged, but in rough weather they would doubtless be of the greatest service to the tender little ones, or even in fine weather afford them the means of sunning themselves or of drying their scanty plumage.

Much has been said as to the effectual way in which the old bird covers her eggs on leaving the nest, but, although I have examined over a period of many years a large number of nests, I never saw one so completely covered that the eggs were entirely hidden. In the scanty use of hiding material this species is a decided contrast to the little grebe, which covers her nest so effectually that it has all the appearance of a casual heap of dead vegetable matter.*

Mr. J. H. Gurney, jun. ("Zoologist," 1885, p. 436), states that on the 15th June, having climbed into a tree, by the use of binoculars he saw a loon quit her nest without covering the eggs, but on the 25th he saw the same bird perform the act of covering them before leaving the nest, and describes the action as almost momentary, "a couple of tugs with the beak and that was all." Mr. Gurney opposes Mr. Seebohm's opinion that the eggs are covered to protect them against cold; in this I quite agree with him, as I have repeatedly seen a single freshly laid egg (very conspicuous in its whiteness), which there would be no need to protect against cold, covered as carefully, or perhaps I should say as slightly, as a full clutch in the last stage of incubation. If part of the eggs are taken the old bird will lay others to supply the deficiency, but I am informed that if all are removed they make a fresh nest not far distant from the old one, though of this I cannot speak from personal observation.

The number of eggs is generally three, occasionally four, and rarely five. I have only seen five on one

---

* I may state that my observations of this bird have been confined to the county of Norfolk and part of Suffolk, but I imagine their habits in other localities do not differ greatly from those of the birds found in this county.

occasion; from this nest one egg had been taken, but the bird replaced it with another, and on a subsequent visit I found the unfortunate fifth nestling trodden flat at the bottom of the nest. Mr. F. Norgate mentions in some notes he was kind enough to send me on this bird, several clutches of four eggs, two of five, and one of six, but none of the three last were seen *in situ* by himself. Mr. R. W. Chase informs me that on one occasion on Ormesby Broad he took a clutch of six eggs. The eggs when fresh laid are white, but, as Mr. Stevenson and others have remarked, soon become stained by contact with the wet materials of the nest, small fragments of which often adhere to their chalky surface. Both birds take part in the business of incubation, and at this time they are rarely seen on the wing, seeking safety by diving, and if not molested are far from shy.

Not long since, through the kindness of Mr. H. E. Buxton, I watched from a path in his shrubbery, along the margin of the lake at Fritton, a grebe seated on her nest in a small clump of reeds not many yards from the shore: she was quite aware of our presence, and though watchful, was apparently not alarmed. While we were looking at her the male bird approached with a great croaking, and after some preliminary greetings the female slid off the nest and he took her place.

Mrs. Buxton witnessed a curious circumstance with regard to this pair of birds. She had been watching the grebe on the nest and was walking away when suddenly she heard a loud noise which she compared to the bleating of a sheep in distress; she asked Mr. Buxton who was close by to wait and see what it could mean, when suddenly there was a splash and loud cries which the grebe on the nest answered and then left the nest, disappearing in the reeds. A great deal more noise followed, and soon, emerging from the reeds they saw two grebes fighting desperately, "rolling one over the other like balls, each having hold of the other's neck; they went on," says Mrs. Buxton, "fighting till we again lost sight of them, then the noise grew less and less till after a time we saw a grebe swimming off to another part of the lake and the other two were left in quiet possession of the bay in which their nest was placed. I am not sure whether it was the male or female which was on the nest when the noise was first heard. My idea is that the fight may have been occasioned by another grebe attacking the female, in which case the male would at once go to the rescue. It was a wonderful sight, and I could not have believed grebes possessed the power to make such a noise. This fight happened in the middle of May." There was another grebe's nest at the corner of the shallow bay, at the bottom of which the nest referred to was situated, and probably Mrs. Buxton's explanation of the affair is the correct one, but the great crested grebe has always been, and rightly, I think, regarded as very sociable in its breeding habits and it seems difficult to account for this sudden *fracas*, which was not repeated.

My experience, as I have said in a note in the second edition of his "Fauna" (p. 127), differs from Mr. Lubbock's with regard to this bird's supposed habit of hiding when its nest is approached. On Hoveton Broad and other places I have often been directed to the site of the nest by the anxiety of the old birds, and on the 19th of May, of the present year (1889), two lovely birds remained anxiously swimming up and down within half-a-dozen yards while I was inspecting their nest, which contained three eggs and a freshly hatched young one. Nothing but the desire to ascertain the temperature of the nest could have induced me to disturb them at so critical a period of their existence as the birth of their first-born, and they loudly expressed their dissatisfaction at my intrusion, erecting their crests and swimming up and down in a state of great excitement. The young bird was still wet from the egg, but it struggled vigorously in my hand; it was a pretty little striped thing, with the triangular bare patch of skin on the forehead a brilliant scarlet.

Notwithstanding the number of eggs produced by the loon I do not remember (possibly owing to the depredations of the pike) ever to have seen more than two young ones following the parents. When very young they are easily attracted by any strange object, and it is very interesting to watch the anxiety of the old female as she hurries backward and forward just within the shelter of the reed "bush," endeavouring by her warning cries to recall them to a place of safety; or when undisturbed to see the proud and beautiful bird bearing her prettily striped little ones on her back, or at a later period still, instructing them in the accomplishments necessary to their after well-being.

The food of the great crested grebe, of course, consists chiefly of fish, but they are said to supply their young with tadpoles and small frogs, and aquatic insects and their larvæ as well. I have often watched them carrying food to their young, but was never near enough to perceive of what it consisted. The fact of their stomachs almost invariably containing masses of their own feathers, and even of the very young birds having swallowed those of their parents, has been noticed by almost every authority who has written upon the subject, from the time of Buffon (who says he got his information from the elder Baillon), and, perhaps, even earlier still, to the present, and is too well-known to need repetition, but the object for which those feathers are swallowed does not appear to me to have yet been satisfactorily explained.

Only those who have watched this splendid bird when in full enjoyment of its life and liberty can form any conception of its beauty on the water or the stately gracefulness of its motions. Gliding along like a frigate in full sail, it will stop, and, careening over so as to expose the glistening white of its under surface to the rays of the sun, with one foot extended backwards, it will preen its lovely breast, smoothing and apparently caressing its silky plumage with its not less beautiful head, then resuming an usual position for a moment it seems to be lost, so great is the contrast from the gleaming whiteness of the breast to the sombre hue of the upper parts. Soon it dives away, and as we follow it up, to use the words of one who gloried in this beautiful bird:—"The stately loon rises before the advancing boat, and shakes the drops from his crest, looking back with arched neck at the intruders, and plunges again to his unseen course in life and joy." ("Fauna of Norfolk," 2nd edit., p. 123, note.)

On the approach of autumn the grebes, with few

exceptions, take their departure to the sheltered bays along the sea coast, and are there met with till severe weather sends them south, to be no more seen till the returning spring.

Mr. A. W. Partridge, writing to Mr. Stevenson, says that he has met with these birds in some numbers in Heacham Bay in the end of August and beginning of September, and it is possible that they may often be mistaken for other divers. Forty years ago, Mr. Dowell writes that great crested grebes were not uncommon at Blakeney during the winter months, and that he had killed several in that neighbourhood, all of which were young ones. The late Mr. Cresswell, of Lynn, told Mr. J. H. Gurney, jun., that from a gun punt, when shooting in the Wash, he once winged a great crested grebe, which "came at him like a fury" and would have attacked him had he not knocked it on the head with a paddle. See also a similar instance of the pugnacity of this bird quoted in the "Rambles of a Naturalist in Egypt" (p. 284) from "Land and Water." The same pugnacity has been observed in a red-throated diver, which, left by the tide, charged at his assailant till stopped by being caught between his legs.

PODICIPES GRISEIGENA (Boddaert.)*

RED-NECKED GREBE.

Mr. Stevenson thus writes in 1860 of this species:—
"A regular, though not very numerous, visitant late in autumn and early in spring, appearing on our broads and inland waters between the beginning of November and the middle of March."

Scarcely a winter passes without a few of these birds (not unfrequently adults) being met with, but the months of February and March, 1865, witnessed a remarkable influx such as has never been repeated. Mr. Stevenson communicated the results of his observations

---

* The specific term *rubricollis*, by which this bird was long known, has to give way to that of *griseigena*, bestowed previously.

in this neighbourhood to the "Zoologist" for 1865, pp. 9574-5, and I here quote them in full:—"A few specimens invariably visit us in the winter months, though but rarely remaining late enough in the spring to have acquired the rich red throat of the breeding season; but their numbers of late, judging merely from the specimens brought to our bird stuffers for preservation, or exposed for sale in our markets, have far exceeded anything I have previously witnessed, or of which any local record, to my knowledge, exists. I have myself examined, or have heard of on reliable authority, at least five and thirty examples brought into Norwich alone, a large proportion of them between the 18th and 28th of February, when from eight to ten were seen in a week, and these have been brought, with but few exceptions, from the immediate vicinity of the coast, as at Yarmouth, Salthouse, and Blakeney. These would appear to have formed a part, and probably a considerable part, of a large flight, which, from some cause difficult to arrive at satisfactorily, has visited our eastern and south-eastern counties during the late severe winter, as, from previous notes in the 'Zoologist,' I find they were simultaneously met with in Yorkshire and Lincolnshire, and others were observed at the same time in the markets of Cambridge and London. Of those killed in Norfolk, the chief portion appear to have been adult birds in full winter plumage, with perceptible tufts on either side of the head, and their throats greyish-brown. Here and there a bird showed slight traces of the red patch, but not more amongst the early than the later specimens, as of the two I noticed myself as having most red on the throat, one was killed on the 10th of February, the other on the 28th; and an old male shot on the 25th March, showed no indications whatever of this nuptial tint. The stomachs of these birds, as usual, contained a mass of long curled feathers closely matted together, and stained bright green from some minute vegetable substance, apparently *confervæ* from the surface of the water. Mixed also with these were small flinty particles; but, in such at least as I examined myself, no further indication of their usual food than a strong fishy odour."

This species has been met with on the broads in full nuptial dress, but it has never been proved to breed in Norfolk. The Rev. T. J. Blofeld saw three on Hoveton Broad, on the 14th April, 1845; and a very beautiful specimen in full breeding plumage, in Mr. Gurney's collection, was shot at Yarmouth, about the 2nd of April, 1848; another at Scottow, on the 22nd of the same month; and Mr. Gunn had a male in perfect plumage, which was shot in this county about the 20th May, 1889.

Mr. J. H. Gurney, jun., tells me that he saw a very young bird, belonging to Mr. Haycock, of Cley, which was shot on August 10th, 1886, on a little broad inside the sea wall, at Salthouse. In this example the black stripes on the throat were very strongly marked, and the bird was not full grown. Mr. Booth shot a red-necked grebe on Breydon, in August, 1875 ("Catalogue of Birds," p. 217). Mr. Gunn records a female killed on the 30th September, 1871, at Winterton, and a second a few days after at Beeston, both birds with red necks. An adult, in nearly full summer dress, was killed on Breydon, on 30th October, 1873 ("Trans. Norfolk and Norwich Nat. Soc.," ii., p. 213). In fact, this species has been met with in Norfolk in every stage, from the immature plumage just mentioned to the most perfect nuptial dress.

Hunt states, in his list of "Norfolk Birds," that a pair of red-necked grebes were once killed near the Foundry Bridge in this city. One of Mr. Dowell's notes says that Overton, an experienced and very observant professional gunner, at Blakeney, told him that he has repeatedly met this species, with which he was well acquainted, as far at sea as the Dudgeon light.

In the stomach of one of these birds, dissected by Mr. Stevenson, he found a large number of long bones, which he took to be those of the frog, and the wing cases of *Notonecta*, but in this instance no feathers were present. Mr. Gunn describes the iris of the young bird of this species as a pale straw colour, and that of the adult as white. Mr. Booth states that in a freshly killed adult in summer plumage " the iris was perfectly colourless, a pearly white with lines like crystal radiating from the pupil and giving a very brilliant appearance to the eye;" and Mr. J. H. Gurney, jun., tells me the iris of the bird he saw at Mr. Gunn's (p. 247) was also white.

PODICIPES AURITUS (Linnæus.)*

SCLAVONIAN GREBE.

This bird is a regular and common early spring migrant to the Norfolk waters, but is rare in breeding plumage, thus differing from the next species, with which it is sometimes confounded. It is also occasionally met with in the late autumn, but by far the greater number occur in January or February, especially the latter month, so that it is by no means rare in the fully adult winter plumage.† These occurrences are much too numerous to particularise, but several fine examples were shot in February, 1870, and again in 1871; as well as on the coast in the winter of 1872-3. Mr. J. H. Gurney, jun., gives a singular instance of one being killed with a stone on the 15th February, 1879, as it swam in a horse-pond at Sidestrand, near Cromer, close to the public road, about a quarter of a mile from the sea; and on the 27th January, 1881, Mr. Stevenson records the capture of another in an exhausted state on the rocks at low water between Runton and Beeston, near Cromer.

The occurrences later than the month of March have been comparatively few. Yarrell mentions a very fine specimen, which was killed near Yarmouth, in May, 1826; and the Rev. T. Berney, of Braconash, has one in his collection, shot by himself on Breydon, on May 14th,

---

* It is now known that the *P. cornutus* of later writers is the real *Colymbus auritus* of Linnæus, and, therefore, his specific name has to be used for this species.

† In a communication to Dresser's "Birds of Europe" (viii., p. 646), Mr. Stevenson gives the proportions of those which have come under his notice in this county as one in October, five in November, one in December, nine in January, twelve in February, and four in March.

1845, which is in perfect breeding plumage; the next in order of time with which I am acquainted is recorded in Mr. Stevenson's note-book as follows:—"Of the few examples taken in Norfolk in full summer garb, I am fortunate enough to possess a magnificent pair [now in Mr. J. J. Colman's collection], which, together with a young male in Mr. Gurney's collection, were killed on Sutton Broad, on the 16th April, 1862. Having the opportunity of examining them in the flesh, I found them on dissection to be an old male and a young male and female about the same age. The first, a magnificent bird, with a rich crest of black and orange, resembles exactly the bird figured by Yarrell. The colours of the two younger birds are less vivid, and the crests much smaller, with a few white feathers still visible on the chin and throat, being apparently birds of the previous season. In the old male, evidently in full breeding vigour, the testes were large and pure white, the same parts in the immature male being smaller and dark in colour. The female contained a large cluster of eggs, but none larger than good sized pins' heads. On examining the contents of the stomachs I found them in in each instance crammed with a compact green mass, which on closer inspection proved to be nothing but feathers mixed with and stained by the green *confervæ* from the surface of the water, the only atom of real food discoverable being a small brown beetle in the stomach of the female. With the exception of the minute beetle, I found nothing whatever capable of sustaining life, although the stomachs were in each case greatly distended, the contents being closely matted together and at least half-an-inch in diameter. The stomach of the old male was extremely muscular, indeed, a true gizzard, the inner surface rough like a file, and the coats extremely thick. The same parts in the younger birds were also, though in a less degree, indicative of strong digestive powers. It is not improbable that these birds, if undisturbed, would have remained to breed, as a fourth example was shot at at the same time, but escaped by diving amongst the weeds."

Mr. Stevenson mentions a Sclavonian grebe killed in Norfolk in full summer plumage in 1864 as being in the

Dennis collection.* The Rev. C. J. Lucas informed Mr. Stevenson that a fine specimen of the Sclavonian grebe, in full plumage, was shot on the 27th August, 1869, near Acle bridge, by Mr. Mollett, of Acle, in whose possession it was in 1870; and Mr. Lucas bought one in full summer plumage at the sale of Mr. Alfred Master's collection, which was said to have been killed at Wroxham, but in what year I do not know. Mr. Gunn records (Zool., 1870, p. 2225) his receiving a male of this species on the 12th, and a female on the 24th of April, as also another on 28th of same month, which had been killed at Hoveton, and Mr. Booth ("Rough Notes," p. 99) states that in June, 1870, he noticed a Sclavonian grebe in full plumage on one of the Norfolk Broads, but on making inquiries it was ascertained that during the previous winter the keeper had knocked down a bird of this species, and, discovering shortly after that his prize had recovered from the effects of the shot, it was pinioned and then turned out on the broad. I know of no more recent occurrence of this bird in summer plumage.

Mr. J. H. Gurney, jun., informs me that the iris of the Sclavonian grebe before-mentioned as having been picked up alive on the shore near Cromer in January, 1881, was golden yellow, without any inner rim.

PODICIPES NIGRICOLLIS, Brehm.†

BLACK-NECKED OR EARED GREBE.

The present and preceding species being somewhat similar in their winter plumage, may be easily confounded by persons not well acquainted with them,

---

* See vol. i, p. 28, foot-note, for some remarks on this collection.

† As already stated (p. 218) the name *auritus*, which had been commonly given to this species, really belongs to the preceding, *nigricollis* must, therefore, be used; and I have adopted the English equivalent as more descriptive and in harmony with that applied to its near relative, the red-necked grebe.

and thence some records of the appearance of either species may be uncertain; but, as has often been pointed out, the black-necked grebe may be readily distinguished at any season by the slight upward curve of its bill, which is also sharper at the tip than that of the Sclavonian grebe. Another and perhaps even more conspicuous difference is to be found at all seasons in the disposal of the white feathers in the wing; the Sclavonian grebe has only the nine secondaries white, whereas, in the eared grebe, not only the secondaries but also the four inner primaries are more or less white. Unlike the Sclavonian grebe, which is essentially a bird of the north, the black-necked grebe has a decidedly southern habitat, and consequently occurs with us much more frequently in the early summer than in the winter months, many having been met with in April and May in full breeding plumage, whereas, in its winter dress, it is exceedingly rare, and the single example mentioned below as having been obtained in December, must be regarded as a very exceptional occurrence.

Messrs. Sheppard and Whitear ("Norfolk and Suffolk Birds") say, "We received a specimen of this bird from Yarmouth in the autumn of 1817. It was caught alive, and was remarkably tame, pluming itself with great composure soon after it was taken.* Mr. Sabine had likewise a bird of this species, which was also sent him from the same place." In the catalogue before referred to, Hunt gives the following additional information with regard to this bird:—"A beautiful specimen of this species was taken from among the rushes by a spaniel belonging to the late Rev. Wm. Whitear;" and in the same author's "British Ornithology" he states that "the portrait of this elegant species was taken from a beautiful specimen presented by the Rev. Wm. Whitear, of Starston. It was taken by a water dog in the neighbourhood of Yarmouth." Mr. Whitear also mentions the same bird in his diary (Cf. "Trans. N. and N. Nat.

---

* This curious tameness has been more than once exhibited by the Sclavonian, as also by the little grebe.

Soc.," iii., p. 260), but in neither instance is the precise date given. Graves, in his "British Ornithology," states, "We have observed it [the "eared grebe"] on some of the extensive broads near Yarmouth, in April, and received a fine living specimen, which was caught in a net in the river Yare in May, 1820." Messrs. Gurney and Fisher, writing in 1846 (Zool. 1381), observe, "In the month of April last [1845] no less than five specimens of the eared grebe [one of which, in full breeding dress, is now in the collection at Northrepps Hall] were killed within a week at Wroxham and other places in the county, and it is somewhat remarkable that these have all proved upon dissection to be male birds."

"On the 8th December, 1846," Mr. Dowell says, "I killed an eared grebe at Stiffkey Freshes in winter plumage; I take it to be an old bird. The same day I wounded another grebe near the same place, but left it, supposing it to be a Sclavonian grebe, to chase a little auk."

From this time to the year 1862, Mr Stevenson has the following notes: "A fine specimen, in full breeding plumage, was shot at Sutton, in April, 1849, and in the 'Zoologist' for 1851 (pp. 3116 and 3175) I find two notices of eared grebes from the neighbourhood of Yarmouth, being purchased in the London markets. The first, killed on the 14th April, was sent to London with some great crested grebes, and passed into the hands of Mr. Richard Strangways, whose account of the uses to which their skins were applied I have already quoted (p. 234). A fine male and female shot on the 17th were also purchased by Mr. J. Green, 'naturalist,' of the City Road. The latter individual also received another pair, in the following May, 1852, from the same locality, and the females in both instances contained eggs about the size of small marbles. One killed at Blakeney, on March 9th, 1853, passed into Lord Leicester's collection, and Mr. Newcome possesses one killed by himself on the 9th of May in the same year, in Hockwold Fen, which also exhibited the full summer plumage. In 1854, about the 18th May, a very beautiful specimen was killed at Filby, which is now in the collection of the Rev. C. Lucas.

"In 1861 a pair assuming summer plumage were shot at Kimberley, on the 30th of March, and on the 24th of April of the same year a perfect example at Martham, and one in change on Hickling Broad. The following summer of 1862 was, however, even more remarkable for the examples obtained in full summer plumage. Two pairs were killed on Horsey Mere, about the 9th of May,* and of these birds Mr. F. Harmer, of Yarmouth, who had heard the particulars from a friend who was present when they were killed, says ('Field,' May 31st), 'Two were shot at the first discharge, one soon afterwards, the fourth four days later. When first shot at they were diving amongst a small patch of weeds and water lilies, not far from the shore; the fourth bird was shot quite out in the open water, at the end of the mere. The man who shot them said he could get as near them as he wished, in fact, quite close to them.' A fifth specimen, an adult male, equally rich in plumage, was obtained near Yarmouth (I believe on Rollesby Broad), on the 28th, and this bird, which was sent to Norwich to be stuffed, I saw in the flesh, and had the chance of comparing as to the contents and character of its stomach with that of the Sclavonian grebe. Besides being far less stout and muscular, I found the stomach of this species differ greatly in the smoothness of its interior surface, the almost file-like roughness observable in that portion of the Sclavonian grebe being entirely wanting. The same peculiar habit of swallowing feathers

---

* Two of these birds were in Mr. Rising's collection, and the following note in his copy of "Yarrell" probably gives the correct date on which they were killed:—"Eared grebe. Three shot on Horsey Mere on 6th May, 1862, and a fourth the next day, two of which were stuffed and cased." They were purchased at his sale by Mr. George Hunt. Mr. Rising also sent Mr. Gould some fuller particulars, which will be found in the "Birds of Great Britain," vol. v. He says, "These birds, two males and two females, had been seen continually on the mere through the winter, and up to the time they were shot, and would most likely have been killed long before, had they not been luckily mistaken for dabchicks. . . . I can scarcely resist the conviction that they had already nested there, as the water had become so entirely their habitat during all this time."—T.S.

was also plainly discernible in this species; but in this instance these formed only a small portion of the stomach's contents, the greater mass consisting of the half-digested fragments of insect food. Two or three entire feathers were stained brown by contact with the actual food, and many remains of others, at first sight having exactly the appearance of hairs, were blended with the mass. The appearance of these portions of feathers thus operated upon by the action of the stomach, would seem to imply that, however innutritious, they are entirely disposed of through a digestive process. A careful examination of the various fragments of insect food, consisting mainly of the *elytra* or wing cases of Coleoptera, resulted in the identification of claws and scales from the back of the larva of a *Dytiscus*, or great water beetle; several bodies of some smaller species, probably *Noterus sparsus*, found abundantly in the marshes and stagnant waters; two or three bright metallic green Coleopterous wing cases, from a species of *Donacia*, generally found upon aquatic plants; and heads of both species of water boatman *(Notonecta)*, besides also a minute fragment of bone, probably swallowed unintentionally with other portions of food.

"One of these females is said to have contained a quantity of eggs, and there is little doubt that from their late appearance on our broads in summer, this grebe, like the Sclavonian, would occasionally remain with us to breed, if undisturbed, but, unfortunately, though little observed in the sombre garb of winter, the very brilliancy of its nuptial plumage ensures instant persecution."

Mr. Stevenson subsequently heard of a specimen which was killed at Lynn in November, 1857, and seen by Mr. Gurney, who tells me it was in full winter plumage; he also records at p. 263 of the "Zoologist" for 1866 the occurrence of one of these birds on Breydon water on 10th April. The Rev. H. T. Frere, of Burston, writes me that he has a pair of these birds in his collection which were killed on Diss Mere "thirty years ago." An immature bird was shot on Breydon on 15th September, 1883, which is now in the possession of Mr. J. H. Gurney, jun. On the 3rd November, 1884, the Rev. Julian Tuck received one in the first year's plumage from Hunstanton.* On the 21st August, 1888, an individual, still in summer plumage, was killed at Salthouse.

It will be seen that the eared grebe is almost entirely a spring visitant to this county. Out of the thirty-three specimens enumerated above three occurred in March, twelve in April, ten in May, and one, to which no precise date is accorded, was evidently in summer plumage, thus twenty-six out of a total of thirty-three were spring birds; August, September, and "autumn" each produced one, and the only winter occurrences of which I am aware were, one on the authority of Mr. Dowell, killed at Stiffkey on the 8th December, 1846; a second, reported to Mr. Stevenson as having been killed at Lynn, in November, 1857; a third, at Hunstanton, in November, 1884; and a fourth example, which killed itself by coming in contact with a vessel in which Mr. Booth was steaming through St. Nicholas' Gat, on the 9th November, 1879. It will also be observed that since the spring of 1862 this species has been met with less frequently than appears to have been the case in previous years.

No instance of the nest of the black-necked grebe having been found in Norfolk is on record, but the balance of probability seems to favour the opinion expressed by Mr. Stevenson (*ante* p. 254) that this species, and, perhaps the Sclavonian grebe also, if not disturbed, might occasionally remain to breed on our broads, which appear to be situated near the northern limit of the breeding range of the one species and the southern limit of the other. Most of the early writers are of opinion that the "eared grebe" did occasionally remain to nest, and at a time when the broads were more extensive and less frequented it is not at all improbable that such was the case. Hunt, speaking of this species, says, "we have no doubt it sometimes

---

* Mr. Tuck informs me that on a subsequent examination the the bird recorded in the "Zoologist," 1885, p. 480, as an "eared grebe," proved to be a Sclavonian grebe. I may also mention that by a MS. note in Mr. Stevenson's copy of the "Zoologist" (1868, p. 1127), two "eared grebes" there recorded as killed at Bacton and Reedham on 11th January, 1868, should also have been Sclavonian.

breeds in the county of Norfolk." Messrs. Gurney and Fisher also favour the possibility of its " sometimes nesting in the county," but the only direct evidence on the subject is to be found in Mr. Booth's " Rough Notes " (part xii.), where, speaking of the probability of this species having been a not unfrequent visitor in the breeding season " in days gone by to several of the broads in the eastern counties," he states that, although he had never seen it alive in summer, " a full-plumaged adult and a couple of downy mites " were brought to him by a marshman. In reply to further enquiries, Mr. Booth informed Mr. Gurney, jun., and myself that he did not remember either the date or the precise locality of this occurrence, but that it was certainly in the county of Norfolk. In several of the females referred to in the above list eggs have been found in various stages of development up to the size of a small marble.

Mr. Strangways describes the eyes of his bird, killed on the 14th March, as of a " rich, deep orange, the eyelids edged with the same colour."

PODICIPES FLUVIATILIS[*] (Tunstall).

LITTLE GREBE.

The dabchick or didapper, by which names the little grebe is generally known in this county, is not only the smallest but also the only species of the genus which remains with us the whole year round. Each returning spring sees a considerable number of pairs resort to our smaller lakes—often to ponds of insignificant size—very few being found in the broad district proper, and then only in some " pulk hole," or marsh drain. Here they establish themselves and breed, quitting their summer home when their young are full grown. The numbers thus nesting in the county are, however, surpassed by accessions received late in autumn; and these seek

---

[*] This specific name has priority of *minor*, which has hitherto been commonly used.

the running and more open waters, less liable to be frozen over, where in summer they are seldom seen—appearing, as Mr. Lubbock puts it, "as if by magic, popping up and down upon our rivers and small streams, and the marvel is how they got there." At this time the dabchick, like the water-hen, may be easily caught by the hand when squatting under the overhanging banks of streams. A favourite resort of the little grebe is the drains and small rivers of fresh or brackish water which meander through the broad expanses of marsh bordering on the sea at Heacham, Holme, Blakeney, Salthouse, &c. I am assured that they remain in these localities throughout the summer (at Holme Mr. Le Strange tells me he has taken a nest), but it is in the autumn and winter that they are found there in the greatest numbers; a marshman at Holme told me that he has seen twenty or thirty at one time on a piece of water there in severe frost, and that at such times he has frequently caught them by hand.

As the weather becomes more severe, and the water is gradually covered with ice, they still cling to their restricted haunt so long as any portion of it remains unfrozen, and at such times many individuals may occasionally be found congregated in a very small space, instances of ten or twelve of these birds assembling in one small swan hole not being unfrequent, and several are generally to be seen in winter in the pools of water into which the river Yare expands just below Earlham, Cringleford, and the Harford bridges, near Norwich, as well as at many other pools along the courses of the Wensum and upper Yare, which rivers appear to be favourite winter resorts of these birds. When the water is entirely frozen over, then their fate becomes a hard one, only the sea-shore and estuaries of the tidal rivers are left for them, and those that do not find their way to such situations suffer great privations, and have even been known, like water-hens, to consort with domestic fowls (cf. "Trans. Norfolk and Norwich Nat. Soc.," iii., p. 401), while their frequent appearance in the poulterer's shops tells the tale of their persecution.

Mr. W. E. Fisher, writing in the "Zoologist" for 1843, p. 247, refers to the regular appearance of little grebes at Yarmouth in spring and autumn, although they are not known to nest in that immediate neighbourhood.* Miss Gurney records in her note-book ("Trans. Norfolk and Norwich Nat. Soc.," ii., p. 20) the occurrence of a little grebe at sea off Cromer, on the 13th November, 1830; in January, 1880, one was picked up on the shore at Cromer; and in April, 1884, Mr. J. H. Gurney, jun., received the remains of one of these birds, which had struck the Happisburgh lighthouse with great force. Mr. Lubbock, in a manuscript note on the margin of his copy of "Bewick," says, "this species often appears in autumn and spring in small companies of four to eight on the broad," they doubtless being fresh arrivals. Mr. Dowell tells me that when he used to shoot at Salthouse, little grebes were found in winter in some numbers, as mentioned above, on the brackish waters near the coast known as Salthouse Broad, where they assembled about Michaelmas. They received some sort of protection from the gunners as being useful to decoy other fowl. From this locality Mr. Stevenson obtained a pair of these birds in December, 1862. Other evidence of the migratory habits of this bird on the east coast of England will be found in the Reports of the Migration Committee of the British Association.

Messrs. Gurney and Fisher, in the "Zoologist" for 1847, record an instance of what may possibly have been only a local movement on the part of some individuals of this species, but which proves them to be possessed of considerable power of flight, as well as illustrates their nocturnal habits. "More than one specimen of the little grebe," says Messrs. Gurney and Fisher, "was taken about the 14th inst. (December, 1846) in the streets of Norwich, and on the 23rd, a gentleman, who was passing about eleven o'clock at night along the street, was surprised by a bird of this species suddenly striking

---

* A remarkable instance of the sudden occurrence of little grebes in considerable numbers on the Sussex coast will be found in the "Zoologist" for 1868 (p. 1482). It is there stated that between Lancing and Shoreham they appeared on the 6th November of that year in great numbers, and were said to be "in every ditch," but on the following day not one was to be seen.

a gas-lamp, which was above him, and immediately afterwards falling upon his head. The bird was picked up alive, but died on the following day. The lamp to which it appeared to have directed its course is fixed to a wall facing the south-east, from which quarter a sharp gale was blowing at the time." In January, 1880, when the weather was very severe, and "the channel filled with veritable icebergs," Mr. J. H. Gurney, jun., met with this species at Cley, and about the same time many were received by the local bird stuffers.

Whilst speaking of the flight of the little grebe, I will offer no apology for quoting still further from Mr. Lubbock's note before-mentioned (p. 257). Writing in 1824, he remarks, "It has often been a source of perplexity to me to account for the immigration (as it may surely be termed) of these little active aquatics. They appear to possess the power of flight in a more limited extent, therefore do not seem likely to shift their quarters in any great degree, yet during the whole summer a most accurate observer would be puzzled to produce an instance of one remaining in their hibernal haunts. Having myself generally fished a good deal in the spring and summer, and almost always being accompanied by a water-dog, I have had every opportunity, but have never seen one. Yet no sooner does the weather begin to be severe than they are seen enlivening every stream; indeed, what renders this more remarkable is their apparent preference to the smaller rivers. If they only appeared on the larger pieces of water, we might suppose that extensive reed-beds, etc., effectually concealed them during summer from the prying eye of the ornithologist, but on the smaller streams nothing can resist the searching nose of the water-spaniel. What are we to think of this appearance and disappearance? I know not. If accomplished by flight their power of wing must be far greater than is generally supposed. I have more than once taken this little bird with the hand in shallow drains leading from the stream it frequents with us in winter; and on carrying it to some distance from the element and giving it liberty, it could never achieve above sixty or seventy yards, its flight then became a run, its toes touching *terra firma*."* On the 30th October, 1834, Salmon writes in his diary, "Saw some little grebes flying, being the first time I ever saw any upon the wing. They appeared to fly very sharp and strong." Mr. J. H. Gurney, jun., tells me that on the 3rd September, 1871, he saw a dabchick with a fish in its bill fly a few yards when chased by another bird of the same species.

In March the little grebes resort to their breeding stations, and are then only seen in pairs, but they are still somewhat gregarious, and in a favourable locality more than one pair may often be found nesting at short distances from each other. The nest is constructed, like those of the other grebes, of a large quantity of dead vegetable matter, built up from the bottom, in rather shallow water, and generally somewhat concealed by the coarse water-weeds growing at the edge of the pond, but I have seen them in a small piece of water close by a much frequented road, and perfectly bare of cover. In the "Zoologist" for 1869 (p. 1803) I gave a description of the nests as observed at Scoulton, where they formerly nested. The eggs are usually four to six in number. Mr. Salmon, however, mentions in his diary finding a nest of the little grebe, at Stanford, on the 27th May, 1837, containing seven eggs; they are laid about the middle of April, but I have met with fresh eggs late in June, and Mr. Norgate tells me of a nest at Brandon containing two fresh eggs on the 29th July. When first laid the eggs are pure white, soon to become stained by their wet bed, they are much more carefully covered than those of the great crested grebe; the nest is, however, if possible, even wetter and more uncomfortable in appearance than that of the last-named species, and in some instances which I have seen

---

* Mr. Cordeaux thus describes the flight of a little grebe which was brought to him alive at Great Cotes (Lincolnshire), and which he liberated the next day:—"The little bird went off at once in a direct line, flying along the surface of the water, its wings moving rapidly, and its feet at the same time working alternately like paddles, the tip of each foot catching the water at every stroke—exactly the same motions as it would have used in diving. It is seldom we have an opportunity of seeing the little grebe fly."

I am convinced the weight of the parent bird must have brought the eggs to the level of the water, and probably even submerged them. If suddenly disturbed they will leave the eggs uncovered, but the old bird will watch an opportunity of returning to cover them as soon as the coast is clear. These very uncomfortable-looking nests are probably not so devoid of warmth as their appearance would seem to indicate (*cf.* p. 240, great crested grebe), and it is not unlikely that a low degree of temperature may be sufficient to bring the embryo to maturity.*

Scoulton Mere was formerly a regular breeding-place of the little grebe, as well as of wild fowl of various descriptions. In the year 1869 I found the little grebes still nesting there, but on my next visit in 1873 they had quite disappeared, and I believe that only an occasional pair have been found there at intervals since. The reason is not far to seek. In the year 1864 a number of half-pound pike were placed in the water; ten years later they were found to have so increased in size (and, doubtless, in number also), that many were obtained weighing 18 lbs. each. It is obvious what had been the fate of the little grebes! The presence of the pike may also account for the readiness with which the grebes took wing in 1869, for I see from my notes made at the time that "they rose freely and skimmed over the water, their feet dip, dipping like a coot," a mode of proceeding, as before stated, very unusual with this species at any time, but especially so when nesting. It may be, experience had taught them that the hungry pike

---

* Colonel Legge, in the "Zoologist" for 1867, p. 603, gives a remarkable instance of the vitality of the "chick" of this species. He says, after describing the nest of a little grebe, "a proof occurred here of the small amount of warmth required to hatch and keep alive the young 'chick.' I took away two eggs, one of which was addled. On arriving at my rooms I laid the eggs in the bath, and was surprised to hear periodical chirpings coming from one of the eggs. Not having time to open the egg then, I put it away, and on taking it up the following evening, thirty hours after I had taken it from the nest, I was still more surprised to hear the chirpings again. Assisted by a friend, I liberated the hardy little monster, and wrapped him up in flannel. He departed this life, however, on the following day. The egg was not kept in a warm room."

were more to be dreaded below the surface than any enemy above it.* Doubtless vast numbers of the young gulls also are annually entombed in the greedy maws of the pike. I fear the bulk of the wild fowl bred on the Wretham meres meet a similar fate, and in some waters it is even doubtful whether any escape. As a contrast to the behaviour of the Scoulton grebes, I may quote from my note-book the following experience in another locality where no pike are. "Their powers of hiding are truly wonderful. I cautiously approached a little pond on Wretham Heath, and, in concealment, watched several couples of little grebes popping up and down in ceaseless motion; no sooner did I show myself than down they all went, and not a vestige of a grebe was to be seen so long as I stayed. The pond is, or was then (for it varies according as the season is wet or dry), barely larger than a good-sized horse-pond, the turf sloping down to the water's edge, with no hollow banks and one little clump of rushes only, and yet the birds completely concealed themselves, and remained hidden till I was tired of waiting for their re-appearance."

I have had very few opportunities of watching the little grebe when accompanied by its young, but in Mr. Stevenson's journal for 12th July, 1873, occurs the following interesting passage; the event took place on Surlingham Broad: "Saw an old dabchick and one young one; the old bird caught a small eel for it, but the chick kept dropping it again and again, and the mother seized it as often and gave it to the young one. They were sometimes not more than fifty yards off us." Mr. Cremer also told Mr. Stevenson that at Salthouse he used to find numbers of little grebes in summer, and had shot them with their young under their wings, a mode of protecting their offspring which appears frequently to be resorted to both by this and other species of grebe. The many instances recorded of this bird having been found choked by attempting to swallow miller's thumbs *(Cottus gobio)*

---

* A correspondent in "Land and Water" (June 22nd, 1889) states that he saw a 4 lb. pike swallow three young moorhens in succession. It may readily be imagined what mischief would be effected by 18 lb. monsters.

appear remarkable, and I know of several having been found dead in their breeding plumage without any apparent cause. On the 10th of April, 1884, Mr. F. Norgate had one brought to him which was thus found dead in the Little Ouse, near Brandon; and on the 16th April, 1885, I found one under similar circumstances on Surlingham Broad.

In addition to small fish the food of the little grebe consists of larvæ, water-beetles, tadpoles, and small aquatic snails.

I have mentioned the disinclination shown by the little grebe to seek safety in flight,* but no bird, perhaps not even excepting the great northern diver, is a more expert diver, or more skilful in effecting concealment when alarmed, availing itself of the slightest cover, and remaining submerged for a length of time which is quite surprising. Probably the head is not completely under water, but as before mentioned (p. 262), in a small, exposed pond several pairs of these birds remained so completely concealed that I was quite unable to discover how and where they hid. As to their mode of progression under water, Mr. F. Norgate has favoured me with the following notes:—"February 11th, 1878, I saw in the river Wensum, at Sparham, two little grebes diving without using their wings. This seems to be their usual way of diving when scared, but I believe they can and do use their wings under water, passing through it and the air as if the two elements were one and the same to them;" and again, on the 26th February, 1880, he writes, "At Sparham I saw a little grebe which *suddenly* appeared *on the wing* from *under water*, near the middle of the river Wensum, and flew a few yards and dived as suddenly, just as if it had continued its flight from underwater, through the air, into and under the water again; it did not re-appear." Mr. Norgate does not tell us in what way and to what extent the feet or wings were used in this instance in the progress through the water. On the 19th March, 1834, Salmon writes as follows:— "Saw a pair of little grebes (*P. minor*) swimming at the

---

* See also on this subject remarks by Lord Lilford and Mr. Cordeaux, "Zoologist," 1883, pp. 466 and 503.

bottom of the river. I observed they did not make use of their wings, only their legs; they could not remain under water long, although I endeavoured to keep them under; they went under the opposite shore, and just put their head above the water, keeping their bodies quite immersed." Mr. J. H. Gurney, jun., also tells me that on the 4th of May last he had a live dabchick brought to him which he put into an aquarium for the purpose of observing its mode of diving; he found that the legs only were used when progressing under water. He then transferred it to a pond, where it could dive as deep as it chose, with the result that there also he could not detect any use of its wings when diving.

Professor Newton has favoured me with the following interesting note on the mode of progression of a very young individual of this species:—"A newly hatched little grebe will use its fore limbs as instruments of progression. A friend of mine brought me one which could not have been more than twelve hours old, and when laid on the table it crawled about and completely across it, not actually sustaining its weight, it is true, by its wings, but dragging itself forward by their means quite as much as it impelled itself by its legs. The resemblance of its actions to those of any slowly moving reptile was very remarkable." On holding a newly hatched great crested grebe on my open hand one day recently, I was struck by this remarkable mode of propulsion. Its body remained prone, and I could distinctly feel the pressure of its wings on my hand as it partly raised itself, followed by the thrusting of its largely developed extremities, each pair of limbs distinctly taking part in the forward motion. On placing it on the edge of the nest it rapidly gained the centre in the same manner, and this it repeated more than once. At that time I was not aware of Professor Newton's observation, or I should have watched the action even more closely. Mr. J. H. Gurney, jun., informs me that Mr. Hancock had one of these birds alive, which he believes he kept in a bowl for goldfish, so that he could see beneath it when it was floating on the water. Mr. Hancock could not see the bird's legs, and found that it tucked them up under the wings, and in this position he stuffed one. This habit of disposing of one foot is common to all the grebes, but it seems difficult to conceive how the bird could maintain its balance on the water with both feet in such a position.

The young dabchick shows the stripes on the side of the head even when it is full grown, and a Norfolk killed specimen in that stage of plumage is in the collection of Mr. J. H. Gurney, jun., at Keswick.

COLYMBUS GLACIALIS, Linnæus.

GREAT NORTHERN DIVER.

Mr. Stevenson, writing in 1863, expressed his belief that this species "is decidedly the most rare, in any plumage, of our three British divers," adding that he had known of only four or five examples during the previous twelve years. In the adult breeding plumage there can be no question as to its extreme rarity off the Eastern Counties, and, so far as I am aware, it has not been met with on the Norfolk coast in that stage; but in immature and winter plumage it cannot, I think, be considered so rare as the black-throated diver, though both species are far from frequent at any time. Sir Thomas Browne doubtless refers to this diver in the following passage:—"As also that large & strong-billd fowle, spotted like a starling, Clusius nameth *Mergus maior Farroensis*, as frequenting the Faro Islands, seated above Shetland; one whereof I sent unto my worthy friend Dr. Scarburgh;" and, in a letter to Merrett, he says that he has thrice met with this bird, which "were taken about the time of the herring fishing at Yarmouth. One was taken upon the shore not able to fly away."

Messrs. Sheppard and Whitear refer to two young great northern divers having been killed on the river at Yarmouth "in the beginning of last winter" [early in the winter of 1823-4, as I learn from a note of Mr. Lubbock's], one of which they state "is in Mr. Sabine's collection." Hunt, in his list of "Norfolk Birds," published in Stacy's "History of Norfolk" (1829), evidently referring to the adult state only, says that this bird "is very rarely taken. Last year (1827) one was caught in a net with the herrings." He was not aware that in the mature plumage there is no difference between the sexes, and under *Colymbus immer*, supposing the adult plumage of the females to resemble that of the young males, he adds, "seldom a season passes without one or more being killed at Yarmouth." Messrs. C. J. and J. Paget state that this species is "occasionally shot on Breydon," remarking that "the young bird, colymbus immer, is the more common."

In Mr. Lubbock's interleaved copy of "Bewick" the following entry occurs:—"Imber, 1838, December 28th. One shot upon Quidenham water—good specimen"—a locality very far inland. A fine old bird was killed at Thornham, in December, 1851, and an adult male, on Rollesby Broad, about Christmas, 1863. Mr. J. H. Gurney, jun., has a female, killed at Salthouse on December 15th, 1866. Mr. Gunn records one from Surlingham Broad, in the "Naturalist" for 1867, p. 83. One was killed near Lynn, on December 10th, 1871. An immature specimen, in Mr. Booth's collection, was shot on Hickling Broad, in January, 1872; and Mr. Stevenson has recorded the occurrence of one of these birds, on Tomston Mere, on 7th November, 1875 ("Trans. Norfolk and Norwich Nat. Soc.," ii., p. 216). Mr. Pashley, of Cley, received three great northern divers for preservation in 1888, two of which occurred on the 26th and 27th of November, and the third on the 4th of December. It will be seen from some of the above instances that this species is by no means restricted to the sea, but occasionally, even in winter, wanders far inland to fresh water lakes.

Mr. Dowell has kindly furnished me with the following notes:—On December 1st, 1846, an old great northern diver was seen in Blakeney harbour. On December 9th of the same year Mr. Dowell chased a young bird in Wareham Hole, but it dived so indefatigably and quickly that he, although an excellent shot, could not kill it. On the 19th of the same month he succeeded, after a long chase, in securing a young bird of this species in the "narrows" of Blakeney harbour. Mr. Dowell says that in the winter of 1846, he watched a young northern diver during the process of swallowing a flounder the size of a man's hand, which it took some minutes to accomplish. That the bird did not resort to the pecking process to enable it to "roll it up into a cylinder," as asserted to be the custom in the "Zoologist" for 1847 (p. 1907), was proved by the fish which was cast up by the bird after being shot, remaining whole and uninjured. Mr. Dowell states that he never observed this species to dive, in the manner so often ascribed to it, by gradual submersion, the head being the last to disappear, but that, like all other divers, it goes down head foremost, showing the tail last.

Professor Newton, referring to the mode in which this and other species of diving birds submerge themselves, remarks, in a letter to the writer, that, so far as his observation goes, "divers, *when fishing*, go under water in much the same fashion as other birds, with a more or less decided plunge head foremost, but when striving *to escape pursuit* they seem to have the knack of submerging the body, leaving the head alone above the surface; this is, at the critical moment, dipt, and the bird disappears. It does not seem to me that this difference, which obtains not only in the divers proper, but also in other diving birds—razor-bills, guillemots, grebes, and ducks, has always been recognised."

At the present time Mr. Monement tells me the great northern diver is occasionally met with at Blakeney, but he has never shot a mature specimen.

The great estuary of the Wash seems to be a favourite winter resort for the various species of divers; here among the many channels of deep water intersecting the numerous sandbanks, they find an abundant supply of food in the shoals of sprats and other fish which frequent the sheltered waters, and if, under stress of weather, they are compelled to enter the rivers the same abundance still awaits them in the shape of eels, flounders, and other fish frequenting the mouths of the tidal waters, while still further inland the numerous lakes and broads teem with fresh water fish which form an easy prey.

The following remarks on the plumage of this species

from Mr. Stevenson's notes may be considered as supplementary to Mr. Gurney's excellent papers on the same subject in the "Zoologist" for 1850, p. 2775, and 1851, p. 3301:—"December 13th, 1865. On examining the great northern diver at Sayer's [a Norwich bird stuffer], killed last week, and one shot last year about the same date, I find the beaks in both birds small and more green than yellow. The feathers about the head and throat have a downy appearance, and the white on the sides of the head and throat is powdered as it were with grey. In each also the back is free from white spots, but each feather is dark in the centre, with a light grey edging, giving a variegated appearance, and this I have also observed in black-throated divers, having small beaks and downy head-feathers. I believe these in both species to be young birds that have never assumed the black throat. I have never seen the great northern diver with a black throat, or any appearance of it in Norfolk; they occur only in the winter months."

On the Norfolk coast this species is sometimes known as the "herring loon."

ADAMS'S DIVER *(Colymbus adamsi)*. An example of this form of the great northern diver, now in the collection of Mr. J. H. Gurney, at Northrepps, given to him by the late Mr. Abraham Scales, of Pakefield, which had partially acquired the breeding dress on the back and wing coverts, was killed off Pakefield, near Lowestoft, in the early spring of 1852, and was exhibited at a meeting of the Zoological Society ("Proceedings of the Zoological Society," 1859, p. 206). Looking to what is known of the home of this bird it could scarcely have reached the place where it met its fate without passing over Norfolk waters. But this is probably not the only example of the form taken in the eastern counties, for, on the dispersal of the contents of the Sudbury Museum, the late Dr. Babington became possessed of a specimen which there is strong reason to suppose was obtained at the mouth of the Stour or the Orwell

("Birds of Suffolk," p. 246). After Dr. Babington's death this specimen passed into the collection of the Hon. Walter Rothschild. Mr. Seebohm states ("Zoologist," 1885, p. 144) that another British specimen exists in the Museum of Newcastle-on-Tyne, which was shot on the coast of Northumberland. Professor Newton informs me that in 1864 he recognised an example of this species in the Museum at Bergen, which he was assured had been obtained in Norway, while Professor Collett has recorded ("Nyt Magazin fur Naturvidenskaberne," 1877, p. 218) one shot in Flekkefjord, in the south of Norway, in November, 1875. It is obvious, therefore, that this form, conspicuous by its large ivory-coloured beak and greater expanse of wing, as noticed by Sir James Ross in the Natural History appendix to his father's "Second Voyage in Search of a North-west Passage" (p. xlii.), visits north-western Europe not very unfrequently, and it may well be expected to occur in Norfolk.*

COLYMBUS ARCTICUS, Linnæus.

BLACK-THROATED DIVER.

This species occurs in Norfolk between the months of November and March, most of the specimens having been met with in the first three months of the year. The greater number even at this season are immature,

---

* Professor Newton, in his article "Diver," contributed by him to the later issue of the ninth edition of the "Encyclopædia Britannica," adds to his notice of *Colymbus glacialis*:—"In this connexion should be mentioned the remarkable occurrence in Europe of two birds [supposed to be] of this species, which had been previously wounded by a weapon presumably of transatlantic origin. One had 'an arrow, headed with copper, sticking through its neck,' and was shot on the Irish coast, as recorded by Thompson ('Natural History of Ireland,' iii., p. 201); the other, says Herr H. C. Müller (' Vid. Medd. nat. Forening,' 1862, p. 35), was found dead in Kalbaksfjord in the Færoes, with an iron-tipped bone dart fast under its wing." This article is not to be found in the earlier issues of the edition, a fact which may justify the above quotation.

it rarely appearing in fully adult winter plumage. It has been met with in change, but, so far as I am aware, never in perfect summer plumage, although one, shortly to be mentioned, procured by Mr. Gurney, near Lowestoft, had all but assumed the nuptial dress. Hunt, in his list, speaks of the black-throated diver as very rare in this county. Sheppard and Whitear merely catalogue the species without remark, and neither Gurney and Fisher, nor Lubbock give any definite information with regard to it. A note in Mr. Lubbock's copy of "Bewick" is as follows:—"In 1832 saw a very fine pair of these birds preserved by Smith; they were shot in Postwick Reach, about four miles from Norwich, in the winter of 1831-2. Stuffed for Mr. Penrice. A good many divers were killed and exposed for sale in Norwich after a furious gale from the north-west in the beginning of December, 1836." Mr. Blofeld has in his collection a fine adult male, which was killed on Barton Broad, in the winter of 1849-50, and in his notes he also mentions an immature bird, which was taken in a net on Wroxham Broad. Mr. Dowell has a good adult specimen, which was killed at Wells early in January, 1849. Mr. Stevenson has notes of a young bird killed in Scottow park on 10th November, 1851; one on 7th December, 1852, at Blakeney; a young male at Colney, 17th February, 1855; one also, immature, at Ranworth, 11th December, 1856; a fine old male on Barton Broad, 20th January, 1860; and a young one, also killed at Barton on 1st March of the same year; in 1861, 12th January, an example assuming the black throat; 2nd February, an adult at Hickling, with white spots on the wings; and on 16th November of the same year, a young bird. Mr. J. H. Gurney has a beautiful specimen in his collection killed about 1857, near Lowestoft, in nearly full breeding plumage, slight remains of the winter dress still shewing on the cheek. There is also a specimen in Mr. Lucas' possession, killed at Yarmouth, which he purchased at the sale of Mr. Overend's collection in 1876; and another in the Norwich Museum, presented by Mr. Gurney, which was killed at Yarmouth about Christmas. Both of these have, to a considerable extent, assumed the breeding plumage. Mr. Gunn purchased at Mr. Rising's sale a black-throated diver, killed at Horsey, which had assumed quite half the breeding plumage. In 1868 several immature examples were obtained, and on the 14th of January in the same year, Mr. Gunn received a male, apparently assuming the adult plumage ("Zoologist," 1868, p. 1221). In March, 1871, one was killed at Horning, and two on Breydon ("Zoologist," p. 2828). In 1877 a bird of the year was sent into Norwich about the first week in January, and an adult in winter plumage was shot on the mill pond at Hempstead on the 3rd of the same month; an immature bird was also killed on the 13th December, as far up the Yare as Surlingham Broad ("Trans. Norfolk and Norwich Nat. Soc.," ii., p. 479).* Mr. J. H. Gurney, jun., was offered a black-throated diver, at Cley, which had been killed there on the 26th November, 1884.

In the "Field" for May 17th, 1856 (see also "Zoologist," 1856, p. 5159), will be found a communication to the effect that after a north-easterly gale, which occurred on May 7th, the coast was strewed with thousands of dead puffins and gulls; and a subsequent communication on the 24th of the same month, states that among them "one or two black-throated divers" had been found. The writer, however, was evidently not an ornithologist, and may possibly have been mistaken as to the species; it is much more probable that they were red-throated divers.

From the foregoing records it will be noticed that many examples of the black-throated diver have been killed in this county in inland situations, and on the various broads, indicating a predilection for fresh water; and, although the same has been observed of the great northern diver, it has not happened to such an extent as with this species; in fact, it may be that the frequent records of its occurrence are to some extent due to its venturing into such dangerous localities. The shore gunners seldom meet with this species, and at Blakeney

---

* In January, 1879, a considerable number of divers of various species, chiefly immature, were observed off Yarmouth during the severe weather which then prevailed.

Mr. Monement tells me it is much the rarer of the two large divers.

With regard to the assumption of the breeding plumage in this species, I find the following note amongst Mr. Stevenson's papers:—" The black-throated diver in one instance has been killed in December* with the black-throat coming on, therefore they must assume it very early. I never saw any others but with white throats here, old and young, but some have spots on the back; they also never appear before November, and mostly between January and March—surely the one in December was an exceptional case only."

COLYMBUS SEPTENTRIONALIS, Linnæus.

RED-THROATED DIVER.

The "sprat loon" is by far the most numerous representative of this family occurring on the coast of Norfolk. It is found from September or October till March, and, occasionally, as late as the month of April, not unfrequently in the breeding plumage, but more often in the winter dress, young birds being the most common. Although numerous at times all round the coast, and occasionally entering the harbours and tidal rivers, it appears less frequently on fresh waters than the preceding species. Exceptions, however, have occurred, as in January, 1838, when one was killed near the New Mills at Norwich; at Hickling Broad, in November, 1852, and again in October, 1854; at Rockland, January, 1855; one shot at Bawburgh, and another at Wroxham, about the same year; at Hickling, in October, 1865; and perhaps in a few other instances, but the proportion killed on fresh water is far less than with the black-throated or even the great northern diver. Occasionally this species is quite numerous, as appears to have been the case in October, 1862, when some fine specimens, recorded as retaining their full

---

* This is probably the specimen in the Norwich Museum.

summer plumage, were procured. Again, in October, 1865, when between the 5th and 16th of that month several others in a like state of plumage were obtained; the years 1867 and 1868 also appear to have produced an unusual number of these birds. On 1st October, 1880, an extraordinary migration of red-throated divers was witnessed off Cley by the Messrs. Power; the divers were a quarter of a mile from the shore, and were going south-east, following the trend of the shore, in an almost constant stream for nearly four hours. ("Trans. Norfolk and Norwich Nat. Soc.," iii., p. 349). Only one was obtained, and that had a perfect red-throat. In Mr. Dowell's notes he states that on the 5th October, 1846, he saw the first "mag lowan" (*i.e.*, "large loon," a name by which all three divers appear to have been known at that time on the Blakeney coast), which had entered Blakeney harbour that year, and continues: "These birds are common there all through the winter, six or seven pairs frequenting the harbour. Killed a male in full summer plumage at Blakeney, on April 15th, 1848; Overton saw another a few days afterwards. A 'mag' in the harbour, April 7th to 14th, 1849, had assumed full dress. October 23rd, 1851; shot two in the harbour, one in summer plumage, the other just lost it, *i.e.*, with the white feathers predominating over the red. September 25th, 1863, bought a red-throated diver in summer plumage, with the red mark well developed, at Hunstanton, of a boy who had caught it alive on the beach." Mr. Booth states that "the summer plumage is occasionally retained till late in the year. During the last week in October, 1872, while steaming with the herring fleet outside the Cross Sands, off Yarmouth, I noticed some hundreds of red-throated divers; many were in full, though the majority showed intermediate, plumage, their necks being much speckled with white. The whole of the birds were collected within a short distance of the herring-boats, evidently attracted to the spot by the abundance of fish." ("Rough Notes," p. 111). "Before these birds dive," says Mr. Dowell, "they may be observed to sink about half an inch lower in the water, so that a person who knows this habit may always tell the moment before a bird of the kind means to dive as surely as a duck's flying is prefaced by turning head to wind."

Mr. Stevenson has a pretty full note on the plumage of the red-throated diver, as follows:—" From frequent opportunities of examining this bird, I cannot help concluding that the summer dress is both retained later and re-assumed later than in either of the other species, and that the specimens observed by Audubon [on the coast of Massachusetts] with red throats, in February, had not then lost the plumage of the previous summer. Whenever these birds appear very early in autumn, that is to say, from the first to the third week in October, some few birds are sure to exhibit the red patch on the throat as perfect as it is during the breeding season, and others in every state of change appear at the same time, but I have never observed any trace of red in specimens killed in November or any later period. It is only, however, occasionally, as before stated, that these divers appear here early enough to present their full summer dress, and this was particularly the case in the autumn of 1862 [*vide supra*], when a most unusual number of these birds appeared off our coast, accounted for by the extraordinary shoals of herring at the time. Several very beautiful specimens were sent to a bird stuffer in this city, from whom I purchased one now in my collection, as perfect an example of this species in full nuptial plumage as I ever saw in collections from high northern localities. More than a dozen were shot in the course of a week or two off the Sheringham beach, one of which being held up by the legs disgorged sixteen young herring from its capacious throat." At a much later date, also, occurs the following note on the same subject:—"The red-throated diver comes as early as September, often in October, and many of these have perfect red throats, and no appearance of change in some specimens, whilst others killed at the same time have lots of white feathers among them. Again, the white feathers appear first in the red patch in *some*, but more usually in the cheeks and neck first, the patch remaining pure. Some with perfect red throats, &c., have no spots on the back; others with as good red patches have many spots on the back. May not the oldest birds lose their back spots altogether and not resume them?"

On the 23rd of March, 1883, I found one of these birds washed ashore dead on the beach, near Cromer, after a very violent storm (see also p. 271); and Mr. J. H. Gurney, jun., tells me that he has more than once seen them dead under similar circumstances. To Mr. Clement Reid we are indebted for the discovery of the fossil remains of this species in the post-glacial deposit known as the "Mundesley River Bed" ("Geological Magazine," March, 1883, p. 97).

URIA TROILE (Linnæus).

COMMON GUILLEMOT.

No member of the family of Alcidæ, although some species are found at times in considerable numbers off the Norfolk coast, can now be claimed by us otherwise than as a bird of passage, more or less abundant as the food supply is greater or less. In summer the almost cliffless shores, so unattractive to these rock birds, offer them no inducement to stay,* and, although guillemots and razorbills are to be found *at all seasons,* fishing in the deeper waters, it appears certain that those seen in summer are either birds which for some reason have not betaken themselves to the great breeding-places on the rocky headlands of our island, or are merely engaged in procuring food with which they return nightly to their homes, the nearest of which is Flamborough Head, on the Yorkshire coast, a distance of some ninety-eight miles.†

---

* Mr. J. H. Gurney, jun., suggests that the old name of Foulness for the highest part of the Lighthouse Hills, at Cromer, seems to imply that they may have bred there at one time.

† There is a vague tradition, it can hardly be called more, that the guillemot formerly nested in Hunstanton cliff, say at the commencement of the present century; although this is by no means impossible, the only evidence on the subject I have been able to

Mr. Stevenson says, "these birds are still met with on our coast throughout the summer months, and both young and old together during autumn and winter, and at times in considerable numbers. I have seen the old birds off Cromer in full summer plumage, both in May and June, and, in the latter month especially, have observed small flights some three or four miles out at sea, for the most part towards the close of day, flying north-westward like the larger gulls. All the fishermen

obtain is in an unpublished MS. work left by the late Rev. George Munford, entitled "Outlines of the Natural History of Hunstanton," which contains, in a chapter on the birds, the following paragraph :—" Of the rock birds, it is said the Foolish guillemot (*Uria troile*), the puffin (*Fratercula arctica*), and the razorbill (*Alca torda*) occasionally breed in the cliff, each of these laying a single egg." I am indebted to the Rev. E. E. Montford, son of the writer of the above, for a sight of this MS.; it was written about the year 1856 (dated 1858), and in a very popular style. I am not aware that the writer, although an excellent botanist, made a special study of Ornithology, and the fact that the passage was not reproduced in Mr. P. Wilson's guide to Hunstanton, published in 1864, which contains a list of the birds from Mr. Munford's MS., I think is conclusive evidence that the authority for the statement was not considered of much value. There is no mention of the guillemot or any of its allies breeding at Hunstanton in Sir Thomas Browne's letters, although he writes to Merrett, "that fowl, which some call a willick, we meet with sometimes; the last I saw was taken on the sea-shore," but says nothing of its nesting at Hunstanton. Mr. le Strange tells me this bird is not mentioned in the le Strange "Household Book," and that he has questioned an old gamekeeper on his estate, eighty years of age, who had no recollection of any such bird. This is only negative evidence, but it is all I can obtain. On the other hand, these cliffs afforded nesting shelter to the peregrine falcon as late as the year 1820, and, although they are now deserted even by the jackdaws, which till within the last few years bred there in numbers, and only inhabited by sparrows and swifts, they are in every way suitable (or were when less disturbed) for the nesting of this species, and there is nothing improbable in the story; it is not, however, likely that the puffin ever nested there. Mr. J. H. Gurney has a very ancient specimen of the guillemot in full breeding plumage, which was picked up at Hunstanton below the cliffs, perhaps about the year 1824, by the Rev. Edward Edwards (whose daughter the Rev. George Munford married), and given to him; and it is possible the idea that this species bred there may have originated from that circumstance. I think, however, that unless further information can be unearthed the case must be considered at least not proven.

I have conversed with on this point believe that both guillemots and razorbills, during the breeding season, come down from the Yorkshire coast, daily, as far as the Yarmouth Roads, whence they are seen to return apparently laden with food. It is only after heavy gales that these birds are observed close in shore, riding like corks among the crested billows, and, though here and there some luckless examples are found washed up on the beach, as soon as the storm subsides they are no longer visible from the shore, having returned again to the ' bosom of the deep,' the true home of these perfect divers."

At the close of the breeding season old birds, accompanied by their young, appear off the Norfolk coast, and Mr. J. H. Gurney, jun., has met with the half-grown young ones at Blakeney as early as the end of July.

On the approach of winter, guillemots occasionally seek the shelter of the harbours and bays along the coast, and are sometimes driven far inland. Mr. Dowell has known one taken alive on Kelling Heath, and Mr. Frank Norgate tells me that one was caught in the rectory garden, at Sparham, on the 2nd November, 1876, as it was " flapping along the gravel path, probably too tired to fly. It did not seem to have any wound or disease." Mr. Newcome has a specimen in winter plumage, killed on the river near Hockwold, some years ago. On December 1st, 1869, a guillemot, a razorbill, and a grebe were caught in a bird-net on the shore of the Wash, which was intended to intercept the nocturnal flight of waders, but must have been submerged by an unusually high tide.

The following extract is from Miss Anna Gurney's note-book, under date of June 12th, 1839 :—" A Foolish guillemot's egg was found on Cromer beach, and another the following day " (" Trans. Norfolk and Norwich Nat. Soc.," ii., p. 22), and other eggs have been found in the shrimpers' trawl-nets off Lowestoft and Lynn.*

---

* No doubt many eggs are dropped at sea, and Mr. J. H. Gurney, jun., tells me that when on the Yorkshire coast some years ago, on the 21st March, he was told of six eggs being brought into Bridlington by a trawler who dredged them up.

Occasionally very large numbers of this and kindred species are found cast up dead upon the beach; these wholesale destructions occur immediately after protracted storms, and doubtless arise from the birds, through stress of weather, and consequent inability to obtain food, being buffeted by the breakers until they perish from exhaustion induced by their involuntary fast. A remarkable instance of this sort occurred in 1851, and is thus recorded by Mr. J. H. Gurney, in the "Zoologist" (p. 5159): "On the morning of Sunday, May 11th (after some severe north-easterly gales), a very large number of sea birds, recently dead, were observed on the beach in the neighbourhood of Cromer: they were washed up, mixed with seaweed, and were found lying near the edge of the water in considerable numbers, so much so that a lady counted 240 in the space of not more than two miles; many were gathered for manure, one man collecting four cartloads, partly composed of seaweed, but principally of dead birds. I have ascertained that they extended along the beach for full six miles; I am informed that many were washed up at Caister, near Yarmouth, and I have no doubt that others may have been found on other parts of the coast, respecting which no information has reached me. I have had some difficulty in ascertaining the exact species of the birds thus destroyed; but, as far as I can learn, they were chiefly foolish guillemots, intermingled with razorbills, puffins, and gulls." Mr. Dowell says that about Blakeney there were vast numbers of guillemots and razorbills washed up, but he saw no gulls; he dissected several of the birds but found no trace of disease. A writer in the "Field" (May 17th, 1856), who examined many of the birds, also states that he could find "no marks of external violence nor any indication of starvation. After similar gales, on the 22nd March, 1875, in the spring of 1878, and again on the 23rd March, 1883, I was witness to like mortalities, but on a much smaller scale, on the same coast.*

---

* In the "Field" for the 9th October, 1859, will be found an account of a similar mortality amongst the sea-birds on the west coast on even a larger scale. It is said that vast numbers of dead

The curious variety known as the RINGED GUILLEMOT has been procured on the coast of Norfolk several times. The first example recorded was killed at Yarmouth on the 9th October, 1847. Messrs. Gurney and Fisher, in announcing it as "new to the Norfolk List," say ("Zoologist," p. 1965), "It was noticed that the mark on each cheek which forms the 'bridle' is not merely a line, but an indentation or groove in the feathers throughout its length." One of these birds, in Mr. F. d'A. Newcome's collection, was stuffed by the late Mr. E. C. Newcome; there is, unfortunately, no date or locality preserved, but a note in Mr. Stevenson's writing (who probably had his information direct from the late Mr. Newcome), states that it was taken at Feltwell. Another, in Mr. le Strange's collection, at Hunstanton Hall, was killed in that neighbourhood; and Mr. Stevenson possessed one which was taken alive on the beach at Cromer about the middle of May, 1863. Mr. Dowell also has a ringed guillemot killed at Blakeney, in November, 1850. "Of these examples," says Mr. Stevenson, "all but the last are in full summer plumage, with the white ring round the eye, and the 'bridle' on the cheek very pure in colour and clearly defined." On the 31st of January, 1872, Mr. J. H. Gurney, jun., purchased two of these birds in the Norwich Market, which he found amongst old and young guillemots and razorbills sent from Cromer, and said to have been shot some few miles out at sea. Mr. J. G. Overend's Yarmouth collection contained a ringed guillemot, killed prior to 1875. An example was also shot at Yarmouth in the last week of February, 1881; and a very immature specimen, the beak of which was not fully grown, was killed at Cley, in November, 1885.

---

and dying birds were found in the Firth of Clyde and Belfast Lough, Dublin Bay, and Waterford Harbour, and along the Wexford shore, and many dead birds were found on the coast of North Devon.

URIA GRYLLE (Linnæus).

BLACK GUILLEMOT.

The first example of this species known to have occurred on the Norfolk coast is, I believe, one which Mr. Dowell tells me was brought to him on the 16th November, 1850, by Mr. Brereton; it was an immature bird, and had been shot at Blakeney by Mr. Loads, a shore gunner there, who noticed it in company with some ducks. Mr. T. E. Gunn recognised in the breast and wings of a bird shot in November, 1866, and sent from Wells to be made up for a lady's hat, an immature black guillemot, and he was told on what he considered good authority that a second example was shot at the same time; about the same date, also, Mr. J. H. Gurney, jun., obtained another from Blakeney. On January 21st, 1887, an immature male black guillemot was picked up on the beach at Salthouse and sent to Mr. Gunn ("Zoologist," 1867, p. 710). The Rev. R. H. Tillard has a young black guillemot, which was shot at Blakeney "some twenty years ago," but, unfortunately, he does not remember either the year or month in which he received it. An adult male in change, which passed into the possession of Mr. J. S. Dawson, chief officer of the coastguard at Yarmouth, was picked up by one of his men on the beach near Caister, early on the morning of the 22nd March, 1875. It was washed ashore dead with other birds, such as common guillemots, razorbills, &c., during the easterly gales which prevailed about that time. When at Yarmouth, in October, 1879, Mr. J. H. Gurney, jun., ascertained that a black guillemot had been killed there in the winter of 1878-9. Dr. Whitty, of Hunstanton, has a young bird of this species in his collection, which was procured at that place in 1884, and Mr. Gurney saw on the 20th December, 1886, at Mr. Pashley's, of Cley, an immature black guillemot, which had been sent to him a few weeks before to be preserved.

Messrs. Sheppard and Whitear include this species in their catalogue, but without comment.

MERGULUS ALLE (Linnæus).

LITTLE AUK.

Mr. Stevenson has left a note on this species apparently written in 1863, which, though long, I have no hesitation in quoting in full: "This curious little sea-bird has hitherto been described as only an occasional and storm-driven wanderer to our coasts, but I think I shall be able to prove, from my own notes on the species, that it has undoubted claims to be considered an annual visitant, though in small numbers. I find that since 1852 I have seen in the hands of our Norwich bird-stuffers from one to six or eight specimens in each succeeding year, the total number amounting to between thirty-five and forty examples, of which at least half have been picked up in a dead or dying state far inland. Nearly all have appeared with singular regularity between the first week in November and the end of December, two or three chance specimens only being obtained in February or as late as the 18th March.*

"Judging, therefore, from these statistics of many successive writers, I believe that the little auk appears

---

* In confirmation of Mr. Stevenson's conclusions, I append the following list of the recorded occurrence of the little auk in Norfolk, with dates. Hunt ("British Ornithology") mentions one of these birds having been taken at North Walsham, on 4th or 5th November, 1821, the only precise date given by him. Messrs. Gurney and Fisher, as referred to further on, mention the occurrence of great numbers in October, 1841. In Mr. Dowell's notes he gives December 1st and 9th, 1846; November 24th, 1849; November, 1850; "early part of winter," 1849-50; November 11th, 1853; and he also mentions that Lord Leicester procured one at Holkham or Wells during the hard weather in February and March, 1853. Mr. Norgate has notes of one November 9th, 1861; and another November 18th, 1874; others occurred in October, 1867, and on the 11th and 16th of November, 1870 (both in the Keswick collection); one was found dead on 4th December, 1872; and many were obtained between the 5th and 9th November, and the 3rd and 18th December, 1878; also one on the 15th January, 1886, at Cromer. Mr. Pashley, of Cley, had one killed on the 4th and another on the 25th February, 1889. It will thus be seen that Mr. Stevenson's observations with regard to the frequent occurrence of this species in the last two months of the year are fully borne out.

regularly on our coast, towards the end of autumn, on its passage southward from the Arctic regions, where it breeds in summer in countless myriads. That they are rarely seen except in stormy weather is accounted for by their strictly marine habits, preferring "with their wondrous powers of swimming and diving" the open sea, being well able, like the guillemots and razorbills, to get their living in the deep waters. The prevalence, however, of autumnal gales at the very time of their arrival on our eastern coast accounts for the large proportion of specimens obtained inland, which have been borne helplessly onward with the fury of the storm, to be dropped exhausted in some strange locality far from their natural haunts. Amongst other places at a distance from the coast, I have known the little auk to occur at North Walsham, Fakenham, Stalham, Stratton Strawless, South Walsham, Scottow, Reymerston, Hevingham, and Buckenham; also at Eaton, Lakenham, and Thorpe, near Norwich, between twenty and thirty miles from the sea; in several instances in the city itself. Many of these birds were alive when first picked up, but, either from previous injury or want of proper food, soon died. A curious instance of the boldness of one of these birds soon after capture was related to me by Mr. Rising. In the autumn of 1862 a little auk was taken alive in a turnip field by a labourer at Horsey, who carried it home with him, and having supplied it with a 'keeler' of water, it dipped and washed itself most happily, and arranged its feathers afterwards with the most perfect indifference to its strange position. I regret to say that this most interesting bird, though apparently doing well, was afterwards killed in order to be stuffed. It would not, I imagine from the above instance, be very difficult, with moderate care, and provided it has been in no way injured, to keep one of these auks alive, at least for some time, and I would strongly advise any one fortunate enough to obtain a living specimen to forward it at once to Mr. Bartlett, at the Zoological Gardens, who would, I am sure, devote every attention to so novel a visitant.

"Little auks were particularly numerous in the latter part of October, 1841, and again in December, 1846, and

November, 1862. On the 25th May, 1857, a little auk in full summer plumage, having a rich black throat, was killed at Wells. This bird, now in my own collection,* was shot whilst skimming over the waves close inshore; it was covered with sand when brought to me in the flesh. It is difficult to account for its appearance on our coast at such a season. So rare is the occurrence of this arctic breeder in England during the summer months that I know of but one other recorded instance, an adult bird, said to have been obtained at Downham, in Norfolk, during the second week in July, 1846. Mr. C. B. Hunter, who communicated the fact to the 'Zoologist' at the time, says 'it was in an extremely emaciated condition,' and, therefore, most probably from accident or disease was unable to migrate. Mr. Hunter, however, does not state whether the throat of this specimen was black or white."

In addition to the two instances of the occurrence of this bird in summer plumage given above, Mr. J. H. Gurney, jun., has at Keswick a specimen said to have been obtained about five miles from the entrance to Lynn harbour, on the 15th July, 1872; another in complete summer plumage, in the same collection, was killed at Surlingham about 1870; Mr. Gurney also informs me that Mr. Tillard, of Blakeney, has one, killed at that place, which had about half assumed summer plumage.

Mr. Stevenson made an analysis of thirty-five occurrences of this bird in Norfolk up to the end of 1862, with the following result:—Seventeen were killed inland; four on Hickling Broad; and fourteen on the coast. Of these, thirty-one occurred in November and December; two in March (13th and 19th); one in February (23rd); and one May (25th).

Mr. Dowell tells me the little auk was (and perhaps is now) known to the Blakeney gunners as the "King John."

---

* Purchased at Mr. Stevenson's sale for the Norwich Museum.

FRATERCULA ARCTICA (Linnæus).

PUFFIN.

This species, although met with occasionally off the Norfolk coast, is by no means so frequent as might be expected; possibly it may be more numerous some distance out at sea, where I have occasionally seen a few old birds in summer; but it is certainly best known to us as a storm-driven stranger, and as such has been picked up far inland. Sir Thomas Browne says of "*Anas arctica Clusii*, wch though hee placeth about the Faro Islands is the same wee call a puffin . . . sometimes taken upon our seas;" and Hunt thus writes of this species: "We have had them sent to us in the months of February and December, and we have two specimens now before us taken alive from off the beach, at Cromer (after a violent storm) on the 4th day of November, 1819. There were great numbers cast on shore at Wells and other parts of the Norfolk coast."

Mr. Dowell says the "sea parrot" is "found off the Norfolk coast during the latter end of summer and autumn, but is never common," and mentions one having been caught alive on the shore at Blakeney, on 1st December, 1846, during rough weather, with a northerly wind; and the remains of another found on the 18th of the same month. Mr. Stevenson has notes of three specimens procured in February, 1860, at Winterton, Palling, and Wells, all of which, he says, judging from the size and form of their beaks, were young birds.* A

---

* Puffins, with remarkably small bills, are occasionally met with on the Norfolk coast. Mr. J. H. Gurney tells me he had such a bird sent to him in the flesh, which was shot at Yarmouth, on the 26th December, 1884; he has also one which was picked up dead at Cley, in September, 1888; and Mr. Pashley, of Cley, has stuffed several presenting the same peculiarity. Professor Newton tells me that one in the Newcome collection was bought at Lynn as a "greenland dovekey." These examples were, I believe, all young birds of the year; indeed, so rare is the adult puffin on the Norfolk coast in winter that I do not remember ever to have seen one which was procured at that season, and it appears that the process of development of the singular form of bill in this species has not attracted the attention which so

fine old bird in Captain Longe's collection was picked up after a heavy gale, in April, 1858, on Winterton beach, and about the same time another was found at Yarmouth. It will be seen also by reference to p. 278 that a few of these birds were observed amongst the mass of sea birds washed up on our coast in May, 1856; others were obtained in 1867, 1868, and 1869. An immature puffin was picked up inland, at Cawston, about the middle of March, 1876; and another on the coast, on the 29th of the same month; many were found dead on the shore, at Cromer, in the spring of 1878; and two were sent to Norwich to be preserved, on the 4th December of the same year. In May, 1879, Mr. Stevenson records ("Trans. Norfolk and Norwich Nat. Soc.," iii., p. 133) several examples sent to Norwich for preservation in the second week of that month: "They were, for the most part, in very poor condition, though rich in the colouring of the bill; and, in the absence of gales, or hard weather, their emaciation was probably due to some disease, as with the razorbills and guillemots." I have no doubt many other examples have been met with of which I have no record; probably few years have passed without such occurrences, but never in any numbers except under stress of weather as before explained, and sufficient examples have been mentioned to indicate at what seasons, and under what circumstances, they are generally met with.

---

remarkable a feature deserves. Mr. Newman, in the "Zoologist" for 1862 (p. 8004), and Mr. Blake-Knox shortly after (p. 8331), referred to the subject, but no results followed; the examples figured by them were also, I believe, birds of the first winter, and their bills were quite different in appearance from that of the adult after the shedding process so well described and figured (Fig. 2) by Dr. Bureau (cf. Mr. Harting's translation of his paper, "Zoologist," 1878, p. 233), some very interesting examples undergoing which change I had the opportunity of examining last August; none, however, had fully completed the transformation. According to Mr. Cordeaux and Mr. Hancock, as quoted by Mr. Dresser in his "Birds of Europe," the puffin is found by the fishermen off the north-east coast all the year round, and as this species probably does not become fully adult till the end of the second or third year, it would be interesting to obtain examples showing the various stages of development, and at various seasons of the year, of that most singular organ.

ALCA TORDA, Linnæus.

RAZORBILL.

"With the exception of being far less plentiful than the common guillemot," says Mr. Stevenson, "the same description will apply equally as to the habits of either species, some few razorbills appearing on our coast throughout the summer as well as in the spring and autumn. I have seen the old birds taken alive off Yarmouth, in full breeding plumage, on the 16th of June, and very fierce they are in confinement, biting at everything within reach with their formidable bills, and as clamorous as young jackdaws, with a very similar note. By the 17th August I have seen both old and young birds off Cromer* already come down from their northern breeding stations, the nearest to us being Flamborough Head, and the cries of the young birds, if separated by accident for a time from their parents, may be heard for a long distance at sea in still weather. At this time the old birds still retain their breeding plumage, but the white line from the beak to the eye is becoming less distinct. Examples killed in February and March exhibit the white throat of the winter plumage, but without the white line, which, so far as my own observation goes, is confined, as a rule, to the nuptial dress. It is wanting entirely in the young bird killed in its first autumn or winter, but appears indistinctly by the following March; yet, strange to say, a nestling from Flamborough Head, in the Norwich Museum, only just emerging from its downy state, has this very mark distinctly shown,† although an older bird by

---

* Mr. Dowell mentions having met with old and young birds in Brancaster harbour as early as July 20th and 22nd, in 1851 and 1853 respectively. On July 28th, 1880, Mr. Gurney had one brought to him, at Northrepps, which had been caught on the shore at Runton; it was an adult with full black neck. On August 14th, 1874, Mr. J. H. Gurney, jun., captured one of these birds, which he found left by the tide at Blakeney; it was three-quarters of a mile from the water, but quite uninjured.

† The white line is also present in a bird of the year caught at Hunstanton, on 4th of August, 1870, in the collection of Mr. J. H. Gurney, jun.

some months is entirely without it. The young birds, also in their first winter, have no grooves on the bill, but these are acquired in the following spring, and the white groove once obtained appears to be permanent in the adult specimens."

In the open sea off the north coast of Norfolk, razorbills, both old and young, are found in considerable numbers after the breeding season, and in rough weather the young birds seek the shelter of the harbours, especially during north or north-east gales; at such times Mr. Dowell says the Blakeney harbour is sometimes filled "with these and such-like visitors," but, so far as my experience goes, the adult birds are rare near the shore at any time, save under the exceptional circumstances mentioned. During the early part of March, 1847, the same gentleman says, "I picked up several specimens which seemed starved to death, and killed others in nearly as bad a plight, and yet the weather was not at all severe. The last found, March 25th, had nearly assumed the black throat of summer."

Mr. F. d'A. Newcome has a young razorbill in his collection, which was picked up alive with a broken wing, on the heath at Wilton, in July, 1884.

In August, 1889, a number of razorbills were shot off Cley by a Mr. Bagster, and skinned by Mr. Pashley; some were adults in full moult, others only half grown. On September 18th of the same year, Mr. F. Barclay and Mr. J. H. Gurney, jun., shot two of these birds off Cley, about three miles from the shore, one of which, an adult, had no primaries, and therefore could not have flown.

PHALACROCORAX CARBO (Linnæus).

CORMORANT.

Two hundred years ago the cormorant was a resident in Norfolk and the border of the adjoining county of Suffolk. It must now be regarded as an occasional and uncertain visitant to our coast, and less frequently still to some of the inland waters, almost invariably occurring in the spring and autumn months. Turner, writing in 1544 of this bird under the name of "*mergus*," says that he had seen it nesting on the Northumbrian rocks at the mouth of the Tyne, as well as among herons in tall trees, in Norfolk. His evidence, quoted by Aldrovandi, in 1603, as that of "a certain Englishman," was repeated by Willughby in 1676; but Sir Thomas Browne, in his account of the birds found in Norfolk a few years earlier, gives more precise information; and speaks of "cormorants building at Reedham, upon trees from whence King Charles the first was wont to be supplyed." There is no evidence, I believe, to show at what date the cormorant ceased to breed at Reedham, but at Herringfleet, on the banks of the beautiful Fritton Lake, they continued to do so far into the first half of the present century. Mr. Lubbock states that "cormorants have in some seasons nested in the trees around Fritton decoy in some numbers; in other years there has not been one nest. These woods used to be, perhaps are, their favourite resort during the time of low water upon Breydon."* There is, I believe, no record of their having bred at Herringfleet since the date mentioned by Mr. Lubbock, and it is probable they ceased to do so about that time. Hunt does not mention their breeding either at Reedham or Herringfleet, although an incidental remark when writing of the shag ("British Ornithology," vol. iii., p. 47) would lead one to suppose that he had the fact of their doing so present in his mind. "This species" [the shag], says he, "is never known to visit our fresh-water rivers, which the cormorant frequently will, and in some places make their nests in trees, on which they frequently perch by the sides of rivers."

---

* In Mr. Lubbock's copy of "Bewick" occurs the following more precise and interesting note:—"These birds seem fond of occasionally exchanging salt and fresh water; perhaps, like epicures, variety of fish is the inducement, but during the autumnal months they are to be seen at all hours sailing like wild geese with slow and steady motion, in their trips from Breydon to Herringfleet, where in some years there have been fifty or sixty nests (in 1825, for instance); in the present year, 1827, not a single one."

The late Mr. Howlett informed Mr. Stevenson that probably about the year 1820 a flock of cormorants suddenly appeared in the neighbourhood of Bowthorpe, near Norwich, at a time when the meadows were unusually flooded. Mr. Gurney also tells me that, he thinks in the year 1828, a similar flock was seen one morning near the pond in the park at Earlham, and remained thereabouts all day; one of these birds, which he still possesses, was shot in the afternoon just below Earlham bridge, and was stuffed by Hunt; there were probably thirty to fifty of them. The ground was covered with snow at the time, but Mr. Gurney does not know the exact date. It is not impossible that these two accounts may refer to the same flock.* Mr. Gurney has also a very fine plumaged bird, with a white mane and pure white thigh patch, which he obtained in the Norwich Fish Market many years ago. Mr. Rising had a cormorant in his collection, which was shot at Horsey, in April, 1843; an old male bird was killed on the river Yare, at Thorpe, in February, 1855; an immature bird was sent from Yarmouth, in January, 1861, a season remarkable for the intensity of the cold; a fine old male in full breeding plumage was shot on Breydon, in April, 1862; and, on the 9th October, 1864, an immature female was shot at Coldham Hall, on the Yare, from "a tree on which it usually settled when driven off the river by the swans, who attacked it fiercely." ("Zoologist," 1865, p. 9405). Mr. J. H. Gurney, jun., records the occurrence of three cormorants, at Cley, between the 10th and 18th February, 1889 ("Zoologist," 1889, p. 334); these, like some others already mentioned for the same reason, are worthy of note, as having been met with unusually early in the season. As might be expected in a maritime county like Norfolk, there are many records of the occurrence of this bird, only a few of which I have given; it is not unfrequently met with in inland localities, where, in addition to those

---

* Mr. Gurney also informs me there is a tradition that early this century a flock of "black swans" appeared on the Keswick meadows. They were chased about for two or three days by the late Daniel Cooper, of Keswick.

already mentioned, one was shot on Salhouse Broad, in April, 1865; another from the Church tower at Necton, near Swaffham, in March, 1883—like the shag it shows a predilection for lofty perches;—others at Hickling, Kimberley, and Ormesby. Occasionally they are seen during the early summer perched on the stakes which mark out the channel in Hickling Broad, or on other inland waters; in May, 1867, Mr. Stevenson and Mr. Blofeld saw one flying over Hoveton Broad; and on the 16th May, 1889, Mr. J. H. Gurney, jun., and myself saw two others flying over the same piece of water. Other flocks of seven and three have been seen on the coast in the same month. Of forty-five individuals either killed or seen, only one occurred in January; five in February (including Mr. Gurney's three); three in March; six in April; fourteen in May; four in August (seen by the Rev. Julian Tuck off Hunstanton); seven in September; four in October; and one in November; thus the spring months of March, April, and May contributed twenty-three towards the total; the autumn months of August, September, and October fifteen, and January, February, and November seven only.

The January bird must be regarded as an accidental straggler, met with in exceptionally severe weather, and those in February were probably the advance guard of the spring migration, which would indicate the cormorant to be essentially a spring and autumn visitor to the county of Norfolk, never numerous but by no means rare.

PHALACROCORAX GRACULUS (Linnæus).

SHAG.

Unlike the preceding species, the shag, or crested cormorant, must be regarded as very rare on the Norfolk coast, where it has only been met with in winter, between the months of October and February. Hunt makes no mention of it, and the Messrs. Pagets merely remark it is "very rare." Messrs. Gurney and Fisher state that it is "occasionally met with, the specimens, which chiefly occur in autumn, being mostly immature." Mr. Lubbock also speaks of it as "very uncommon here."*

Mr. Stevenson has the following notes of the occurrence of the shag:—"An immature bird killed at Yarmouth during severe weather, in November, 1854; an old bird from Blakeney, in February, 1855; and another exhibited about the same time in the Norwich Fishmarket. These were no doubt stragglers driven to our coast with other rare fowl by the intensity of the cold that prevailed at that time, but I could not hear of a single example having been met with in the still more severe winter of 1860-1. An immature female was also killed at Hingham, on 9th October, 1856." In the "Zoologist" for 1868, p. 1128, Mr. Stevenson writes, "Amongst other effects of the heavy gales, which, commencing on the 1st February, lasted for several days, was the appearance of a fine adult female of this rare visitant on our eastern coast. The bird was sent up to Norwich on the 8th, having been picked up dead on the beach in the neighbourhood of Cromer, and was in most beautiful plumage, the rich bottle-green of its feathers contrasting with the bright yellow round the gape of the beak." On the 6th October, 1880, an immature shag, now in Mr. H. M. Upcher's collection, was caught in a fishnet off Beeston. Mr. Gunn received an immature bird, killed on the 22nd of February, 1883, while perched on the spire of Attleborough Church; and another, killed at Yarmouth, on the 1st March of the same year, described as a "second year bird," passed into the pos-

---

* The annotator to Wilkin's edition of Sir T. Browne's works appears to be wrong in supposing the following passage in the "Birds found in Norfolk" is applicable to the crested cormorant or shag. The words are: "besides [*i.e.*, in addition to the common cormorant already mentioned] the rock cormorant, which breedeth in the rocks in northern countries, and cometh to us in winter, somewhat differing from the other in largenesse and whitenesse under the wings." Sir T. Browne's idea evidently was that the "rock cormorant" was distinct from the bird which he knew to breed in trees in Norfolk, and was larger and whiter than that bird; but this description certainly will not suit the shag.

session of Mr. Stevenson. The late Mr. Alfred Master's collection contained a very fine adult bird, which was was killed at Yarmouth.

In Mr. Lubbock's copy of "Bewick" occurs the following note on this species:—"1824, November 9th A striking instance of the voracity of this bird occurred to me to-day. I shot one, which fell from a broken wing, but, the wing being fractured nearly close to the body, the impetus of its fall stunned it; on recovering it very gravely raised itself on its latter end and vomited *eleven* flounders, or butts, as they are provincially termed at Yarmouth; the one first swallowed, at least last ejected, was hardly acted upon perceptibly by the gastric juice, so rapid had been the work of destruction."

SULA BASSANA (Linnæus).

GANNET.

Mr. Stevenson has left the following notes on the gannet, which, although written about the year 1862, are equally applicable to the present time:—

This species "occurs on the coast nearly every autumn, and occasionally, also, in spring, their movements being chiefly regulated by the shoals of fish which at certain seasons frequent our shores. At these times they attend upon the herring boats, in the roads off Yarmouth, and most of the specimens taken are either shot from the smacks or captured with a hook baited with some tempting morsel. Captain Longe has seen several killed on Breydon during the last five years, and one or two at Hickling. He says, 'the usual times of their appearance on the Norfolk coast are between the months of September and December, and again in March and April,' although, as he suggests, the fact of examples being obtained in January and February may probably be accounted for by the fishermen being employed for the most part during those two months. One or two of the finest old birds I have seen (killed in April) resembled a specimen examined by Mr. Longe, at Yarmouth, which had the top of its head a beautiful yellow, with a patch of the same colour on the breast similar to that on the spoonbill. Young birds occasionally follow the course of the rivers, and have been killed as far inland as Upton, Rockland, and Surlingham, and, in one instance, according to Messrs. Sheppard and Whitear, at Pulham, near the south-east boundary of Norfolk, a still more unlikely locality for such a bird.

"In December, 1860, some twelve or fourteen gannets, all killed off the coast, were brought to Watson, the game dealer at Yarmouth, besides several others sent to Norwich, for preservation about the same time, preceding only by a few days the first severe frost of that memorable winter. The Messrs. Paget also speak of their appearance in some numbers after a heavy gale in 1827."

Judging from the numbers of these birds which have been found far inland, or on the beach either in a dying condition or washed up dead upon the shore, it would seem that this species must be very susceptible to the influence of continuous stormy weather; it appears not unlikely that, owing to the fish retiring to the less disturbed waters at greater depths during stormy weather those birds which search for their prey on the wing are the first to succumb to the scarcity of food thus produced, and either perish or fly inland, when, in their exhausted condition, they fall easy victims. Sir Thomas Browne records two such instances: "A white, large, and strong-billd fowle, called a Ganet, which seems to be a greater sort of *Larus;* whereof I met with one kild by a greyhound, neere Swaffham; another in marshland, while it fought, and would not bee forced to take wing; another entangled in an herring-net, which, taken aliue, was fed with herrings for a while." Two hundred years later a similar scene was enacted at Bodney Field, also near Swaffham, and probably not far distant from the scene of the first encounter, when, in November, 1854, a gannet stoutly defended itself from the attack of a sheep dog, and was killed by the shepherd's boy with a stick. Miss Gurney records in her journal that, on the 25th May, 1834, "during a heavy gale a gannet was taken at Trimingham asleep under a bank near the cliff—it died a few days after it was caught." Mr. Dowell mentions one picked up dead on the Blakeney "wash," in October, 1846; on the 2nd June, 1853, the year after the flood in the fens, the late Mr. Newcome saw a gannet flying over Hockwold fen; on the 27th June, 1863, Mr. F. Norgate found an adult gannet dead at Blakeney; in October, 1865, an adult was killed near Holt; Mr. Stevenson records in the "Zoologist" for 1868, p. 1126, the capture alive of a fine adult bird, on the 7th December, 1867, at Harford Bridge, near Norwich; it was driven inland by a gale, and after its capture exhibited in a fishmonger's shop as a "wandering albatross." A young bird was caught alive in a field at Dunton, near Fakenham, about the 10th October, 1870; in this instance there had been no stormy weather to account for its appearance so far inland; about 16th January, 1877, an adult, storm driven, was taken at East Ruston; a male bird, also adult, was caught at sea off Blakeney, resting on the water, apparently asleep or exhausted, on 2nd September, 1880; Mr. J. H. Gurney has another which was taken under similar circumstances off Runton, on 15th October, 1887. An unusual number of these birds were brought into Yarmouth by the smacksmen in the following winter, many of them fine adults. An adult gannet was found dead on the beach, at Wells, by Mr. F. B. Middleton, in the middle of September, 1887.

Mr. J. H. Gurney, jun., also, says that after protracted storms he has found birds of this species washed ashore at Cromer, which he presumed had perished for want of food. One of these, a male, found on the 14th October, 1885, was in the piebald stage of plumage, which is very uncommon here; and, on the 14th September, 1889, a gannet, in speckled plumage, was taken alive in a turnip-field, near Cley-next-the-Sea.

The above are only some of what may be styled the abnormal occurrences of this bird, the number of which, on looking through my notes, I confess surprised me. As a rule the gannet keeps far out at sea, following the shoals of surface-swimming fish, more particularly of sprats and herrings, which habit has obtained for it the name of "herring gant" among our fishermen; it thus happens that it is seldom seen from the shore, and generally passes unnoticed. Occasionally a small flock makes its appearance in the summer months (cf. "Zoologist," 1872, p. 3226), fishing off our coast, at which time they are doubtless wanderers in search of food from their nearest breeding station, the Bass Rock, or possibly barren birds.

PELICAN *(Pelicanus onocrotalus?)* It may be well to mention here that both in his "Account of Birds found in Norfolk," and in his letters to Merrett, who had included this as a British species, Sir Thomas Browne mentions the pelican as having been killed in Norfolk. In the list of birds we learn that " An *onocrotalus* or pelican shott upon Horsey fenne, 1663, May 22 wh, stuffed and cleansed I yet retaine it was three yards and a half between the extremities of the wings the chowle and beake answering to the vsuall description the extremities of the wings for a spanne deepe browne the rest of the body white. a fowle wh none could remember upon this coast about the same time I heard one of the king's pellicans was lost at St James perhaps this might bee the same." In the letter to Merrett dated September 13th, 1668, he writes as follows: "In your Pinax I find *onocrotalus*, or pellican; whether you mean those at St James or others brought ouer, or such as have been killed here, I know not. I haue one hangd up in my howse, which was shott in a fenne ten miles of, about four yeares ago; and, because it was so rare, some coniectured it might bee one of those which belonged vnto the King, and flewe away." Whether such were the case or not it is impossible to say; but, either as an escape or a genuine "straggler," the Norfolk broads of those days would have proved a very suitable resting-place for the wanderer. That the pelican, however, in former days frequented the fens of the Eastern Counties is proved by a *humerus* of this bird, given to the Zoological Museum at

Cambridge, by Mr. J. H. Gurney, jun., which was found in the peat, at Feltwell Fen, some few years ago (cf. "Proc. Zool. Soc.," 1871, p. 703.)*

STERNA CASPIA, Pallas.

CASPIAN TERN.

Mr. J. H. Gurney, jun., has at various times taken considerable trouble to investigate the reported occurrences of this fine bird in England, and from his latest communication to the "Zoologist," 1887, p. 457, I copy the following list of Norfolk specimens:—

1. Yarmouth, or Breydon Broad, October 4th, 1825. "Mag. Nat. Hist.," iv., p. 117; Babington's "Birds of Suffolk," p. 247.
2. Yarmouth, 1830. "Zoologist," 1856, p. 5035. In the Norwich Museum.
3. Cromer, 1836. On the faith of a letter to T. C. Heysham, dated November 21st, 1836. Lent by Rev. H. A. Macpherson.
4. Yarmouth, April 16th, 1839. Received in the flesh by my father [Mr. J. H. Gurney].
5. Yarmouth. Female. June 2nd, 1849. Is or was in the possession of Captain Barber.
6. Yarmouth. Male. June, 1850. In Bury Museum.
7. Yarmouth. ——— July 16th, 1850. Mr. Gurney was informed that others were seen at the same time.
8. Yarmouth. Male. August 11th, 1851. Preserved at Northrepps.
9. Yarmouth. Male. May 2nd, 1862. Stevenson, "Zoologist," 1862, p. 8093.

---

* It may be mentioned that this was the second pelican's *humerus* from the peat of the Bedford Level; the first, however, was found in the Isle of Ely ("Proc. Zool. Soc.," 1868, p. 2), and the condition of the bone shows that it must have belonged to a young bird, and one, therefore, almost certainly bred in the district. It is worthy of note that in 1856 one of these birds was found dead on the shore at Castle Eden, in Durham, according to Canon Tristram ("Zoologist," p. 5321), but so many foreign birds are brought home by merchant vessels, that it would be rash to affirm it had not died on shipboard and been thrown into the sea to be cast ashore at Castle Eden.

As will be observed, all these specimens, with one exception, were killed at or near Yarmouth, and, since the last named, I am not aware of any more recent occurrence. This is probably owing to the gradual decrease in numbers of this species, which has now been going on for many years in its old breeding haunts on the Frisian coasts, whence we may infer that the birds visiting us were wanderers.

Nos. 1 and 2 of the foregoing list probably both refer to the bird mentioned in Hunt's "List" (1829) as having "been recently killed near Yarmouth." Messrs. Paget (1834) record two specimens—one the 1825 bird, and another "in the Norwich Museum, which was shot here;" this latter is doubtless No. 2 of Mr. Gurney's list; it was presented to the Norwich Museum with other birds, by the Rev. G. Steward, in 1831.

No. 3. I can discover nothing of the history of this bird, and if really a Caspian tern it is probably not in existence.

No. 4. This example came into Mr. Gurney's possession soon after it was killed, and was given by him to Mr. Heysham; at the dispersal of that gentleman's collection in 1850, Mr. Gurney re-purchased it, and it is now at Northrepps; it had a curious history attached to it, as I am informed by Mr. Gurney; some one went into a Yarmouth gunsmith's to buy a gun, and chose one which suited him, but before completing his purchase he asked the gunsmith to allow him to try it, and, the latter consenting, he took it just outside the town and shot the first bird which passed over him, and which proved to be this Caspian tern.

No. 5 was shot on Breydon by Mr. H. J. Barber, for whom it was preserved, and has recently passed into the possession of Mr. P. B. Bellin. (There is a note in Mr. Stevenson's MS. book as follows: "Mr. C. S. Preston says he owns this bird.") It was recorded in the "Zoologist" by two different correspondents (pp. 2500 and 2529) as measuring in length $22\frac{1}{2}$ inches, and from tip to tip of wings 4 feet $3\frac{1}{2}$ inches, but they differ as to the sex.

Nos. 6 and 7. It seems probable that these two refer to the same bird which was shot on Breydon; it was discovered by Mr. Stevenson in the Dennis collection, at Bury, marked Breydon, June, 1850; it is a male in perfect summer plumage; the sex was not noted (see "Zoologist," p. 2915). It is doubtless this bird also to which the following note in Mr. Stevenson's MS. refers:—"Young Harvey, of Yarmouth, says one of Mr. Gurney's birds was shot by a Mr. Goodwin, of the North End [Yarmouth], thirty years ago [there is nothing to indicate the date of this memorandum], on Breydon Wall. It was never in Mr. Gurney's possession, although he recorded it."

No. 8. An adult male in full summer plumage, is in Mr. Gurney's collection, at Northrepps. (See J. O. Harper, in Morris's "Naturalist," 1852, p. 128.

No. 9, and the last example known to have occurred in this county, is an adult male, now in the collection of the Rev. C. J. Lucas, of Burgh; like the earlier specimens this was also killed from the Breydon Wall.

The above particulars are gleaned mainly from Mr. Stevenson's note books, and with the assistance of Mr. J. H. Gurney, jun.

STERNA SANDVICENSIS, Latham.*

SANDWICH TERN.

This species has occurred much more frequently on the Norfolk coast than the Caspian tern, although perhaps not so frequently as might have been expected.

Messrs. Sheppard and Whitear state that it had been killed at Yarmouth; Hunt also, in his "List," says that "several have been killed on Yarmouth beach;" and the Messrs. Paget say it is "not uncommon." Of late years their remark, as will be seen, certainly would not apply; but as it formerly bred in Suffolk (Latham), Essex (Parsons), and Kent (Plomley), it is quite possible that

---

* "General Synopsis," Suppl. I. p. 296 (1787). In his "Index Ornithologicus" (vol. ii. p. 806), published in 1790, the same author named it *S. boysii*, after its discoverer; and in the meanwhile Gmelin (1788) had called it *S. cantiaca*.

it may also have bred in this county within the memory of the Pagets' contemporaries.

The first recorded example of which I am aware is mentioned in the Hooker MS. as "taken on Yarmouth beach, in September, 1827." In Mr. Rising's MS. list of birds in his collection, the Sandwich tern is mentioned with the affix "Birds, 1829, R. R.;" also "Horsey, 28th September, 1830," both undoubtedly records of their having been obtained there. Neither Messrs. Gurney and Fisher, nor Mr. Lubbock, give any instances, although the former say that it "sometimes occurs on the coast in early spring and autumn;" in the "Zoologist" for 1849, p. 2353, they, however, record the occurrence of a Sandwich tern, at Lynn, in the end of September, 1848. Dr. Babington mentions one in the Dennis collection, killed at Yarmouth, on the 4th May, 1849; one was killed at Salthouse, in the summer of 1849, as I am informed, by Mr. Dowell; on the 6th September, 1851, an example was shot at Hunstanton; Mr. Dowell also informs me that in the autumn of the same year Overton saw several of these birds at Blakeney, and shot some young ones. On July 22nd, 1853, Overton saw two in Blakeney harbour; on April 28th, 1854, he also saw one in the Cley channel; and on August 18th, 1857, he shot two at Stiffkey freshes. An adult and young bird of the year were killed at Yarmouth, on October 5th, 1857, the former, according to Mr. Stevenson, just losing the black head of summer, and the latter in the "striated" plumage, with also a little black about the head. In 1862, Mr. Ellis, the bird preserver, at Swaffham, showed Mr. Stevenson a fine specimen shot by himself, at Hunstanton, some two or three years before, in the month of July; three or four others were seen at the same time, but not knowing their rarity he obtained only one. A male was shot at Blakeney, on the 13th of August, 1862, which exhibited a state of change between summer and winter plumage ("Zoologist," 1863, p. 8332); an immature bird was shot at Yarmouth, on 20th October, 1875; a female, one of a pair seen on Breydon, was killed on the 8th September, 1880; and on the 24th August, 1881, an adult was shot on Breydon.

It will be seen by the above records that the greater number of these birds occur on the Norfolk coast towards the end of summer or in autumn, beginning to appear in July, reaching their greatest numbers in August and September, and that after October they are no more seen till the following spring, at which season they have been met with in the months of April and May, but are then decidedly rare.

As these birds merely visit us in passing, of course there is nothing to be said of their habits here, except that they are found in some of the same localities as those frequented by the common tern, but appear never to leave salt water. It is not improbable that they often pass unnoticed.

STERNA DOUGALLI, Montagu.

ROSEATE TERN.

The roseate tern is included in Hunt's "List" (1829), where it is stated to be "very rare." The Messrs. Paget also (in 1834) say, "Mr. Youell has known this to have been shot here." Following these two authorities, this species had been retained in all the successive lists since that time, but no specimen had ever been produced to substantiate its claim to the position until the one about to be mentioned. As this species was undoubtedly more frequent on the east coast many years ago, of course it is quite possible that both Hunt and Youell may have had sufficient grounds for their remarks but no precise evidence was adduced to verify their statements.

Since that time the roseate tern is said to have been seen on the Norfolk coast more than once, the most precise account being perhaps that of Mr. Booth, who states ("Rough Notes," pt. xi.) that one of these birds "flapped slowly past the punt on Breydon mud-flats, on the 26th May, 1871," but both barrels being empty at the time, the bird was out of shot before another cartridge could be inserted. All doubt, however, was removed in 1880, when an adult male was shot at Hunstanton by Mr. George Hunt, on the 12th July.

This beautiful specimen passed into the possession of Lord Lilford, who very generously presented it to the Norwich Museum, where it forms part of the rich collection of local rarities; it is, of course, in full summer plumage, and is said to have been accompanied by another bird of the same species, Mr. Hunt's attention being first attracted by their peculiar note.*

STERNA FLUVIATILIS, Naumann.†

COMMON TERN.

The common and lesser terns are the only two species of this genus which now breed in the county of Norfolk, and the numbers of both have for many years past steadily decreased. With regard to the species under consideration there can be no doubt that such is the case to a very serious extent; all the older writers speak of the tern as exceedingly common. Sir Thomas Browne refers to the "Hirundo marina or sea-swallowe, a neat white and forked tayle bird butt much longer than a swallowe;" which may safely be taken to be the common tern. Hunt mentions it as breeding on "many parts of our coast," the Pagets as "very common," and Messrs. Gurney and Fisher also state that it is "very common in spring, summer, and autumn, and breeds at Salthouse and some other shingly parts of the coast." The first

---

* Mr. G. F. Frederick was kind enough to inform me that the roseate tern, in the collection of the late Mr. Rising, was killed on the Sussex coast, near Eastbourne, about the year 1848. I think it well to mention this, as Mr. Rising's collection consisted so largely of local rarities that it might be supposed, no locality being given in the catalogue, that this was a Norfolk killed specimen.

† It being certain that Linnæus did not recognise the distinction between the two nearly allied species of tern (this and the next), many recent writers have discarded the use of the name *S. hirundo* applied by him, in favour of the more definite appellations of Naumann.

alarming note occurs in Mr. Lubbock's copy of Bewick, where he states that they " used to breed in large numbers on the islands in Hickling Broad; at present but few scattered nests are there." From this neighbourhood I need not say it has long since disappeared in the nesting season, as also from Ormesby, Hemsby, and Winterton, where the late Mr. Rising remembered their breeding; and in the place last named Whitear found this species, as well as the lesser and black terns, breeding in 1816 and 1818. Salthouse, once a favourite resort of this species, now knows it no more as a nesting species; and a former breeding place, on the marshes at Holme, owing to drainage and cultivation, is also quite deserted. Such being the case, it becomes necessary to speak with caution respecting the few remaining haunts of this species on the Norfolk coast; and I am sure all true lovers of this beautiful bird will forgive me if I refrain from particularly mentioning the sites of those breeding-places to which I shall have to refer in what follows.

A correspondent writing to me in 1883, with regard to a locality where twenty-five years ago I remember the common and lesser terns breeding in very large numbers, says that the colony of the former amounted to about one hundred pairs; but that the eggs are very much taken, and in August a great many of the young are shot. Fortunately for the birds that are left, the sand-hills blown up on what have once been "binks," or gravel-ridges, that surround the locality, are let for the sake of the rabbits which frequent them, and not only are persons with guns warned off, but even indiscriminate wandering over the hills is not permitted, so that the terns have some protection. But this does not go far, for he informed me that in a walk of five miles along a line of coast where hundreds upon hundreds of common terns used to breed so lately as ten years before (1873), he did not count above six pairs, and his companion assured him that at that time five hundred pairs were breeding at a place where they now did not see more than twenty. The man added that he had taken with his own hands 170 eggs in a day, and that he was only one of many who gathered them. My correspondent believes that the diminution in numbers, which will lead to the ultimate extinction of the common tern on our coast, is due partly to indiscriminate egging, and partly to the comparatively few broods which hatch off falling victims, in the early part of August, so soon as the close time is over, to parties of persons who imagine that they are sportsmen, and, coming from a town in the neighbourhood, shoot all the terns they can. It is well known that at this season a wounded tern, by its screaming, attracts every other of its species within hearing, and the gallant gunner shoves in his cartridges and blazes away until it is time for him again to seek the tap-room, whence he had set out in the morning, there to recount to an admiring circle his glorious deeds of skill. According to a well known professional gunner, who of course regards such behaviour with scorn, " the —— gents," as he called them, naming the town whence they came, in one August " nearly cleared off the big terns."

I could mention more to the same effect, but enough has been written to show what I fear must be the fate of this charming species at no distant time unless all lovers of birds exert themselves in its behalf. There can be no doubt that the establishment of a " close time " has had the effect of retarding its extermination, which otherwise would have probably been accomplished by now, so far as this county is concerned; but more remains to be done to ensure its safety, and this can only be effected by the prolongation of the protected period for perhaps another fortnight, so as to allow the late hatched broods and their parents to escape. I am indeed assured that in one of its haunts it has not decreased in number, and that in another until 1886 it had actually increased of late years, the people in the neighbourhood being under the impression that its eggs were protected by law. I am pleased to hear that the Earl of Leicester has informed his tenants on the portion of his estate on which the two terns and the ring dotterel breed that he does not allow the eggs to be taken: consequently all pillagers are warned off the ground, and the result was a very fine show of terns, both old and young, last August, frequenting the mouth of Wells Harbour, and the colony referred to has certainly held its own, and probably increased in numbers this year; that it may continue to do so is sincerely to be hoped; but where robbed of their eggs, and both old and young cruelly shot at a time when they are so unsuspecting, their banishment cannot be long delayed. To ruthlessly destroy these lovely summer visitants, by taking their eggs and shooting them, seems little short of infatuation; and when once the race which has bred on our shores for perhaps thousands of years has been extirpated, it is extremely doubtful, even under the most rigid system of protection, whether others would ever come to occupy the vacant breeding site. It is wrong that we should deprive our descendants of the pleasure afforded by the sight of these exquisite creatures, and commercially it is short-sighted policy, for the presence of a carefully protected breeding-place of terns, in close proximity to a sea-side town, must certainly prove a great attraction to visitors, and therefore a source of profit.

"On their first arrival in spring, towards the end of April," says Mr. Stevenson, the common terns "are frequently seen on the broads and rivers; but, in autumn, when in company with their young broods, they keep almost entirely to the sea-coast, and rarely remain with us later than the first week in October." Mr. A. Patteson, who has observed this species at Yarmouth, tells me that when the wind blows from a westerly direction, especially south-west, he has found the terns most numerous, "flock succeeding flock playfully, busily, but leisurely, working southward; at such time herring fry, which swims inshore and near the surface, offers them great attraction, and they seem literally to gluttonise on them."

The common terns arrive at their breeding stations in the month of May, and the 24th of that month is the earliest date on which in a particular season an egg was found, and in that instance it was a single one; on the 26th many were laid, and by the 1st June the deposition of the eggs was general.* I have found fresh eggs as

---

* On their nests and nesting habits Colonel Feilden has contributed some interesting notes to the "Zoologist" for 1889, p. 264.

late as the end of July, but probably the old birds had been robbed more than once, and were making a last effort to rear a brood. Mr. J. H. Gurney, jun., and myself found a young one unable to fly, on the 7th August, 1885. So soon as the young can fly they follow their parents on their fishing expeditions; and, as has been already said, early in October they take their departure.

Mr. Gurney has a common tern, which is white with the exception of a shade of brown on the back of the head; it was killed at Lowestoft on August 12th, 1853.

In Norfolk the common tern is variously known as the "darr," "perl," "stern," or "big mow."

STERNA MACRURA, Naumann.*

ARCTIC TERN.

There is no reason to suppose that the arctic tern has bred in Norfolk, although from its general similarity to the common tern it might easily escape notice. For many years Norfolk naturalists have been on the look out for it, but neither Mr. Stevenson, Mr. Gurney, nor myself ever met with, or heard of, an instance of its nesting here. In the month of April small parties of these birds pass on their way to the more northerly breeding-stations; probably the bulk proceed direct to their destination, but, at that season of the year, such as are seen here follow the course of the inland waters, very few being met with along the coast line; on their return migration, however, in August and September, they are almost invariably met with along the coast, and at that time are by no means rare. Some few make their appearance in the middle of July, but the bulk do not appear till August, at which time they are frequently met with.

As this species is difficult to distinguish on the wing, and is probably often passed over in mistake for *S. fluviatilis*, it may be useful to some if I quote from Mr.

---

* See foot note at page 301.

Dresser's "Birds of Europe" the following characters by which they may be distinguished from the common tern :—

1. The tail is much longer, and in the adult bird in summer plumage reaches beyond the wings. Hence Naumann's name of *macrura*.
2. The tarsus, both in old and young birds, is much shorter than in the common tern.
3. The bill is almost entirely crimson, without any black on the culmen;* it is also more slender.
4. *Young birds* are more difficult to determine; but the tarsus [which is only half an inch in length] is a sure sign of distinction; and, moreover, there is a less amount of greyish black on the outer primaries.

The under surface of the body is also as grey, or nearly so, as the back, whereas in the common species it is, as a rule, much lighter or white.

STERNA HYBRIDA, Pallas.

WHISKERED TERN.

Mr. J. H. Gurney possesses the only Norfolk speci- of this rare tern, which was shot on Hickling Broad, on the 17th of June, 1847, by the late Mr. J. Sayer, of Norwich. Mr. Gurney, in recording its occurrence in the "Zoologist" for 1847, p. 1820, remarks, "It proved to be an adult female, and contained ova in an advanced stage, the largest being apparently almost ready to re-

---

* Professor Newton tells me that in Iceland, in the year 1858, in company with Mr. Wolley, he shot from amongst a flock of undoubted arctic terns an over year bird which had the bill *black*. The conclusion he arrived at was that the bill in this bird for some reason or other had not attained the full colouring, and he thinks it possible that occasionally, though very rarely, such a variety may be met with.

ceive the shell. In the stomach were found the remains of about twenty of the *larvæ* of the broad-bodied dragon-fly."*

STERNA ANGLICA, Montagu.

GULL-BILLED TERN.

The only record of this species as a Norfolk bird of which I am aware previous to the year 1849, occurs in Messrs. Gurney and Fisher's list, in the following guarded terms:—"We have seen a specimen of this bird, which was said to have been killed in West Norfolk." This note refers, I believe, to a bird said to have been killed at Hunstanton in the spring of 1839. The appearance, therefore, of several examples of this rare British tern, in the three successive years of 1849, 1850, and 1851, as will be seen from the following extracts from Mr. Stevenson's notes, is certainly very remarkable.

On the 14th April, 1849, an adult male in full breeding plumage, now in the Dennis collection at Bury, was killed on Breydon, and, according to the statement attached to the specimen, was 16 inches in length and weighed 8¼ ounces. On the 31st July of the same year an adult male, in full breeding plumage, was also killed on Breydon by Mr. P. B. Bellin, as recorded in the "Zoologist" by Mr. Gurney and Mr. John Smith, of Yarmouth (pp. 2569 and 2653). On the 1st of the following September Mr. Gurney ("Zoologist," 2592) also notices the occurrence of two other specimens at Yarmouth; they were "male and female, both adults and beginning to assume the winter dress, the change having progressed somewhat further in the female than in the male." This pair were shot on Breydon, and are now in the Northrepps collection.

---

* It may be well to mention that the whiskered tern, which was sold in the late Mr. Rising's collection, was shot on the river Swale, at Hornby Castle, in Yorkshire, by one of the Duke of Leeds' gamekeepers, in 1842, as I am informed by Mr. G. F. Frederick, in whose collection it formerly was.

In the following year, 1850, on the 24th May, a fine male bird in breeding plumage was killed at Yarmouth, which, as Mr. Gurney remarks ("Zoologist," p. 2853), was probably on its way towards its breeding-place, in the Frisian islands, as were other spring birds of passage obtained on the same day. In 1851, early in the month of July, an adult male was killed also at Yarmouth, which is now in the collection of the Rev. H. T. Frere, of Burston.

I know of no other occurrence of this bird in the county of Norfolk till the year 1878. On the 8th of May of that year, as recorded in the "Zoologist" for 1880, p. 53, by Mr. Gunn, two gull-billed terns were shot near Yarmouth, one of which, an adult bird, was too far gone to be preserved, but the other, also adult, and a female, was preserved by him. And, lastly, in July of the same year, one now in the collection of Dr. Whitty, of Hunstanton, was killed near the railway station of that town, and taken to its present owner in the flesh.

Mr. Dack, bird-preserver, of Holt, assures me that he had a gull-billed tern about the year 1865, at the end of August. Mr. Dack is well acquainted with all the terns usually found on our coast, and I have every reason to credit his statement. Mr. Gurney gave one of these birds, which was killed at Hunstanton, to Mr. Heysham, but whether one of those already recorded it is impossible to say; he also gave a Norfolk-killed specimen to the Wisbech Museum, where it now is. Beside these he saw another at a poulterer's at Peterborough, believed to have been killed somewhere in the Wash. It is unfortunate that the dates of all these three birds are now lost.

It will be seen that of the nine recorded examples of this bird which have occurred in Norfolk, eight were killed at Yarmouth, or perhaps Breydon, and one at Hunstanton. One was killed in the month of April; three in May; three in July; and two in September.

STERNA MINUTA, Linnæus.

LESSER TERN.

I am sure the reader cannot fail to derive pleasure from the following notes which Mr. Stevenson has left on this charming little bird; and, although rather out of date in some particulars, they exhibit so truthfully the loving spirit of the writer, as well as his keen powers of observation, that it would be a positive loss to condense them.

" Long may it be ere these exquisite little birds cease to frequent our coast during the summer months, and yet, when considering their extended range in former days, and the contracted area within which they are still found breeding, one can but contemplate the worst result from the combined effects of shooting and egging. Too often, I am sorry to say, these delicate little creatures are slaughtered in their nesting season for the mere sake of *sport*, their pretty forms being left to rot upon the beach to which each action of their harmless lives had lent a further charm. The collector, even though with some excuse in the desire to obtain specimens for preservation and study, will feel but small delight in the possession of his prize, if, as once happened to myself at Salthouse, his victim's mate, with plaintive cries, comes hovering round. Heedless of danger to itself, this widowed bird called on its dead companion with every accent of distress and grief, and finding still no answering note, it gently seized its partner by the beak and tried to bear it off. What would I not have given to recall that luckless shot! and oft as I remember that touching instance of animal affection, the thought recurs—

> Have they no feeling? or does man pretend
> That he, alone, can make, or mourn a friend?*

---

* Mr. Stevenson told me that he was so grieved to see the distress thus plainly evinced by this bereaved one, that, to end its sorrows, he felt compelled to shoot it also, not, however, thereby extinguishing his regret at the result of his thoughtless act. In order to commemorate this touching incident he made a rough

"I know few prettier sights by the seaside in autumn than watching a flock of terns slowly passing along the coast, and fishing as they fly. Each bird in turn, pausing for an instant, with its wings closed over its back, drops like a shot into the splashing waves, and rising again in an instant with its finny prey, pursues its onward course. Occasionally a shoal of sand eels* close in shore, or round the piles of some jetty or breakwater, attracts the birds' notice, and for hours the same little game of 'ups and downs' goes on continuously, whilst, off the breeding-stations in the summer months, some few are seen throughout the day, dipping out at sea. Now poised an instant over the waves with beaks pointing downwards; now with sure aim and upraised wings darting upon their prey; or, with a fish protruding from their bills, hastening back to some expectant partner on the shingly beach.

"The note of these 'little mows,' as they are called by our beachmen, to distinguish them from the 'big mows,' or common terns, is very pleasing as they chase one another in amorous flight, and not unlike the soft babbling notes of the swallow, whose quick turns and graceful sweeping flight are likewise so closely imitated by *la petite hirondelle de mer*.

"Mr. Lubbock states that he has found the nests both of this and 'a larger species of tern formerly upon an island in Hickling Broad,' and that 'they used also at times to breed at Horsey.' At the present time, however, I know of no nesting-place between Yarmouth and . . . † but the latter beach is still, as it always has been, the chief haunt of the lesser terns, although

---

sketch of the position of the birds, and carrying away the eggs and the shingle which did duty for a nest, with his assistance, the late Mr. John Sayer reproduced the scene, in a charming group still in the possession of the family, and on the case was painted the couplet quoted above, which he wrote at the time.

* Sir Thomas Browne, in his "Account of Fishes, &c., found in Norfolk," mentions the abundance of sand eels "about Blakeney and Burnham," and says they were "taken out of the sea sands with forks," and are "a very dainty dish."

† As this locality is still frequented by the lesser tern, although in sadly reduced numbers, I hope I may be excused for withholding its precise situation.

their numbers are but small compared with former times. They usually arrive about the middle of April and commence laying in May, but the first batch of eggs being invariably gathered by the beachmen, fresh ones may be found up to the middle of June, and the young are hatched by the middle of July. No nest is made, but a slight depression in the sand or shingle contains the eggs; never more than three in number, and generally two; these, from their extraordinary similarity to surrounding objects, are by the novice more likely to be found by accident than otherwise, though the practised eye of the 'egg gatherer' detects them at once, and dogs are also trained to assist in the search." To the west of the place to which Mr. Stevenson refers they were at the time he wrote still found "in some few localities in marshy spots between the sandhills, in company with the ring dotterel and the common tern, the eggs of all three birds being occasionally sent to Norwich for sale every year from different places. As soon as the young are able to fly, at which time they present a very pretty appearance from the mottled tints of their plumage, they join their parents in their search for food, and continue on our coast until the end of September or beginning of October,* when they again betake themselves to more southern quarters. A singular instance, however, of one being exhibited in Norwich Market, in the third week of December, is recorded by Yarrell on the authority of the Rev. William Howman."

I have little to add to what Mr. Stevenson has so charmingly told us of this species, except that it has suffered rather less from egging and considerably less from shooting than has its larger relative, with which it is so much associated during its stay on our coast that much which has been said of that species will apply equally to this. They are decidedly more numerous than the "big tern" as a species; and although, perhaps, there is no

---

* I think very few lesser terns remain so late as the 1st of October; they leave us earlier than the larger species. Mr. J. H. Gurney, jun., tells me he has noticed that they continue to feed their young on the wing after they are full grown and strong flyers.

breeding-station of that bird which is not also frequented by a few lesser terns from some of the still existing colonies of the latter, the larger species has altogether disappeared. Mr. Stevenson considers that this bird has an advantage also in being more difficult to shoot than the common tern, but its fatal gift of curiosity, although exceedingly pretty to witness, renders it an easy victim. I have seen them leave their fishing in a body to examine any strange object floating in the sea, and a boat is almost sure to attract them.

On the Norfolk coast the names of this bird are various but expressive; "sea swallow," "small perl," "chit perl," "dip-ears," and at Holme this year an old man called them "shrimp-catchers," an occupation in which they were busily engaged in a backwater on the salt marsh as we were watching them.

STERNA NIGRA, Linnæus.

BLACK TERN.

The black tern is one of the birds which in former days must have been very characteristic of certain districts of the county, especially in its most eastern and most western parts, and peculiar interest attaches to it as a species formerly very abundant, but which has, perhaps in the memory of some now living, been lost to us as a summer resident. It can hardly be any other species than this which is mentioned in a letter by Sir Thomas Browne to Merrett, dated 29th December, 1668, as the "sterne," and said by him to be "common about broad waters and plashes not farre from the sea," since Turner,* to whom in the preceding paragraph Browne had referred, almost unquestionably meant this bird, as usual

---

* In Wilkin's edition of Browne's works the word in the MS., abbreviated "Turn.," is stated to be illegible, whereas Professor Newton tells me that it is perfectly plain, and the difficulty in reading it must have arisen through ignorance of the name of the author cited by Browne.

latinising the English word, which he writes "*sterna;*" but particularising its small size and blackish colour.* Turner [p. 64] spoke of its excessive clamour during the breeding season, which was enough to deafen those who lived near the lakes and marshes it frequented, and as he was personally acquainted with the fen district of the Eastern Counties, his experience may not impossibly have been derived from that portion of it bordering upon or comprised in Norfolk. Pennant, in his regrettably short description of the East Fen, which he visited in 1769, writes† in almost similar terms:—"The pewit gulls and black terns abound; the last in vast flocks almost deafen one with their clamors." What the East Fen of Lincolnshire then was, the south-western corner of Norfolk continued for some years longer, and in February, 1853, William Spencer, of Feltwell, whose name has appeared before in this work (vol. iii., p. 3), told Professor Newton and his brother, Sir Edward, that he as a lad recollected " starns," as he pronounced the name, being very numerous, and especially common at Poppylot in that parish, on the border of Southery. Their eggs, he said, were of no use except to put in lapwing's nests, as the latter would "lay to" them. This statement probably refers to the first fifteen or twenty years of the present century, as it was made when he was fifty-three years of age. On the opposite side of the county we have evidence of its breeding much about the same time, for Whitear, in his diary, says ("Trans. Norf. and Nor. Nat. Soc.," iii., p. 243), under date of 2nd July, 1816, that he went shooting with a friend "in the marshes between Winterton and Horsey," and among the birds killed he mentions the black tern, adding that they found an egg said to have belonged to

---

* Prof. Newton, however, points out that it must be admitted that Willughby (Lat., p. 268; Eng., p. 352) took Turner's *sterna* in the general sense of tern as we now use it. He imagined it was derived from the north-country term tarn (a small lake), not knowing that it existed with but little change in cognate tongues, and with absolutely none in Dutch. Turner's latinised form, having been repeated by Gesner and other writers, has come to be recognised by all ornithologists as the well known generic name.

† See his first "Tour in Scotland," ed. 5, p. 12.

that species.* Again, in 1818, on the 26th of May (tom. cit. p. 247), he "went into the marshes at Winterton," and saw (he does not say that he was this time shooting), among other birds, common, lesser, and black terns. Considering the season of the year, I think we may safely infer that this species was then nesting in those marshes. Moreover, Lubbock, in a note, probably made in that same year, states that it "breeds in myriads at Upton," near Acle. Subsequently, but before 1843, he sent Yarrell word that it "used to breed in Norfolk in great abundance, but that the great breeding-place in a wet alder carr at Upton, where twenty years back hundreds upon hundreds of nests might be found at the end of May, has been broken up for some years,"† adding that the "blue darr," as the species was called in that district, "can hardly be said at present to breed regularly in Norfolk, a few straggling pairs only still nest here." Sheppard and Whitear merely catalogue its name, while Hunt, in his list of 1829, simply says of it, "on most of the broads." The Pagets, in 1834, speak of it as being "sometimes in plenty on the beach" at Yarmouth. Mr. J. H. Gurney tells me that he visited Winterton decoy, he thinks in 1838, and recollects being told that black terns had quite recently nested in that neighbourhood, though they had then ceased to do so regularly. In 1846 the species was regarded by Messrs. Gurney and Fisher as a merely passing migrant, "common" in spring and "occasional" in autumn. More precise data for determining the extinction of this once abundant species as a native of the eastern part of Norfolk are wanting.

In the western part of the county, the Fen district, it possibly held its ground longer, though there is no direct evidence of the fact. On the 19th of May, 1832, as shown by an entry in his diary, Salmon visited Crowland Wash, which is indeed in Lincolnshire, though con-

---

* The Rising collection at Horsey contained two eggs, with the birds which had no doubt been taken there; they were purchased at his sale by Mr. Holmes.

† An "Upton Drainage Act" was obtained in 1799, and, though not alluded to by Lubbock, the destruction of the breeding-places of the black tern was most likely one of its effects.

terminous with our own county, and there he found, as he informed Hewitson, "immense numbers" breeding; but he makes no mention of its doing so in Norfolk. Prof. Newton tells me that though he believes a pair or more appeared almost every spring or summer, as they still may occasionally do, on the Little Ouse below Brandon, they had long ceased to make any stay in that neighbourhood until 1853, when the great flood of the preceding year, which laid thousands of acres under water, was subsiding. On the 21st May, 1853, he, with his brother and the late Mr. Newcome, saw four pairs on Hockwold Fen; and on the 8th of June following, three nests, two containing three, and one with two eggs, were taken in the adjacent Feltwell Fen. Three of these eggs, one from each nest, are in his collection, and two more, together with a pair of birds shot about the same time, in that of Mr. Newcome. Since then the species has not been known to breed in that district; and the next, and probably the last, instance of its nesting in Norfolk was in 1858, when a pair bred at Sutton, laying two eggs, both of which were taken, and the birds unfortunately shot. These were sent to Sayer, of Norwich, to be stuffed, and from him bought by Mr. Stevenson, at the sale of whose collection both birds and eggs passed into Mr. Colman's possession. In the "Zoologist" for 1869 (p. 1868) Mr. Gunn states that on the 20th of April in that year, "an egg of the black tern was found on a marsh near Yarmouth," which he, however, considered to have been probably dropped by chance.

It remains to speak of this elegant bird, which is lost to us in its original capacity of an abundant native, as a mere passenger in spring and autumn, of pretty constant appearance indeed, and sometimes in not inconsiderable numbers. The arrivals in spring begin as a rule towards the end of April, and continue till the end of May, an occasional bird occurring in June or July. In August the young are now and then met with, but the main body, old and young, pass along the shore in September and October. Mr. Dowell notes that he has never seen these birds precipitate themselves into the water as the other terns do, but that they take insects off the surface or flying just above it. They have on the wing very much the look of swifts, and show great power of flight in beating up against rough weather which appears very congenial to them.

Whether any of the "blue darrs" or "sterns," to use their local names, which yearly come to us, would remain to breed in Norfolk if they were protected may be doubtful. There is no evidence that the species has been wilfully exterminated from our limits. That re-result must be ascribed to agricultural improvements, and especially to the drainage of places that alone were suited to its habits, most likely on account of the insect food which they afforded, for this appears to be necessary, if not to the adults certainly to the young. The persistency with which the black tern recurs yearly to our shores is very remarkable, for nowadays it has no regular breeding-resort of any consequence in the British Islands, or even further to the northward, and the fact seems to deserve more attention than it has received from those who study the phenomena of migration. Meanwhile, the extirpation of our native race of this beautiful and graceful bird remains to be deeply deplored by every ornithologist, though it may convey an instructive lesson to those who know how to use it.

It is worthy of note that the local name of the black tern differed in the eastern and western divisions of the county. On the Broads it was the "blue darr," according to Whitear and Lubbock, whereas in the Fen district "stern" (pronounced starn) was used. In Lincolnshire, on the other hand, it seems to have been "carr-swallow," or "carr-crow"—which last Willughby, on the authority of his correspondent Johnson, writes "scarecrow."

STERNA LEUCOPTERA, Schinz.

WHITE-WINGED BLACK TERN.

The first recorded Norfolk specimen of this beautiful tern was killed from among a flock of fifteen or twenty black terns, on Horsey Mere, on the 17th May, 1853, and, as stated by Mr. Yarrell ("Zoologist," p. 3911), was shown to him in the flesh. It passed into the possession of the late Mr. Rising, of Horsey, by whose keeper it was killed. At the dispersal of that gentleman's collection it was purchased by the late Mr. William Jary. From the information given to Mr. Rising, it appears probable that two of these birds were associating with ordinary black terns, but his keeper supposing it to be merely a variety, and having killed five of the commoner species, did not shoot the other although he had every opportunity of doing so. The 17th of May is the date usually assigned for this occurrence, but in Mr. Rising's own note, in his copy of Yarrell, from which I gather the above particulars, the 18th is the day mentioned.

The next occurrence of this species was on the 27th of June, 1867, when a beautiful adult male was killed on Hickling Broad ("Zoologist," 1867, p 951); this bird was purchased by Mr. Stevenson, and at the sale of his collection was acquired by Mr. Connop.

On the 26th May, 1871, after a very tempestuous night, Mr. Booth killed four of these birds, two males and two females, out of a flock of five, at one discharge of his big gun, soon after daylight, on Breydon; the odd bird was saved by not alighting with the others. The two males were in perfect breeding plumage, but Mr. Stevenson, who examined them in the flesh, states that the females presented a marked difference in appearance; the tail feathers, which in the adult males are pure white, were in the females light grey, and the feathers on the back lighter in tint than in the males. Mr. Booth ("Rough Notes") states that in the spring of 1872 a pair were seen on Breydon by one of the gunners who had been present when the last-mentioned examples were procured. On the 28th May, 1873, Mr. Booth killed five more of these birds out of a flock of seven, on Hickling Broad, and several others are said to have frequented the broad at the same time, which happily escaped. Mr. B. C. Silcock informed Mr. Stevenson that on the 8th June, 1883, while sailing on Barton Broad, he saw a pair of these birds hovering over the water, occasionally settling on a post; Mr. Silcock had excellent opportunities of watching them, and was quite certain as to the species. That he was right there appears little reason to doubt, for, on the 10th of the same month, a beautiful male white-winged black tern, now in the collection of Mr. R. W. Chase, of Birmingham, was shot from a post on Hickling Broad; it was alone when seen, but was probably one of the pair seen two days previously by Mr. Silcock.

Lastly, Mr. J. H. Gurney, jun., recently recognised a young bird of this species in the shop of Mr. Lowne, of Yarmouth, which was killed on Breydon in 1888, and is now in the collection of Mr. W. W. Spelman, of Brundall, who purchased it of Mr. Lowne. Mr. Gurney's determination of the species was confirmed by Mr. Howard Saunders. I know of no other occurrence of this species in Norfolk, but Mr. J. A. Cole assures me that on the 30th May, 1886, while moored fishing at Horning, a white-winged black tern came leisurely beating up the river and passed close to his boat; he saw it very distinctly both in approaching and going away, and is quite certain that he could not be mistaken.

Whether or not this species has visited us of late years more frequently than formerly is of course uncertain, but it will be noticed that all the recorded specimens, with one exception, have been met with in the early summer, and are in full, or nearly full, breeding plumage, a state in which they could hardly be overlooked; but on the return autumn migration, when immature birds would probably occur, and in a stage of plumage not easily to be distinguished from the black tern, it has been recognised only in the single instance given above. The frequent occurrence in the British Islands of these birds in early summer is a curious fact for which ornithologists have hitherto been unable to account; for it may be confidently asserted that this species has no breeding-place further north which it may have been seeking, and it would really seem as though there had been attempts to establish itself in this county—attempts which might have succeeded but for the cruel reception they have always met with. Their wanton destruction therefore under such circumstances cannot be too strongly denounced by all true ornithologists.

LARUS SABINII, J. Sabine.

SABINE'S GULL.

October, 1881, will long be remembered in Norfolk for the severity of a storm which occurred on the 14th of that month. It was purely cyclonic, beginning in the south, and passing successively to south-west, west, and north-west, and was accompanied by a deluge of rain such as has been rarely experienced in this county; the seven following days were calmer but very cold, the mean, 43° 3′, being about 6° below the average temperature. One result of this exceptional weather was the arrival on our coast of several rare birds; among others, Buffon's and Richardson's skuas and little gulls, but the 17th and 22nd were rendered notable by the appearance on Breydon of two immature examples of Sabine's gull, the first known to have occurred in this county. One of these northern strangers was sent to Mr. Stevenson in the flesh, and proved upon dissection to be a female; the other, a male, was purchased by Mr. Connop, of Caister, in whose collection it now is; Mr. Stevenson's bird subsequently found a home in the Norwich Museum.

In a paper by Mr. Stevenson, published in the "Transactions of the Norfolk and Norwich Naturalists' Society" (vol. iii., p. 373), giving full particulars of this event, occurs the following passage, which is worth quoting, as his prediction has to some extent already been fulfilled. He says, "It seems remarkable that this arctic species, which has been procured several times in Great Britain, should not, hitherto, have been recognised on some portion of our extensive sea-board; but, probably, as I have remarked in more than one instance,* of late years, in the case of additions to our list of rare and accidental migrants—the wanderers of 1881 may prove but the pioneers of others of their race;" and thus it happened, on the 6th October, 1889, under similar conditions of wind and weather, and

---

* "The white-winged black tern for example."

accompanied by examples of *Larus minutus*, an immature male of this species was killed at Palling, and came into the possession of Mr. Lowne, of Yarmouth, by whom it was preserved for Mr. W. W. Spelman. This Palling bird, which was consorting with lapwings when it was shot, and is in a less mature stage of plumage than Mr. Stevenson's specimen before referred to, is probably a bird of the first year.

LARUS MINUTUS, Pallas.

LITTLE GULL.

The first mention of the little gull as a Norfolk bird is, I believe, in Messrs. Pagets' list (1834), where they briefly remark that it is "rarely met with." As the species was originally described and figured as British by Montagu* in the Appendix to his "Ornithological Dictionary," in 1813, of course little could be expected to be known of it on the Norfolk coast at so early a period, but it seems probable from more recent experience that Mr. Lubbock's remarks in 1845, that it "has been found several times within the last three or four years," and the still more explicit statement of Messrs. Gurney and Fisher that "a few specimens are generally obtained every year in their vernal and autumnal migrations,"—which has been fully substantiated of late,—would apply with equal force to the earlier date, unless, indeed, this species has deviated from the line of migration formerly pursued by it. Certainly, at the present time, the little gull must be regarded as a regular migrant, more particularly in late autumn, and sometimes occurring in rather considerable numbers.

---

* The specimen examined by Montagu was shot on the Thames near Chelsea, and belonged to Mr. Plasted, of that place. From him it passed through Leadbeater to the late Mr. Lombe, and with the rest of the Lombe collection it is now in the Norwich Museum. Its plumage shows, as indeed does the figure given by Montagu, that it was a young bird in its first autumn.

Probably the earliest Norfolk specimens were a pair in Mr. E. S. Preston's collection, said to have been killed on Breydon, in December, 1829, the same severe winter which produced the Steller's duck. One in Mr. Newcome's collection, which was purchased at the dispersal of Mr. Stephen Miller's birds in 1853, must have been procured several years before that date; Mr. Gurney has also an example, which was killed on a pond near the New Mills, within the city of Norwich, in November, 1844. Dr. Babington mentions a male bird which was killed at Yarmouth, on the 18th January, 1850; Mr. Dowell saw a young one in company with terns and common gulls, at Blakeney, on the 5th August, 1851. It was met with in Norfolk again in 1861, 1867, and 1869, and, from the latter year till the present time (1889), I have notes of its occurrence in varying numbers every year, with the exception of 1874, 1876, 1877, 1878, and 1885. Out of twenty-nine recorded instances of examples of this species occurring in the county of Norfolk (several individuals having occasionally been met with at one time) there are only two in August; one in September; but thirteen in October, and seven in November: again only two in December; one in January; two in February; and one in April; thus showing that it is much more frequent on its autumnal migration, after the breeding season, than at any other time. Mr. Dowell speaks of his August bird as being a young one and very tame, allowing him to pass within a few yards of it after the commoner gulls had taken flight; the other August bird, shot on the 25th, described as a very young one, will be found recorded in the "Zoologist" for 1873, p. 3716. The April example, a beautiful adult male, was shot in 1888 out of a flock of five on Hickling Broad, on the 27th of that month; another was wounded at the same time, but not followed, and the bird killed, which is now in Mr. Connop's collection, narrowly escaped being food for the ferrets. Nothing could exceed the beauty of the pure tints and lovely roseate breast of this exquisite little gull. This is the only Norfolk example in breeding plumage with which I am acquainted.

A remarkable influx of these birds occurred on the

east coast in February, 1870; in Norfolk alone at least sixty individuals are known to have been killed, of which the larger portion were shot between the 12th and 14th, and probably there were many others which were not recorded. Mr. Stevenson gave an interesting account of this event in the "Transactions of the Norfolk and Norwich Naturalists' Society" for that year, pp. 65-70. A remarkable feature in this visitation was the large proportion of adults; out of thirty-five examined by Mr. Stevenson twenty-nine were fully adult, and only six immature, and the sex of those examined showed the females to present "about the same proportion in numbers to the males, as the young to the old."

LARUS RIDIBUNDUS, Linnæus.

BLACK-HEADED GULL.

In treating of this species, which has suffered so severely in Norfolk from the combined effects of drainage and inclosure, I think, with the example of the black tern before us, which has been allowed to pass away as a breeding species almost unheeded, it is most desirable to place on record, so far as possible, before it is quite too late to do so, such facts as are still known with regard to its former breeding-places, and I hope I shall be forgiven if I enter into this branch of the subject as fully as the scanty material at my command will permit.

At a time when the fens were undrained, and vast tracts of heath extended for many miles in unbroken solitude in various parts of the county, there are known to have been several breeding-places of these birds which have entirely disappeared, and probably there were others, of the former existence of which we have now no knowledge. To Sir Thomas Browne, whose letters form so rich a source of information as to the past, we are as usual indebted, but I am not aware that there is any published reference to be found to some of the breeding-places of which I shall have to speak.

To commence with the south-west portion of the county.* Many years ago the black-headed gull nested at Stanford on Lord Walsingham's estate. In Salmon's diary are frequent mentions of this gullery; thus, on the 9th March, 1834, he writes—

"A very fine day. Went to Stanford warren to see if the black-headed gulls *(Larus ridibundus)* had arrived at their usual place of nidification; found about twelve pairs there. They do not appear as if they had been long there, as they were not disposed to alight in the pond. They kept wheeling about at a considerable height above the pond, and when I left they went away. I should hardly think they are settled yet, only reconnoitring their future place, which is by no means a very secure place, the pond being very small and close by the road. From the appearance of the old nests I should think a great many resort hither—the nests are placed upon the tops of small hassocks which abound all round the pond, standing just above the water. I understand these birds had made their appearance at Scoulton mere about ten days ago, and that they were tolerably nume-

---

* Professor Newton has very kindly furnished me with the following notes on an extinct gullery, which, although not actually in Norfolk, was so near to the county boundary that it ought not to be omitted here. "The extinct Brandon gullery was on a small mere perhaps half-a-mile from the Brandon and Mildenhall road, and so close to the Wangford boundary that in one place the Wangford warren-bank may be said to have touched the water—indeed, in a wet season, I have seen the water come through on the Wangford side. On the 9th April, 1853, Gathercole, who had been warrener on Wangford for twenty-two years, told my brother and myself that the 'coddy moddies' left off breeding there several years ago; and the tenant of the warren, Mr. Plaice, on the same day told us that 'Mr. Bliss destroyed the 'coddy moddies' by taking their eggs too close.' Mr. Bliss was the former owner of that part of Brandon which adjoined Wangford and Elveden, and had planted, chiefly with Scotch firs, many hundreds of acres on his property, and among them the ground surrounding the mere. The trees may have at the time of which I speak been about twenty or twenty-five years old. I do not think they had anything to do with the disappearance of the gulls, which I have heard others, beside Plaice, attribute to the same cause that he did. The mere, as I remember it, was not of any particular interest, though we once found a redshank which had its brood, just where the Wangford boundary abutted on it."

rous there on the 7th; still the main body of them had not then arrived." In another place Mr. Salmon says that the black-headed gulls, when at Stanford, visited Fowlmere, on Wretham heath, about two or three miles distant, early every morning and late in the evening during their stay at Stanford, but that they had never been known to breed nearer the latter mere than Stanford (as to this see p. 236). On June 15th, 1834, Mr. Salmon writes, "I walked to Stanford warren this afternoon to see the gulls. There are some thousands of birds; several of the nests had young ones, none exceeding three. I should say they never exceed three eggs in a nest. . . . The men appointed to take the eggs still gather them for sale; they make threepence per dozen of them. I understand they will be gathered all this month."

On March 19th, 1835, he says the gulls had arrived at the warren four days ago, and on a subsequent visit, on the 10th May, he found they had been laying rather more than a fortnight. "I find the person who collects their eggs leaves a few of them to hatch their first eggs, and when that is the case the same pair of birds does not breed any more that season. . . . The old bird commences sitting as soon as the first egg is laid, consequently they place stones, rotten eggs, &c., for them to lay to as nest eggs till the complement is laid." July 12th: "Mr. Chambers informs me that the black-headed gulls have left the pond at Stanford warren more than a week."

March 18th, 1836. "The black-headed gulls have arrived at Stanford warren during the past week, but not to remain for the night."

May 3rd, 1837. "In the evening rode over to Stanford warren. The gulls appear very unsettled, although tolerably numerous. A few pairs have commenced laying. The eggs are not to be gathered this season, at least only a few." May 27th: "Gulls not very numerous at the pit; saw a few young ones." Mr. Salmon left Thetford in 1837, and this is the last mention in his diary of the Stanford gulls. The tenacity with which the birds seem to have clung to their old nesting-place is remarkable, but to account for their eventual disappearance we have only to bear in mind the treatment they received. In 1835, although the gulls did not arrive till about the 15th March, and commenced laying about the 6th April (the practice being to take their eggs till the end of June), on the 12th July they had all left the warren "more than a week!" No wonder that in 1837, on the 3rd May, only a few pairs had commenced laying, and the birds appeared very unsettled; and although only a few eggs were allowed to be taken that year, it is probable Mr. Salmon remained at Thetford long enough to witness the practical destruction of the colony. Of the subsequent history of the colony I know nothing till the notes of its final extinction supplied me by Lord Walsingham, which will follow shortly.

Professor Newton informs me that Spinks, a warrener at Elveden, told him in 1853 that he formerly (before 1829) lived at Stanford, and that near the lodge, on the warren, he might have got a tumbrel load of "coddy moddies'" eggs. From Lord Walsingham I learn that the black-headed gulls have been once or twice re-attracted to Stanford by putting down a few eggs obtained from Scoulton, and he thinks this could be done again. A few years ago some twenty pairs established themselves spontaneously at Stanford, and bred for about three years. Lord Walsingham believes they were driven away by otters, and his attempts to bring them back by putting down eggs were only successful for one year.

It was probably to one of these latter temporary visits to its old nesting-place that Mr. Stevenson refers when he says in his diary that on a visit to Thetford warren, on 16th April, 1870, Smith, the warrener, told him that "the black-headed gull still nests at Stanford."

Formerly, when Lord Walsingham was very young, he remembers the black-headed gull breeding at Tomston regularly, but in small numbers; after they had ceased to do so, by putting eggs from Scoulton in old coots' nests, he induced a few pairs to return and breed, but after a year or two they left the place again. Another small colony frequently bred at Bagmoor, a small pond on the Thetford and Stanford road; there were from five to twenty pairs or rather more. This pond is now often dry,* like many others in the same neighbourhood, which at that time held large pike, perch, and tench, and for some years the gulls have forsaken it, and the Stanford colony of this bird has entirely ceased to exist.

In the spring of the year 1853, after the extensive flood of the previous winter, a considerable flock of black-headed gulls remained at Feltwell, some of which attempted to breed; also on Wretham heath, near Ringmere, but they seem to have been seriously inconvenienced by the rapid subsidence of the water at the latter place. Of the former of these localities Professor Newton has kindly furnished me with the following notes:—

"I find from our 'Register' that on 5th May, 1853, I saw a flock of about 200 black-headed gulls on Hockwold Fen, and a nest of the same, which contained an egg. The next day I saw a flock of about 100 on the same fen; but, to the best of my recollection, not exactly in the same part of it. Our 'nidification list' shows that a nest with two eggs was taken 15th May, 1853, on Feltwell Fen, but I am unable to say whether I saw them or what became of them. On 21st May, a flock of about 100 was seen, in Hockwold, by my brother and myself. On 3rd June, 1853, we again saw a flock of about 200, and found some nests. These were all empty; the water was then fast receding, and the nests which had been built on the margin of the flood were now a long way off it. On the 10th June, 1853, my brother and I found a flock of about fifty at Ringmere."

Again, in the spring of the year 1883, when the meres on Wretham Heath had spread beyond their usual boundaries, some forty or fifty pairs of these birds nested in the heather at the south-west corner of Lang-

---

* Lord Walsingham has favoured me with the following note with regard to these ponds:—"The first time that the so-called 'Bagmoor pit' was known to be dry, about 1858, we took from the last remaining muddy pool a tench weighing over 6 lbs., another over 4 lbs., pike up to 8 lbs., and perch up to $2\frac{1}{2}$ lbs. At another, called 'Westmere,' I have seen pike of 24 lbs. and 22 lbs. taken with a line, and have since shot partridges on the same spot. I think the level of the surface water throughout this district has very much lowered of late years."

mere, where their nests were seen by Mr. Norgate, but they did not return to breed there in 1884, nor have they done so since.

Near the centre of the county is the ancient and celebrated Scoulton gullery. This colony has been so often described, and Mr. Stevenson, as before stated, has left such an excellent account of it in the "Trans. of the Norfolk and Norwich Naturalists Society," vol. i., 1871-72, pp. 22-30, that I must refer the reader to his description. Here the birds in the present day at least meet with considerate treatment, and their laying powers are not unduly taxed, while the nature of their breeding-place, a boggy island of some forty acres in extent, surrounded by a lake of about thirty acres, the whole girt by a sheltering belt of plantation, admits of their being effectually protected from molestation. The "pewits," as they used always to be called, arrive about the 7th or 9th March; the first eggs are generally produced about the middle of March, and none are taken after a certain date, which varies according as the season is early or late. In 1889, no eggs were taken after the 24th May; towards the end of July the gulls take their departure; and on the 1st of August, as a rule, the mere has regained its wonted quiet. About nine or ten thousand eggs are now taken, the number of birds for the last few years has been very equal, and their chief enemies appear to be the rats which destroy the eggs, and the pike which prey upon the young. The pike might be removed from the lake to great advantage, and it is doubtless owing to their incautious introduction in 1864 that the various species of duck, and especially the little grebes which formerly bred there, have forsaken the locality. Part of the "hearth," as the island in the lake is called, is too boggy to allow of the men gathering the eggs, and affords a safe nesting-place for a portion of the gulls as it did for the wild fowl which formerly nested there.

The view of the "hearth" and gulls at Scoulton, which forms the frontispiece of the present volume, was taken by Mr. Joseph Wolf on the occasion of a visit paid by him to the gullery, on the 26th June, 1872, in company with Mr. Stevenson, Prof. Newton, and the late

Mr. E. C. Newcome; and all who know the place will recognise the faithful likeness which that masterly hand has portrayed of the singularly beautiful scene presented, one that can be paralleled in but few other spots in this country.

Professor Newton who, as just stated, was of the party when this drawing was made, has sent me the following notes upon it:—" Mr. Wolf was careful to represent the gullery in a state of disturbance, for disturbed it must be to some extent that all its characteristic features may be brought out. Nearest to the spectator's eye are the downy young, that have not left the nest many days, seeking shelter among the water-lilies, while their anxious parents strive to cover their retreat by continually alighting for a moment on the surface of the water. In the far distance is shown the main body of the birds as they rise from the mere on any one walking along its bank, thick as snowflakes, and as white, when seen against the leafy screen of the surrounding trees. In the middle distance is the hearth, with its rough herbage, that hides half the birds which have settled upon it, while others, perching awkwardly on the stakes driven into its margin, or even on the sallow-bushes that grow here and there, exhibit themselves prominently. Above all hover a few of the more daring or more curious inhabitants, within easy gunshot length of the intruder, and against the bright sky their pure plumage shows dark by contrast. In one respect Mr. Wolf failed, as every painter must fail. It is beyond the power of the pencil to give an idea of the tumultuous noise by which the scene is accompanied."

Many years ago a gullery, probably by far the most extensive in the county, existed at Horsey, on the east coast. The late Mr. Rising pointed out to me the former site of this numerous colony—once a vast tract of wet marshes, now converted into fine grazing meadows —and told me that the breeding-ground extended over 300 acres; he related some very amusing anecdotes of his father's encounters with poachers, and of the rough and ready justice meted out to such robbers in those days; the number of eggs, he said, was enormous. It is of this breeding-place that Sir Thomas Browne says, the "Larus alba, or puets, in such plentie about Horsey, that they sometimes bring them in carts to Norwich, & sell them at small rates, & the country people make vse of their egges in puddings & otherwise." Although this was probably one of the largest gulleries in the kingdom, I am not aware there is any other mention of it in the early authorities, and it also seems curious that, notwithstanding the fact that the young of the black-headed gulls were very generally eaten where they could be procured, no mention of these birds appears to be contained in the le Strange Household book unless, indeed, they be intended by the term "whyt plouer."

Early in the present century, when the late Mr. Rising's father purchased the Horsey estate, it was so isolated by marsh as to be almost unapproachable, and probably had very little communication with the outer world during the winter and spring months; hence, although Sir T. Browne says that "great plentie of these birds have bred about Scoulton Mere, and from thence sent to London;" the young birds and eggs from Horsey do not seem to have travelled further than Norwich.

The gulls continued to breed at Horsey till banished by the drainage and inclosure of the land, the precise date of which I have not been able to ascertain, but I have little doubt the group of smaller gulleries which I shall have to mention now, are tenanted by the descendants of the Horsey birds, small parties from which locality dispersed in various directions and settled in suitable spots as near their old home as possible.

Mr. Lubbock mentions that a small colony took possession "formerly" of the margin of Rollesby broad, and, as his "Fauna" was published in 1845, it is not at all unlikely that these birds may have come there direct from Horsey. In June, 1854, I paid a visit to this breeding-place, and was told that they first made their appearance there "many years ago," but had been so persecuted that they gradually deserted the place; however, about the year 1848 or 1849, some five or six pairs re-appeared, and having been protected and allowed undisturbed possession of the spot, had rapidly increased, and at the time of my visit formed a very flourishing colony. These birds, however, did not long remain undisturbed, for, on the erection, I think in 1855, of the pumping-station of the Yarmouth waterworks, which obtain their supply from this spot, they were again banished. In the meantime the neighbourhood of Hoveton had frequently been visited by parties of these birds, and the late Mr. Blofeld told me that even as early as the year 1817 he used frequently to see eggs of the black-headed gull in the cottages. This was before the inclosure of the common, and of course they received no protection; but it was not till 1854 that the Hoveton gullery was really established, and in that year about thirty nests were hatched off in the sedge-fen. The birds, Mr. Blofeld told me, made their appearance rather late in the season as if they had been disturbed in some other locality, but, finding a safe home, increased year by year. In 1858 Mr. Blofeld took seven hundred eggs; in 1864 two thousand; and about the same number annually for many years subsequent to that date. No doubt at least as many might be taken now, but I am informed that, with a view to encouraging the colony, comparatively few eggs have been gathered of late years. It will thus be seen that the establishment of the Hoveton gullery very nearly coincides with the banishment of the birds from Rollesby; but the Hoveton gulls were not to enjoy peaceable possession of their new home; the rats became an ever increasing pest, and gradually the birds forsook the sedge-fen, where they laid their eggs on the ridges left by the turf-cutters, and betook themselves to the adjoining Little Broad, about a mile and a-half from their original haunt, till at the present time there are none of these birds frequenting the sedge-fen, but all build their nests on the broken-down reeds and bulrushes in the Little Broad, a situation in all ways suited to their requirements, and perfectly safe from all unauthorised intruders, whether men or rats. Not so, however, if the poor birds visit the river, which is all too close, for there the British tourist swarms, and Mr. T. C. Blofeld, who with the estate has inherited all his father's love for the exquisite birds which lend so great a charm to its waters, tells me that he finds many birds dead on and about their nests, which have evidently been shot by the "sportsmen" on the river. There is the satisfaction of knowing that so far as it is possible to protect these interesting birds, both here and at Scoulton, they are perfectly safe from undue molestation, and the result is that at Hoveton they are increasing rapidly in number. Occasionally small parties will wander away from the main body and set up housekeeping on their own account, as at Barton, Hickling, Woodbastwick, Somerton, and some other localities, but none of these offsets have been permanent. Mr. Frank Norgate informs me that in June, 1877, a man told him black-headed gulls used to nest at Brancaster "years ago."

Colonel Feilden tells me that one of the creeks in Wells marshes, leading into the main harbour creek, still retains the name of "Mow creek," as it was formerly the resort of black-headed gulls in the breeding season. Frederick Barrett, the well known shore-shooter, whose experience as a gunner on the Wells coast extends over forty years, does not remember the gulls nesting at Mow creek; but Barrett's father, also a gunner, did so perfectly, and often told his son that, standing on Wells quay, a man with a spy-glass could see the birds sitting on their nests. The date of the gulls leaving Wells marshes as a breeding-place would be at least fifty years ago, but in the spring of each year they return to Mow creek, and frequent it for several days, though they never stay or attempt to make a nest there.

I fear I have not space to add much with regard to the nesting habits of these birds, but so much has been already said by others on this subject that I preferred devoting all my available space to a history of their various breeding-places past and present.

At Hoveton, as I have said, the nests are placed on the broken-down vegetation, and are chiefly constructed of dead reeds and "gladdons,"* lined with the leaves of

---

* In the broad district *Typha angustifolia* is known as the "gladdon," and *T. latifolia* as the "she gladdon." *Scirpus lacustris* is here called "boulder."

the reed or rotting pieces of "boulder," but when the time of hatching approaches the old birds considerably enlarge the base of the structure, adding dead stems of gladdon till quite a floating raft is formed; this, as in the case of the great-crested grebe (p. 240), is evidently intended as a resting-place for the young, which otherwise, the nests being only slightly above and quite surrounded by the water, would have no dry surface to rest upon.

It is very rare to observe any signs of immaturity about the breeding gulls. Occasionally one is seen with a barred tail, the last mark of youth, but this is decidedly exceptional; on the other hand, the birds on the coast are most frequently in immature plumage. The appetite of the black-headed gull seems to recognise very little distinction of food; it has been seen hawking for moths on a summer's evening, flying over the grass like a swallow. Cockchafers, wireworms, aquatic insects, frogs, mice, and even small birds in the nesting-season—probably captured on their nests—have been known to minister to its necessities. It may be seen miles away, following the plough, doing inestimable service to the farmer, and, strangest food of all, in the year 1868, when great numbers of the young perished from want of food, owing to the long continued drought, Mr. Stevenson tells us "the dead fed the living, since the maggots from the bodies of the dead nestlings formed a scanty provision for those hatched later in the year."

There is reason to believe that the bulk of the black-headed gulls which breed here pass south in the autumn. Mr. Stevenson says he has frequently distinguished the harsh notes of this species, apparently in large numbers, forming part of the flocks of migrants passing over this city on dark autumnal nights, but the place of these wanderers is filled by others, probably arriving from more northerly localities, for this species is one of those most commonly met with on our coast during the whole year, even during the breeding season considerable numbers of immature birds frequent the shore, as it rarely nests till it has assumed its full plumage, which requires two or perhaps three years to acquire. In

March it usually assumes the black head, but numerous instances are on record of its appearing in the nuptial plumage much earlier. On the 26th December, 1878, Mr. Norgate tells me he picked up the remains of a gull with a black head, barred tail, mottled wing coverts, red bill, and brown legs; he has also met with specimens more than once which had assumed the black head in December or January; and Mr. Stevenson records in the "Zoologist" for 1871, p. 2599, the occurrence of one of these birds at Beeston, in the very severe January of that year, which exhibited an unusually early state of change from winter to summer plumage.

With us this species is variously known as the "cob," "Scoulton gull," "puit," "puit gull," or "coddy moddy."

LARUS MELANOCEPHALUS, Natterer.
ADRIATIC BLACK-HEADED GULL.

In the "Zoologist" for 1887, p. 69, Mr. George Smith, of Yarmouth, records an example of the Adriatic or Mediterranean black-headed gull, which was killed on Breydon, on the 26th December, 1886, and came into his possession. I am indebted to him for the opportunity of examining the bird in the flesh. It was fully adult, and of course in winter plumage, and proved, on dissection, to be a male. Like so many other rarities, its capture, Mr. Smith tells me, was quite accidental, as the man who shot it was simply waiting for some bird at which to discharge a cartridge he could not extract, and the first to present itself was this rarity. It was exhibited at a meeting of the Zoological Society by Mr. Howard Saunders on 18th January, 1887.

The only other example of this species recorded as having been killed in the British Islands was one obtained by Mr. Whitely, of Woolwich, who, without recognizing it, sold it to the British Museum. This example was a bird of the preceding year, and is said to have been shot in January, 1866, near Barking creek ("Ibis," 1872, p. 79).

LARUS TRIDACTYLUS, Linnæus.

KITTIWAKE.

The following observations on this species, which subsequent experience has only tended to confirm, were written by Mr. Stevenson, probably about the year 1861 :—

" This very elegant species is described by the Messrs. Paget as 'rather rare' at Yarmouth, and recent observations both in that neighbourhood and on other parts of the coast, have convinced me that at no season can it be termed common. Some, no doubt, occur every autumn and winter among the large flocks of small gulls which frequent our tidal estuaries, but in comparison with the common and black-headed gulls, very few specimens indeed find their way into the hands of our birdstuffers, and these, according to my own notes, are obtained in spring and summer. I have known an adult bird shot near the railway bridge, at Thorpe, during the first week in June, and, with the aid of a glass, I have recognised both old and young off Cromer as late as the end of that month, the immature birds still carrying the black bar at the end of the tail."

Unlike the other species of gull, the kittiwake is more frequently met with in the adult than in the immature plumage, in which state it is known as the "tarrock." Mr. Dowell considers it rare at Blakeney, and accounts for the scarcity from its habit of keeping more at sea than most other gulls. Mr. Patterson believes they are not so numerous at Yarmouth as formerly, but that observation, I think, applies to all the species of gull in consequence of the different arrangements for landing the fish (see p. 337). He states that they may easily be recognised by their tumbling, erratic, yet easy flight. Although generally most frequent in spring and autumn, this species is occasionally met with at all seasons of the year, but never, as Mr. Stevenson remarks, in large numbers, and less often inland than the common gull.

In the end of February, 1890, many young kittiwakes and old and young razorbills were found dead along the shore just previous to the severe weather accompanied by easterly winds and heavy snow storms which prevailed from the 28th February to the 3rd March.

LARUS CANUS, Linnæus.
COMMON GULL.

"This species," says Mr. Stevenson, "like some of the larger gulls, is seen on our coast throughout the year; a few stragglers, both immature and adult, which from some cause have not repaired to their northerly nesting-places, appearing throughout the breeding season.

"In spring and autumn, and more particularly at the latter season, they are very abundant, and at times in winter during severe weather immense flocks of small gulls, consisting of the black-headed and common gulls, frequent the Breydon muds and other similar localities, which from the alternate action of the tides afford an inexhaustible supply of food. Here they may be seen at low water dispersing themselves in groups over the sandy bars, mingling freely with, though in strange contrast to, their sable companions, rooks, carrion crows, and jackdaws.

"I know no prettier sight on a clear sunny day than the appearance of these gulls thus dotted like white specks as far as the eye can reach, some wading or pitching into the shallow waters, others pacing up and down or fluttering onwards to fresh grounds, their white breasts glistening in the sun and giving to the whole scene an amount of animation that cannot fail to strike even the least observant. Next to the black-headed species this is perhaps more of a land gull than any other, and in the early spring follows the course of the rivers far inland to frequent the freshly turned soil. In February and March I have seen large flocks upon the high ground about Arminghall and Markshall, near Norwich, following the plough like rooks in search of worms and grubs."

The common gull is by no means confined to the sea-shore, but after a flood visits the drowned lands in

Marshland, in company with the black-headed gull, feeding on the worms and other dead animal matter left by the subsiding waters. It is almost invariably in flocks more or less numerous, and its graceful flight and pure tints render it one of the most beautiful of the smaller gulls.

LARUS LEUCOPTERUS, Faber.

ICELAND GULL.

An exceedingly rare bird on the Norfolk coast; it seems probable that most of the specimens said to have been met with were small and pale-coloured examples of the glaucous gull, a much more common species.

Mr. J. H. Gurney, jun., in his list of Norfolk birds contributed to Mason's history of the county, says, "Mr. Stevenson and I, after numerous enquiries, can only certify one undoubted specimen, viz., a young female, in the possession of Mr. G. Smith, of Yarmouth, which was shot at Caister, near Yarmouth, in November, 1874." This specimen was examined by Mr Howard Saunders, who confirmed its identity. All the other recorded specimens have proved on examination to belong to the glaucous gull in one or other stage of plumage.

The great variation in size to which this "lesser white-winged gull" is subject renders its identification a matter of some difficulty; hence, probably, the frequency with which the larger species (the glaucous gull) has been mistaken for it.

Colonel Irby ("Key List of British Birds") gives the length of this species as 22 inches, and that of the glaucous gull 26 to 33 inches, but states that the wings of the smaller species are relatively much longer than those of the glaucous gull, being nearly as long as those of the last named. Schlegel (in "Mus. Pays-Bas, Lari," pp. 4-5) gives the length of the wing in *Larus glaucus* as from 16 in. 3 lines to 17 in. 3 lines as against from 14 in. to 15 inches 2 lines in *L. leucopterus*.

LARUS FUSCUS, Linnæus.

LESSER BLACK-BACKED GULL.

The remaining species of gulls are especially marine in their habits, haunting the sea and its shore, save under exceptional circumstances; their presence inland being due either to abundance of food or stress of weather. "With strong winds blowing from south, round by west to north-west," Mr Patterson tells me, in the neighbourhood of Yarmouth, they "do not, as a rule, get many gulls on the coast; but, with a heavy wind blowing from north-west to north, a host of gulls of various species may be looked for working laboriously along shore, and a heavy depression, with rain, often induces the gulls to fly inland. An easterly gale seems to upset their cautious self-preservative ways, and they line the coast, especially during the latter part of the year, when, sailing along in graceful flight, they speckle the dark trough of the curling wave, or mount above its white crest just as one imagines the breaking surf is about to swoop down upon and tumble them, wet, bedraggled masses of crumpled feathers, on the weed-lined strand. 'Grey,' black-backs, common, and herring gulls form the bulk of these contingents, the first named being the most numerous by far.* During south-east winds considerable numbers of gulls hang round about the harbour's mouth, and in late autumn, say October, the little gull may very likely be met with, being not so rare as is generally supposed. During the herring season, of course a quantity of offal drifts out of the harbour on the ebb, and collects round about the piers at the entrance to the river, or floats outwards to sea; this draws together at almost any time a number of hungry *Laridæ*. But now that fish offal has a market value for manure, and the facilities for trucking at the fish wharf have rendered it worth preserving, the quantity of fragments falling to the gulls has materially decreased; these, when the boats used to land their 'prime'

---

* The immature greater and lesser black-backed and herring gulls are alike known as "grey gulls."

herrings on the beach, were allowed to float away, and all their refuse was thrown overboard as the easiest way of getting rid of it. In consequence, as I have heard the old beachmen say, gulls at such time swarmed, and no doubt many a rarity was overlooked—there were not so many 'educated' slaughterers as now. Also there are not the same numbers of fish inshore now as then; mackerel, gurnards, and haddocks have almost forsaken us. With heavy north-west and northerly gales we get our rarer gulls, blown, I take it, in this direction out of their usual course by stress of weather."

In confinement nothing in the shape of food comes amiss to this and the next two species of gull; a mouse dead or alive is a favourite morsel; a rat even, after being shaken in water, is swallowed whole, and on one occasion I found the sole survivor of a brood of young French partridges, which disappeared in a mysterious way, on the point of being swallowed by a tame lesser black-backed gull.

Owing, probably, to want of acquaintance with the various stages of plumage which the larger gulls assume ere they reach maturity, the early authorities appear to have regarded this species as much less frequent than it really is. Messrs. Sheppard and Whitear simply catalogue the *Laridæ*, giving no information except with regard to the black-headed gull and the skuas; Hunt, in his "List" appears to regard the lesser black-backed gull as a great rarity, so much so that he remarks, "a specimen of this rare gull was shot near Yarmouth;" the Pagets also evidently regarded it in the same light, for they designate it as "rare," and add that there were "two shot April, 1821;" doubtless both these observations refer to the adult bird only. Mr. Lubbock, I think, errs in the other extreme when he states that this species and the herring gull "are perhaps the most common of the larger species here," but Messrs Gurney and Fisher form a much more accurate estimate, observing that it "occurs on the coast throughout the year, except during the nesting season; but not in large numbers." In the immature plumage it is certainly not rare, especially in autumn, but adults are decidedly uncommon. It is difficult, perhaps impossible, to distinguish between the young of this and the next species when on the wing; my observation must therefore, be taken to apply to the united forces of the two species. Which predominates it is difficult to say, but, judging from the frequency of the adults, *Larus argentatus* should be the more numerous.

LARUS ARGENTATUS, Gmelin.

HERRING GULL.

Much that I have said with regard to the preceding species will apply equally to this; adult birds are decidedly scarce at all seasons, but both old and young, I believe, are more frequently met with than is the case with the lesser black-backed gull.

Mr. Stevenson has the following notes on the regular passing and re-passing of this and other species of gull as observed at Cromer. "Frequent visits to Cromer during the summer months have enabled me to watch the habits of gulls at such times during many successive years, and I have always found a certain number of these birds, both of the larger and smaller species (adults of the smaller gulls always predominating), passing and re-passing that line of coast throughout the breeding season. In the early morning, or even later in the day, their course lies invariably towards the eastward. They fly low over the waves, in small detached parties, rarely stopping to feed, but hurry onward as though bound for some distant point. Before the sun sets they are as surely seen returning in larger companies, all of them taking the same course, which is now reversed, being directly westward or north-westward. The marked regularity of their habits in this respect (observable as well during autumn and winter) has often led me to enquire, Whence do they come and whither do they go? The larger species being represented for the most part by young birds of the previous year, I believe they have not yet sought the more northern or southern breeding stations of their respective kinds, but continue their wandering life for twelve months longer, and most probably when thus seen in transit to and fro, are making during the day for Yarmouth-roads, where the refuse from the fishing and other vessels supplies a large portion of their food,* and at night return to roost on some favourite sand-bar, on the flat shores of the Wash; or, it may be, even further still, on the Lincolnshire or Yorkshire coasts."† The two great feeding-places of the *Laridæ* on the Norfolk coast are the flats left bare by the retiring tide at Breydon, in the extreme east, and the shoals of the Wash in the extreme west. On Breydon "muds," and outside Yarmouth harbour, a constant and abundant supply of food is to be found; but from that point, north-west to Cromer, and thence west to Wells, no special attraction is to be found for these birds. Once the low shores at Wells are reached the succession of flats, mussel scaups, and sands left uncovered at every fall of the tide quite round the north-west corner of the county at Burnham, Brancaster, and Holme, to Hunstanton, Snettisham, and Wolferton, can hardly be surpassed as a feeding-ground for gulls and waders, and all along this line of coast they abound exceedingly. I know no prettier sight than to watch their arrival as the tide recedes. Each slight elevation as it shows above the retiring waters is taken possession of by a party of watchful gulls, and soon the low flat shore is dotted with flocks of these interesting birds busily searching amongst the tangle and shallow pools for the abundance of marine organisms to be found in such situations. Not only are these broad tracts the resort of gulls, but long strings of oyster-catchers may be seen winging their way in the same direction, terns are fishing over the shallows, godwits, redshanks, dunlins, and ring-

---

* Sir T. Browne, in his "Account of the Birds found in Norfolk," without attempting to discriminate the species, which indeed were little known in his day, says, "Many sorts of Lari, seamewes & cobs. The Larus maior in great abundance in herring-time, about Yarmouth."

† See also on this subject a paper by Mr. J. H. Gurney, jun., "On the periodical movements of gulls on the coast of Norfolk," "Trans. Norfolk and Norwich Nat. Soc.," iv., p. 326-333. Mr. Gurney demurs to Mr. Stevenson's statement as to the direction taken by the gulls.

dotterels are all busily occupied with the same object, and in the early autumn all are so fearless on their first arrival that with the aid of a glass their every action may be observed without difficulty. Alas, that their confidence should so often lead to their destruction! Some very interesting remarks on this subject from the pen of Mr. Stevenson will be found in the next article; and, having introduced the feeding-ground, I will leave him to describe one of its most distinguished frequenters.

LARUS MARINUS, Linnæus.

GREAT BLACK-BACKED GULL.

"This," says Mr. Stevenson, "is the largest and most abundant species on our coast, and though breeding nowhere in our neighbourhood it is seen in flocks throughout the year. During a recent visit to Hunstanton in May and June, the great black-backed gulls in their immature dress—for none of them had attained their full plumage—were decidedly the common gull of the district. Through the glass I have counted more than a hundred at one time resting in groups upon the sands, waiting for the receding tide to lay bare their feeding-grounds on the scalps. These groups exhibit every variety of immature plumage, from the young brown bird of the previous summer to the nearly adult specimen, with more grey than brown on the back and wings, and the head and under parts of the body pure white. It is a beautiful sight on a bright sunny day to watch unobserved a large flock of these noble birds, some lazily squatting on the dry sand, others dozing on one leg with the head thrown back upon the shoulders, and the beak buried in the feathers of the back. Here and there a more restless bird paces up and down, and with head erect and open bill gives forth at intervals its harsh barking note. Disturbed from their *siesta* by approaching footsteps, one by one they spread their wide pinions, and, running a few paces, launch themselves upon the wing, and then, after a short flight, evidently reluctant to abandon their haunts, they settle upon the sea, just out of gun-shot, and head to windward in extended lines, they rise and fall with the heaving billows like a little fleet at anchor. In winter, particularly in sharp weather, the old birds are plentiful on Breydon, and may be seen on any part of our coast line flying out at sea. I have also occasionally seen them at Blakeney, Sheringham, and other places even in the height of the breeding season. These, I imagine, like others of their class before referred to, as well as various *Tringæ*, &c., are, in spite of their adult state, unpaired stragglers, which from some cause have not sought with the majority of their tribes their usual breeding stations."

There is at present, I believe, no breeding station of this species on the east coast of England, and that at the Bass Rock has long been deserted. Yarrell, in the earlier editions, mentions the marshes on the shores of Kent and Essex, at the mouth of the Thames, as breeding-places, but it has certainly long ceased to nest there. It seems impossible, therefore, that the adult or nearly adult birds which are found on our coast during the summer months, notwithstanding their enormous powers of flight, can merely visit us from their nesting-places on foraging excursions as some species of sea-birds certainly do. I think, therefore, that Mr. Stevenson's explanation of their presence is doubtless the correct one.*

---

* I am informed by Mr. T. F. Buxton that a curious entry occurs in a MS. Natural History Journal kept by himself, Mr. J. H. Gurney, Mr. J. G. Barclay, and the late Mr. Charles Buxton, in the neighbourhood of Cromer. The entry is in the writing of Mr. Charles Buxton, under date of May, 1839, as follows:—"I am informed by James Pearson (a very intelligent man who was our gamekeeper at Weybourne) that . . . he never knew curlews to breed there, but the black-headed gull has, and one pair of the large black-backed gull." Mr. T. F. Buxton writes, "I can quite confirm what my brother says about J. Pearson being an intelligent informant. He was constantly on the Blakeney marshes, and knew the birds extremely well, and gave us a good deal of information as to the habits of the birds and their migrations." The date of the occurrences referred to would probably be between the years 1820 and 1835.

This species is occasionally seen on the larger broads; and on the 6th January, 1881, Mr. Frank Norgate shot an immature *Larus marinus* at Sparham, but its appearance inland is by no means so frequent as that of the lesser black-backed and herring gulls; and Mr. Rising believed that their presence on Horsey mere was always indicative of approaching severe weather.

Miss Anna Gurney, of Northrepps, in a paper contributed to the "Mag. Nat. Hist." for 1830 (iii., p. 155), on the "Natural History of the Neighbourhood of Cromer," writing of gulls, says, "one which we had young in the autumn of 1823, and which proved to be the large black-backed gull, did not acquire his final plumage till the summer of 1827; his bill turned from black to yellow, but the scarlet spot was not perfect till July, 1828. The next spring he died."

Mr. Arthur Patterson tells me that he estimates the relative proportionate numbers of the species of "grey gulls" on the Yarmouth coast as greater black-backed ten, lesser black-backed four, and herring gull one. My much less practical experience, although perhaps more extensive as to area, would have led me to reverse the proportions of the two latter species, but in the immature plumage it would be presumptuous to speak with any degree of confidence, and the relative numbers certainly vary according to season and on different parts of the coast.

LARUS GLAUCUS, Fabricius.

GLAUCOUS GULL.

This fine gull is not unfrequent off the Norfolk coast during the autumn and winter in immature plumage, but in the adult state it must be considered rare. Occasionally examples in full winter plumage have been met with in the offing in considerable numbers; but, as will be seen, the very large majority of those procured are young birds.

This species is not mentioned by either Sheppard and Whitear or the Pagets, but Hunt in his list records "a fine specimen killed at Yarmouth," and adds, " Mr. Norman, of Docking, has another, killed in his neighbourhood." Mr. Lubbock, in his copy of Bewick's birds, has a note written in July, 1831, of one shot off Yarmouth by a fisherman, but the date of the occurrence is not given. Mr. Dowell states that in November, 1847, during a gale from the north, he killed a "cream-coloured gull," which he sent to the late Mr. Yarrell, who pronounced it to be a glaucous gull, and adds, " these gulls visit us [at Blakeney] during the winter, and then only during gales from the north or north-east. At such times, however, several may be seen in a day; they are easily distinguished from the young black-backs by the primaries being light, and the whole bird of a cream colour at a little distance, whence their name here of 'cream-coloured gull;' being much tamer than other gulls they are much more easily procured. None of these birds appeared at Blakeney during the winter of 1848-9, but on November 23rd, 1849, I saw and shot an immature glaucous gull at Salthouse; and, in December of the same year, Overton sent me one out of which Ellis took a whole golden plover. In January, 1862, an immature specimen was killed at Sheringham." A glaucous gull, in the late Mr. Rising's collection, was killed on November 18th, 1847, the same year as Mr. Dowell's first specimen. A fine young male, in the Dennis collection, was killed at Yarmouth in 1848. In the " Zoologist" for 1850, p. 2778, Mr. Gurney has a note of the capture of four of these birds off Cromer, in the month of February of that year, two of which were adults. Mr. J. O. Harper records ("Naturalist," ii., p. 132) the occurrence of a fine adult male, at Yarmouth, on 29th November, 1851; and Mr. Stevenson mentions one killed at Yarmouth in February, 1859; and a second at the same place in January, 1862. Since that time I have many other records, but shall only mention the adults, which form a very small proportion of the whole. Late in October, 1880, a fully adult specimen was killed at Mundesley (not at Knapton, as recorded by Mr. Stevenson, "Zoologist," 1882, p. 375); Dr. Babington obtained a fine male from Mr. Lowne, of Yarmouth, shot off that place on the 16th January, 1881; and, on the 26th of the same month, Mr. George Smith, of Yarmouth, alone had twenty-seven glaucous gulls brought to him by the Yarmouth fishermen and gunners, seven of which were mature birds, nineteen immature, and one in change. It is impossible to say how many of these birds were killed at that time, or what proportion of them could be claimed as Norfolk-killed specimens. Probably most of those brought in by the fishermen were killed on or near the Dogger bank; but, from information collected by Mr. J. H. Gurney, jun., it appears that from forty-five to fifty were offered in the flesh at Yarmouth alone. The weather at the time was of unprecedented severity, a terrible gale blowing from the east-north-east, the anemometer on the 18th registering a velocity of 548 miles in the twenty-four hours, and on the 19th 410 miles, accompanied by a heavy fall of snow and intense frost, a storm which will long be remembered for its terrible severity and the fearful loss of life attendant upon the shipping disasters off our coast. On 19th October, 1887, Colonel Feilden found a dead glaucous gull, on Wells beach.

The carnivorous propensities of the gulls are too well known to need any illustration.* In November, 1847, a young glaucous gull (referred to above) was shot at Horsey in the act of preying upon a dead coot, which had just been shot as it rose out of a "reed-bush," and the gull, which with many others was flying overhead, immediately alighted, and was itself killed while standing upon it. This bird, with its intended prey, was purchased at the sale of Mr. Rising's collection by Mr. Gunn. From the stomach of a young glaucous gull, killed off Cromer on the 11th October, 1886, Mr.

---

* When the operation of flensing whales in the arctic seas is in progress, these birds, in large numbers, are constantly present, and the whalers avail themselves of their rapacious appetites to play off many practical jokes upon them. I am told the size of the pieces of blubber they will swallow is simply astonishing.

Pycraft took the remains of a small bird which he believed to be a lark.

Mr. Booth mentions having seen this species several times off Yarmouth, and once on Hickling Broad, in immature plumage, but only on one occasion did he ever meet with it in mature dress; this was on the 27th October, 1872, off the entrance to Yarmouth harbour; the example being in perfect adult plumage, but too wary to admit of his getting within range for a shot.*

STERCORARIUS† CATARRHACTES (Linnæus).

GREAT SKUA.

This species, which is decidedly rare on our coast, appears accidentally quite independent of weather, but its visits, few and far between, are almost invariably in the autumn months. Sir Thomas Browne, in his manuscript "Account of Birds found in Norfolk," says: "In hard winters I have also met with that large and strong-billed fowle wch Clusius describeth by the name of Skua Hoyeri, sent him from the Faro island by Hoierus, a physitian, one whereof was shot at Hickling while 2 thereof were feeding upon a dead horse."‡

Sir W. Hooker's MS., presumably referring to *S. catarrhactes*, records the occurrence of "four skua gulls" shot off Yarmouth, on October 7th, 1827. Hunt, in his "List," speaks of this species under the name "*Larus*

---

* A gull purchased by Mr. Gurney at the sale of the Miller collection, and long believed to be an IVORY GULL, proved, Mr. Gurney tells me, on careful examination to be a white-plumaged glaucous gull.

† The name *Lestris*, used by Yarrell, has to give place to the older term *Stercorarius*, bestowed by Brisson.

‡ There can be no doubt from the description given by Clusius, though his figure is very rude ("Exotica," p. 367), that Hoier's "Skua" was the bird which the Færoese still call by the same name, though in modern times spelt "Skúir" (See "Zoologist," 1872, p. 3290).

cataractes" as "extremely rare," and mentions that Mr. Norman, of Docking, had a specimen killed in that neighbourhood; Sheppard and Whitear do not include this species in their catalogue; Messrs. Paget merely refer to the four examples already mentioned, while Mr. Lubbock has no information with regard to this bird in particular; Hoy, in the "Magazine of Natural History," for 1837 (N. s., i., p. 117), records one on the coast near Yarmouth, in October, 1836. In the MS. notes in the Dennis copy of Yarrell, "Yarmouth, October 6th, 1849," is given as a date for the occurrence of this species (cf. Dr. Babington's "Birds of Suffolk.") Mr. Stevenson has notes of a pair at Yarmouth, about the 15th September, and one in the same locality, on the 2nd November, 1854; Sayer, of Norwich, had two young and three old birds, all from Yarmouth, on the 12th October, 1855. In the unusually mild season of 1857, two females and one fine male were shot on the 12th October, twelve miles off Yarmouth; Mr. Gould, in his "Birds of Great Britain," says, "On the 19th October, 1857, I saw in Leadenhall Market five great skuas, seven pomatorhine skuas (one adult and six young of the year), and one young arctic (Richardson's) skua, all of which were from Yarmouth, and it was believed had been killed near the lightship after a gale;" on the 21st September, 1858, I find from Mr. Stevenson's notes that four were obtained from Yarmouth; and on the 1st October two others from the same place. In the collection at Keswick is a beautiful dark female, said to have been killed at Yarmouth in the month of October, 1869, and purchased by the late Mr. John Gatcombe, in Leadenhall Market; and in October, 1870, Mr. J. H. Gurney, jun., saw another in the same market, also said to have been sent from Yarmouth; on the 15th July, 1872, a most unusually early date, Mr. Howard Saunders saw a great skua close inshore, at Cromer ("Zoologist," s. s., 3226); Mr. Stevenson records one, on the authority of Mr. George Smith, of Yarmouth, as having been shot off that town on the 3rd October, 1881; and in September, 1887, according to Mr. Pycraft, one of these birds was shot on Breydon. Mr. J. H. Gurney, jun., tells me that Mr. Thomas Berney, of Braconash, has two great skuas in his collection which were killed by himself at sea off Yarmouth, but no date is preserved; there is beside, according to Dr. Babington, a young bird from the same locality, in the Newcome collection, also undated.

I have given, I think, a nearly complete list of the recorded occurrences of this bird, some of them certainly "Leadenhall" specimens, but likely enough to be correct, in order to show its rarity in comparison with the next species. No doubt it is occasionally seen out at sea, but never in numbers like *S. pomatorhinus*, more than a pair together being rarely met with.

Like all the skuas this species sometimes presents an almost whole-coloured form of plumage, although, apparently, much more rarely than two of the three that follow. Mr. Dresser figures in his "Birds of Europe" a remarkable instance of this form from the specimen in the collection of Mr. J. H. Gurney, jun., already mentioned as killed off Yarmouth, in October, 1869.

STERCORARIUS POMATORHINUS (Temminck).

POMATORHINE SKUA.

Notwithstanding the remarkable influx on one particular occasion, presently to be mentioned, of a very large number of individuals of this species, it cannot be considered otherwise than as an occasional autumn visitant, young birds occurring most often. They are frequently seen by the fishermen, far out at sea, following the various species of gull which are attracted by the shoals of herrings and sprats, but only under stress of weather do they come inshore; and as Mr. Stevenson points out, "singularly enough, nearly all the storm-driven specimens picked up far inland have been more or less in adult plumage." It is possible they may pass our coast every autumn on their southern migration in considerable numbers, but far to the eastward and quite out of sight of land.

From various sources I have been able to compile the following list of early examples of the pomatorhine skua which are known to have been obtained in this county. The first in point of date would be the original of a large coloured drawing by Miss Anna Gurney, now at Northrepps, of an immature specimen, which was taken at that place in October, 1822; the bird itself was not preserved. In the "Lombe collection," at the Norwich Museum, is an immature specimen which was shot on October 27th, 1834, at Little Melton, near Norwich, more than twenty miles in a direct line from the sea. In his notes, under date of October 17th, 1848, the wind at the time blowing hard from the north-east, Mr. Dowell mentions seeing four pomatorhine skuas near Blakeney, three of which he shot; and on the 30th October of the same year an immature specimen, now in the Museum of the University of Cambridge, was shot at Elveden (on the Suffolk side of the border); Mr. Stevenson has also notes of two others killed at Yarmouth early in November, 1848, in which year this species appears to have been unusually abundant. Mr. Newcome has an adult pomatorhine skua in his collection, killed on Lakenheath fen (also a short distance into the county of Suffolk), in November, 1850; and an immature example, presented to the Norwich Museum by Professor Newton, in 1860, was procured in the same locality many years ago; an adult bird, also in Mr. Newcome's collection, was shot at Hockwold, in September, 1854. The year 1857 produced several of these birds in the month of October; and in the autumn of 1858, as I learn from Mr. Stevenson's note-book, "when the common and Richardson's skuas were unusually plentiful in the 'roads,' two or three specimens of the pomatorhine were also shot from the herring smacks off Yarmouth. The first of these, killed on the 24th October, was in rapid progress to maturity. A very fine adult male, in the possession of Captain Longe, caught in a rabbit net, on Winterton Common, and a young one in the same collection, taken off the coast, were obtained in November of the same year" (1858).

Others are recorded for 1865, 1867, 1868, 1869,

1870, and 1872. In the year 1874 very many examples of this species were met with; after a heavy gale on the 20th October numbers were brought into Yarmouth, and Mr. Frere informed Mr. Stevenson that one game-dealer in that town had thirty skuas at one time, probably nearly all pomatorhine. The larger number of these birds it will be observed by the above records have been obtained in October, others in November, and a few in September; it will also be seen that many of these were procured in localities far from the sea.

From 1874 till the year 1879, when occurred the remarkable influx of which I have now to speak, I find no mention of this bird either in Mr. Stevenson's or my own notes, and from the most interesting and exhaustive account of this event communicated by Mr. Stevenson to the "Norfolk and Norwich Naturalists' Society," in whose "Transactions," vol. iii., pp. 99-119, it will be found printed, I condense what follows.

"The great ornithological feature of the autumn of 1879 was," says Mr. Stevenson, "the appearance in extraordinary numbers of pomatorhine skuas, with Richardson's and Buffon's skuas in very much smaller quantities, along the entire eastern and north-eastern coast line; not merely, as in most seasons, passing southward on their ordinary migratory course—or collecting far out at sea around the herring smacks on their fishing grounds, or the lightships anchored off the dangerous shoals—but swarming in our harbours, bays, and estuaries. Some few, whirled inland by the force of the wind, have been picked up dead or exhausted far from the coast; whilst others have been observed skimming over the fields in close vicinity to the sea, like ordinary gulls in stormy weather, or tracing back for miles our tidal rivers; the survivors of these northern invaders—for the slaughter has been considerable—returning only after many days, and when the gales had ceased, to their natural element." Mr. Stevenson does not attribute this sudden influx to an extraordinary migration, "but to a succession of violent gales, which compelled the birds while performing their usual autumnal southerly migration, to take shelter for a time," as coasting vessels under similar stress of weather would naturally do, " in our harbours and minor inlets. Once inshore they came directly under the notice of gunners and collectors, who would otherwise have been quite unconscious of their passage; their usual line of flight being probably from thirty to forty miles out at sea."

"That the number of pomatorhine skuas, in particular, should have been so remarkable is due, I imagine, to an equally simple cause and effect, viz., the enormous shoals of herrings on our coast during the past autumn, and of sprats. . . . Such a preponderance of their finny prey would naturally attract gulls and skuas—fishers and 'pirates' alike—to one common feast, and as the direction of the shoals would be indicated by these birds to the smacksmen, their vessels would become the rendezvous of successive arrivals from the north, both of pomatorhine and other skuas. With calm seas and a superabundance of food one can imagine no instinctive force, at that period of the season, sufficient to induce these birds to quit their 'happy hunting-grounds' and resume their southward movement, and thus their accumulated forces within a certain area (instead of passing day by day in their accustomed flocks) seem to have been suddenly exposed in the North Sea to gales of unusual severity, by which they were driven upon our shores—simultaneously almost from the Tees to Lynn Wash, and thence along the projecting coast of Norfolk to Breydon and the mouth of the Yare."

The month of October, from about the 14th, was marked by extreme barometric oscillations, accompanied by gales of wind and driving showers, and during one of these from the north-west, on the 14th, towards afternoon, pomatorhine skuas commenced to pass Redcar, as recorded by Mr. T. H. Nelson, in the "Zoologist," 1880, p. 18, in small flocks of seven or eight coming from the eastward and seaward, until by dusk Mr. Nelson computes that several hundreds, perhaps thousands, had passed. The first Norfolk specimen was obtained by Lord Coke, at Holkham, on the 15th. On the 16th, as observed by Mr. George Cresswell, of Lynn, a great many made their appearance about Wolferton, on the shore of the Wash, and from that time they were found in parties all round the coast of Norfolk, and in many inland localities, the bulk remaining till the end of the month. At Yarmouth Mr. Stevenson was informed that stragglers were seen even later.

From a careful estimate Mr. Stevenson was of opinion that not less than two hundred skuas, mainly of this species, were killed in the county. As may be supposed, the Norwich naturalists had excellent opportunities of examining large series of these birds, of which they availed themselves to the full; and Mr. Stevenson gives some valuable details of plumage and other peculiarities, for which I must refer to the paper already so extensively quoted, merely observing that many excellent examples were present, ranging from very young birds to those which had fully reached maturity, and of both the light and dark-breasted forms; of thirty-four specimens examined sixteen were adult, and six supposed immature examples of the white-breasted form; two fully adult, and six immature birds, probably of the black-breasted form; four others being immature and doubtful. "These four young birds," Mr. Stevenson says, "I can only suppose to be the result of 'mixed marriages,' such as Mr. Saunders refers to in the case of the light and dark forms of Richardson's skua, and are about calculated to drive an ornithologist wild. The more or less adult birds, light or dark, were as eighteen to sixteen young ones."

A marked peculiarity in the plumage of the adult birds is the form of the middle tail-feathers; this was pointed out by Professor Newton ("Proc. Zoological Society," 1861, p. 402), who says "the middle tail-feathers have a *kind of twist* in their shafts, which brings the lower surfaces to meet together towards their extremities in a vertical direction, and this peculiarity gives the bird, when on the wing, a very peculiar appearance." There seems to be no appreciable difference in size, nor any satisfactory means of identifying the sexes short of dissection, and Mr. Stevenson adds that the reliable signs of maturity are—

1. The under tail coverts being a pure sooty black, and the upper of course without bars.

2. Black legs and feet.
3. A more or less vivid yellow colouring on the neck.

Of these three Mr. Stevenson considers the first to be the surest guide. The length of the middle tail-feathers, where perfect, is also a reliable proof of maturity; they project from two inches to three and a half.

I have little more to add with regard to this species, except that in 1880 a few examples were brought into Yarmouth, one as early as the 16th September; and in the following year (1881) a few others were obtained, since which date I have no other record of its occurrence till December 4th, 1889, when, as I learn from Mr. Patterson, an immature example, doubtless wounded, was found alive on the beach at Yarmouth.

STERCORARIUS CREPIDATUS (Gmelin).

ARCTIC SKUA.*

Every autumn brings some birds of this species upon our coast, and occasionally they are by no means uncommon in the immature state; adults are, however, at all times rare.

In Whitear's "Calendar," as printed in the "Trans. of the Norfolk and Norwich Nat. Society" (iii., p. 250), under date of 26th September, 1819, occurs the following entry:—" I am informed by Mr. Sabine that he procured a young arctic gull on a warren, near Brandon [Suffolk], the beginning of this month. About the same

---

* This species was known to Pennant, Montagu, Fleming, Selby, and others as the arctic gull, and it seems desirable to retain that epithet. As pointed out by Mr. Howard Saunders (Yarrell, "British Birds," ed. 4, iii., p. 674, note), "the trivial name, Richardson's skua is certainly distinctive," but it must be borne in mind that " it was originally applied solely to the dark form " of this species, which was at the time thought to be distinct from the arctic skua. Cf. " Fauna Boreali-Americana " (Birds, p. 433, pl. 73).

time an old bird of the same species was killed at Yarmouth, and is now in the possession of Hunt. N.B. The weather was mild at the time." And again on the 19th October of the same year, " Mr. Youell informs me that about a week since a young arctic gull, answering to the black-toed gull of Latham, was killed at Yarmouth." These specimens are also mentioned in the "Catalogue of Norfolk and Suffolk Birds." In his "List" Hunt says, "this species is extremely rare [probably meaning in the adult state] : — Girdlestone, Esq., some few years ago sent the writer of this a specimen alive, which is now in his collection." This may possibly have been the adult bird which Mr. Whitear says was "killed" at Yarmouth and was given to Hunt. Mr. Lubbock, in a MS. note, mentions an arctic skua having been shot at Wells, in the winter of 1830-1. The Pagets say of "*Cattarractes parasiticus.* Both this and its young, the black-toed gull, have occasionally been shot." A Richardson's skua in the Lombe collection at the Norwich Museum was shot at Little Melton, on the 27th October, 1834. Messrs. Gurney and Fisher, in their "List" (1846), state that immature birds in various stages of plumage are occasionally found in autumn, but that the adults are rare ; and in the "Zoologist," p. 1956, they thus notice the occurrence of this species in September, 1847—"We have also to record the capture at different parts of the coast of four specimens of Richardson's skua, two of which were in immature and the other two in adult plumage." In the same year and month Mr. Dowell writes, " Richardson's skua has been uncommonly numerous off the Blakeney coast this month : May brought me two, one of this year, the other of last ; and on the 16th Dolignon killed another ; we saw four or five more the same day." We may, therefore, infer that this species was unusually abundant in September, 1847.* Again, on September 9th, 1848, Mr. Dowell writes, "For the last fortnight these birds have frequented Blakeney Harbour ; two or

---

* It seems not impossible that some of these might have been pomatorhine skuas, which was then thought to be a much rarer species than it has since proved to be.

three might be found any day. September 25th brought a second year bird; October 2nd I refused another;" and in October of the same year an immature specimen was killed at Lynn; on 2nd September, 1853, Mr. Dowell bought two young arctic skuas of a Blakeney gunner. In the middle of September, 1854, Mr. Gurney received one of these birds with great skuas from Yarmouth. In 1858 Mr. Stevenson writes, " October 20th, four at Sayer's shot from the herring-boats off Yarmouth; 24th, two more. Of these six three were in immature plumage of the second year, an uniform brownish tint, with the middle tail-feather projecting about three inches; and the others had attained very nearly their adult state, the under parts being white with the exception of a few dusky markings remaining more or less on throat and breast." In the "Zoologist" for 1861 (p. 7818) is a notice by Mr. W. Winter, of Aldeby, of one killed in the fens at Horning, on the 9th of November, during a snow storm; and Captain Longe has a fine young male procured in November, 1862, on Breydon, after a heavy gale; on the 25th September, 1863, Mr. Dowell saw at Hunstanton a beautiful arctic skua, which chased two terns within ten yards of him, and he says " its power of wing was wonderful." Mr. Stevenson records in the " Zoologist " for 1866 (p. 85) the occurrence of a number of skuas, divers, gannets, &c., off Yarmouth, attracted by the large shoals of herrings, among which, on the 10th November, 1865, was an immature arctic skua. About 4th October, 1867, after severe north-west gales, one of these birds was killed at Lynn, and another on 24th November, 1869, on the coast, a very dark bird in immature plumage (" Zoologist," 1870, p. 2058). In the month of October, 1870, a like abundance of herrings, as in 1865, brought the usual attendance of skuas, and one of this species was killed on the 4th, an immature bird having been previously met with on 27th August, at Blakeney. On the 13th May, 1871, an arctic skua was shot in Brooke wood, near Norwich, a most unusual time and place, and there had been no severe weather to account for its appearance (" Zoologist," 1871, p. 2829); and on the 22nd August of the same year another was sent to Mr. J. H. Gurney, jun., in the flesh, from Blakeney. In the last week of October, 1874, after a very severe gale off Yarmouth, many skuas were brought in by the fishing boats. One game-dealer there, says Mr. Stevenson, had thirty at one time; they were nearly all pomatorhines, but amongst them was one adult arctic skua. Another, an immature bird, was shot at Blakeney, on 18th September, 1877. This brings us to the great skua year of 1879, of which I have spoken somewhat at length in the previous article. I do not know the number of arctic skuas which were obtained, but it was very small (Mr. Stevenson mentions only three), and I think there were even fewer than of the next species. Others occurred in October, 1881, and all four species of skua were met with at various dates in that month. Mr. Dack had a specimen killed near Yarmouth, in October, 1883.

It will be seen from the foregoing rather full list of occurrences that the arctic skua is by no means a rare bird on our coast; indeed, I have no doubt, though it never assembles in such large numbers as are occasionally observed of the pomatorhine skua, that it is a constant attendant upon the herring-boats in autumn, where it takes its toll of the various smaller gulls and terns which follow the fleet to feed on the offal and broken fish which are so abundant, but it rarely shows itself on the shore or comes into the harbours except under stress of weather, and then not in large numbers. I have frequently seen a pair or two under such circumstances years ago beating to windward in Lynn harbour; where Mr. Gurney tells me one of these birds had the temerity to attack a great black-backed gull, which turned and knocked it over; a man, who witnessed the attack, picked it up and found that it had lost one eye; he, however, took it home and it fed readily. Mr. F. J. Cresswell purchased the skua and sent it to the Zoological Gardens.* Mr. Booth says

---

* "Proc. Zool. Soc.," 1868, p. 651. In the 8th edition of the "List of Vertebrated Animals, in the Zoological Gardens," published in 1883 (p. 533), this bird is, however, erroneously entered as a "common," *i.e.*, great skua, *S. catarrhactes*.

he has rarely seen this species attack a gull larger than a kittiwake. The same authority, who gives some beautiful figures both of the light and dark races of this species, is of opinion that it does not assume its mature plumage till the fifth or even the sixth year. The great variations of plumage in this and the next species are certainly sufficiently puzzling to the uninitiated.

STERCORARIRUS PARASITICUS (Linnæus).

BUFFON'S SKUA.

This species is decidedly the rarer of the two smaller skuas on the Norfolk coast, in addition to which it is possible that some of the earlier records, from the bird being imperfectly known, may not be altogether trustworthy.

Lubbock does not mention Buffon's skua, and Messrs. Gurney and Fisher, although they state that the young birds not unfrequently occur in autumn, in a similar way to the arctic skua, do not give any instances. Perhaps the first reliable record of the occurrence of this species in Norfolk is to be found in a communication from Professor Newton to the "Zoologist," p. 2149, where he states that in September, 1847, an immature specimen was found dead at Hockham; in the Newcome collection there is also a young male, which was shot by the late Mr. E. C. Newcome in a turnip field at Methwold, in September, 1854. On the 23rd of November, 1849, Mr. Dowell writes that "whilst walking with Overton by Salthouse broad, a skua pursuing a common gull flew over our heads, which was so small that it could hardly have been anything else than this species." A specimen in immature plumage, lately in Mr. Stevenson's collection, was killed at Salthouse, on October 28th, 1862. The year 1867 produced several of these birds; on the 20th September one was killed at Blakeney; on the 4th October Mr. Gunn records the occurrence of two, one adult male and a bird of the year, at Salthouse ("Zoologist," 1867, p. 992); the stomach of one of these birds contained a beetle. On the 12th, after a succession of severe gales from the north-west, another was procured far inland, at Beechamwell, near Swaffham; on the 30th October, 1869, a Buffon's skua was sent to Norwich to be preserved, from Cley-next-the-Sea. In the remarkable irruption of skuas which occurred in the autumn of 1879, several very beautiful adult examples of Buffon's skua were shot on different parts of our coast, one of which, in fine plumage, taken at Yarmouth, was in Mr. Stevenson's collection (No. 146 of his sale catalogue). Mr. Stevenson enumerates the following in his paper before so largely quoted—" Mr. Cole had one on the 20th of October, picked up exhausted but not dead on the beach at Hasborough, in fine plumage, very yellow on the sides of the neck, but the long middle tail-feathers unfortunately injured at the tips, being scarcely more than $3\frac{1}{2}$ inches longer than the rest of the tail. About the same time as the last Mr. T. W. Cremer obtained an adult bird, which, as he informs me, was knocked over with a stone by a boy on Beeston common, near Cromer. It was in company with another in similar plumage: the central tail-feathers projected $6\frac{1}{2}$ and $6\frac{1}{4}$ inches. Mr. Dack had also an adult and an immature example from Blakeney. The latter (which I presume to be of this species), now in my collection, is an extremely grey-coloured bird, the margins to the feathers of the back, wings, and upper tail-coverts being almost white; the under parts mottled more or less with grey and white, showing scarcely any tinge of brown. The two middle tail-feathers project scarcely half an inch;" this was No. 196 of Mr. Stevenson's catalogue. Mr. Lowne tells me he had two Buffon's skuas, which were killed on the 24th of October, 1879, and that he knew of another obtained about the same time. In 1881 several skuas were procured off Yarmouth about the 21st of October, among them two adult and one immature Buffon's skuas. One of these, No. 223, in the sale catalogue of Mr. Stevenson's collection, was a fine adult male, but, unfortunately, minus the centre feathers of the tail.

The following further extract from Mr. Stevenson's paper may be useful in assisting to determine the species

of the two smaller skuas, which, as I have said, in the immature plumage, are somewhat difficult to distinguish:—
"From the great variation in the plumage at this early stage—in both Buffon's and Richardson's skuas—as well as the similarity in size, Mr. Saunders points out (see 'Proceedings of the Zoological Society,' 1876, p. 327; and the 'Field,' January 17th, 1880), as a certain distinction at any age, that in Buffon's the shafts of the two outer primaries are *white*, those of the remaining primaries being dusky; whilst in Richardson's skua the shafts of *all* the primaries are *white* throughout the greater part of their extent; and even in young birds it is only towards the tips that the shafts are of a shade at all approaching that of the webs." To this may also be added a hint supplied by another well-known authority in such matters, Mr. John Hancock, who states, "Birds of Northumberland and Durham" (p. 137), that the two middle tail-feathers in Buffon's skua are obtuse in the young state, while in the arctic they are always pointed. Mr. H. C. Hart ("Zoologist," 1880, p. 210) states from personal observation that in the male the projection of the middle tail-feathers is from seven to eight inches; in the female, six inches or a trifle over.

FULMARUS GLACIALIS (Linnæus).

FULMAR.

The fulmar petrel, attracted by the herring boats at sea, is not unfrequently met with off the Norfolk coast in the autumn and winter months, the large majority being immature birds, but those occurring inshore are generally storm-driven individuals often in a state of exhaustion. They are not unfrequently brought in by the Yarmouth smacksmen, and Mr. George Smith, in a paper published in the "Trans. of the Norfolk and Norwich Nat. Soc." (iv., p. 223) to which I shall have again to refer, states that between October, 1878, and December, 1885, he had thus obtained fifteen examples, the greater number of which were not quite adult; fully adult birds are very exceptional.

Fulmars usually make their appearance in the month of October, but Mr. Stevenson also records the occurrence of one as early as August, in the "Zoologist" for 1876 (p. 4773).

The earlier Norfolk naturalists seem to have regarded this species as a rare bird. Sheppard and Whitear do not mention it. Hunt, in his "List," merely says that one killed at Yarmouth is in the possession of Mr. J. J. Gurney; in the Hooker MS. is mention of one taken at Yarmouth, on 18th November, 1829; the Pagets say that it is occasionally caught in the Yarmouth Roads; and Messrs. Gurney and Fisher speak of it as sometimes found off the coast in autumn; Mr. Dowell mentions one picked up alive but nearly starved, in Blakeney Harbour, early in the winter of 1855, and another procured at Brancaster, in the winter of 1860-61, since which time many specimens have been recorded in the "Zoologist" and elsewhere.

In Mr. Smith's paper, before referred to, on the habits of this species in confinement, he says that the first he received only lived a fortnight, but he subsequently kept one alive ten weeks, feeding it (for it only took food voluntarily on two or three occasions) by hand on small fish or beef fat. It vomited oil on three occasions, and "certainly attempted to throw it over us; we could see the oil coming up its throat, and when it reached the bill, it shook its head, and threw the oil two feet across the place." It was much troubled with cramp, as was the previous specimen, which Mr. Smith attributed to its treatment in being so long shut up in a hamper or box at the time of its capture, and he thinks that if kept free from cramp they would live longer in confinement. They come so close to the fishing-smacks as to be easily taken by a "didle," a sort of landing net for taking up fish from the surface, and when caught do not bite like the skuas. Mr. Smith's captive "was a good hand at climbing; by the aid of its bill and toes it would get on to a hamper, or any slightly elevated place, and was always delighted to have something to peck at, such as grass or any soft

J.Weir,del.et.lith.

material." Although Mr. Smith's fulmar seized and killed a greenfinch (which it did not, however, eat), he considers it "a harmless bird, and a very tame one."

With regard to the voracity of the fulmar, Mr. Smith states that a fisherman told him that he soaked a piece of oakum in colza oil, and on throwing it overboard into the sea, a fulmar, seeing the track of grease on the water, seized the lump of oakum and swallowed it. This bird was shortly after shot by Mr. Booth, and sent to be preserved to Mr. Gunn, who found the oakum in its stomach ("Rough Notes," p. 218). Mr. T. E. Gunn showed me in November, 1869, a fulmar which he had received from Yarmouth; it was in a very weather-beaten condition, and had been shot resting on the River Bure. On dissection a piece of tarred line was found coiled up in its stomach, which accounted for the inflamed condition of one side. Mr. Pycraft, late of Yarmouth, also obtained in November, 1885, from the stomach of a fulmar, a stout barbed steel fish-hook, two and a half inches in length, or four inches following the curve, and twenty-eight inches of twisted cord attached to it. After taking this hook with the bait, the fulmar actually swallowed a second, thrown from the fishing-smack, and was captured (J. H. Gurney, jun., l. c., p. 225).

Of the fifteen specimens mentioned by Mr. Smith two were examples of the dark form, and were rather smaller than the light-coloured race. At the end of January, 1888, two fulmars were taken at Cley-next-the-Sea, one of which was sent by Mr. H. M. Upcher to the Zoological Gardens, where, however, it did not live long. Others have been obtained from the same locality since; and on the 13th December, 1889, Mr. A. Patterson tells me he found one of these birds dead on the beach at Yarmouth.

ŒSTRELATA HÆSITATA (Kuhl).

CAPPED PETREL.

The first occurrence in Britain of this rare wanderer, and (as will be seen in the sequel) perhaps extinct species, whose true and only home appears to have been the islands of Dominica and Guadeloupe, in the West Indies, was recorded in the "Zoologist" for 1852 (p. 3691), by Professor Newton. In March or April, 1850, a bird was observed by a boy on a heath, at Southacre, in this county, flapping from one furze bush to another, until it got into one and was there caught by him. Exhausted as it was, it violently bit his hand, and he thereupon killed it. The late Mr. Newcome fortunately happened to be hawking in the neighbourhood, and his falconer, John Madden, seeing the boy with the dead bird, procured it, and brought it to his master, by whom it was skinned and stuffed. It still forms the chief attraction in the Newcome collection; and the following note in regard to the accompanying representation of it by Mr. Wolf (plate iv.) is among those left by Mr. Stevenson for publication in this work.

"Being enabled through the kindness of my friend Mr. Newcome to present my readers with a coloured illustration of this singular bird, which, though previously figured in the 'Zoologist' and by Yarrell, required the pencil of a Wolf* to do full justice to its peculiarities, it will be unnecessary for me to describe its plumage. It proved to be a female, and when fresh the irides were deep brown or hazel colour. The following are the principal measurements as given in the 'Zoologist,' *(ut supra)*—'Total length about 16 inches; length of the ulna about $4\frac{1}{4}$ inches; from the carpal joint to the end of the longest wing feather, rather more than 12 inches. The length of the naked portion of the tibia, rather more than half an inch; of the tarsus, rather less than $1\frac{1}{2}$ inches; and of the middle toe, exclusive of the claw, about $1\frac{3}{4}$ inches.'"

Besides Mr. Wolf's plate I am able now to include a reproduction of the original figure of the bird's head from the "Zoologist," which being of the natural size, may aid in the recognition of any future example that may occur.

The occurrence of two other examples has been re-

---

* Since the drawing was made the specimen has been re-mounted, and its attitude changed.

corded in Europe—one preserved in the museum in Boulogne, is said to have been shot near that town; and a second, killed in 1870, at or near Zolinki, in Hungary, was recognised in the Museum at Buda-Pesth, by Mr.

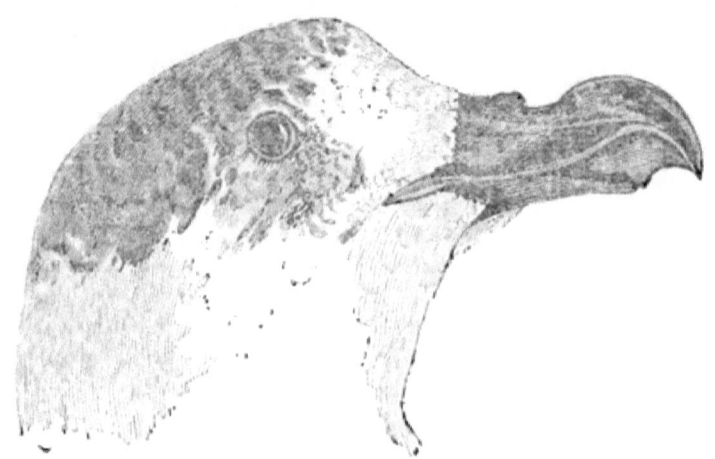

W. Eagle Clarke ("Ibis," 1884, p. 202). Two specimens have been obtained in the United States; but it is certain that the true home of this very rare species is, or was, in the islands of Guadeloupe and Dominica, in the West Indies, where it was undoubtedly once abundant. One of its old breeding-places in the last named of these islands was explored, but without finding a single bird, in February, 1889, by Colonel Feilden, who on the 18th May in that year read an admirable paper on "The Deserted Domicile of the Diablotin"—the name by which it was known to the inhabitants—to the Norfolk and Norwich Naturalists' Society, which will be found printed in their "Transactions," vol. v., p. 24. Therein he gives copious extracts from the interesting accounts of the old voyagers, which are indeed all that is left to us of the history of the species.* From the information he

---

* The authorities cited by Colonel Feilden being mostly rare and difficult of access, it may be convenient here to state that they are :—Du Tertre " Histoire Generale des Antilles," ii., p. 257 (Paris : 1667-71); Labat, "Nouveau Voyage aux Isles de l'Amérique,"

received on the spot, the Colonel has no doubt that its extermination in Dominica has been effected by the introduction of a rapacious marsupial (possibly a South American opossum) as was first announced by Mr. Ober in his "Camps in the Caribbees"—a gentleman who, in 1876, had also searched in vain for the species; but it must be borne in mind that in this island, as well as in Guadeloupe, the young were greatly esteemed as a delicacy, and a great many of them, as well as of the old, were caught in their breeding holes and killed for the table. This destruction of the parent birds, carried on as it was for many years, would naturally reduce their numbers, and, even without this interference of a predatory animal, might in time bring about the extirpation of this petrel.

PUFFINUS GRISEUS (Gmelin).

SOOTY SHEARWATER.

The only Norfolk specimen of this bird of which I am aware is still preserved in the Lynn Museum. At the time of its capture it was recorded by Mr. E. L. King as the "greater shearwater, *Puffinus cinereus*," "Zoologist," p. 3234, and a full description given, but a subsequent examination proved it to belong to this species, probably immature ("Trans. Norfolk and Norwich Nat. Soc.," iii., p. 474). I purchased the bird alive of a boy who had caught it at the mouth of the river Ouse as he was returning to Lynn in a fishing-boat, it being apparently asleep on the water. This occurred on the afternoon of the 25th July, 1851. During the five days which it lived, it passed the day sleeping, but, as evening advanced, became more lively,

---

ii., pp. 349—361 (Paris: 1722); Froger, "Relation d'un Voyage de la Mer du Sud," p. 213 (Amsterdam: 1715); Attwood, "History of the Island of Dominica," pp. 30—32 (London: 1791); and Lafresnaye, "Revue Zoologique," 1844, p, 168; besides Ober's "Camps in the Caribbees," above mentioned.

and readily ate small fish, shrimps, or fresh beef. It probably died of some injury received at the time of its capture. After death it was set up by Foster, of Wisbech, who found it to be a male, and it was deposited in the Lynn Museum.

PUFFINUS ANGLORUM (Temminck).

MANX SHEARWATER.

Sir Thomas Browne doubtless refers to this species in the following passage which occurs in the "Account of the Birds found in Norfolk:"—

"A seafowl called a sherewater, somewhat billed like a cormorant butt much lesser. a strong & feirce fowle houering about shipps when they cleanse their fish. Two were kept six weekes, cramming them with fish which they would not feed on of themselues. The seamen told mee they had kept them three weekes without meat. and I, giuing ouer to fee them found they liued sixteen dayes without taking anything." And, again, in the letter to Merrett, dated December 29 [1668], he says he sends "Also the draught of a seafowle, called a sherewater, billed like a cormorant, fierce and snapping like it upon any touch. I kept 20[*] of them aliue five weeks, cramming them with fish, refusing of themselues to feed on anything & wearied with cramming them, they liued 17 days without food. They often fly about fishing ships when they cleane their fish and throw away the offell."

In the English translation of Willughby's Ornithology, "The sheare-water" is figured (plate lxxviii.) presumably from Sir T. Browne's drawing, as in the text (p. 334), wherein the main facts as already stated are again set forth. Whether the drawing sent by Browne to Willughby was identical with that sent to

---

[*] Professor Newton tells me that the figures 20 certainly occur in the MS., but that they are probably a mistake.

Merrett must remain doubtful, but probably if not, one was a copy of the other, and is conclusive evidence as to the species intended.*

Although of course more frequently met with than the other members of this widely dispersed genus, the Manx shearwater is certainly a rare bird off the Norfolk coast, but it doubtless appears far out at sea, accompanying the fishing boats every autumn, probably going south from its breeding station, such examples as I have notes of having occurred between the middle of August and the middle of October. It is very remarkable that so many individuals of the family *Procellariidæ*, some of them of great rarity, should have been captured far inland, and this species certainly is no exception.

Mr. Gurney has one which was captured at Tivetshall, twenty miles from the sea, on the 9th September, 1859; and, in Mr. Stevenson's notes, he mentions two in his own collection, " taken in a singular manner near Swaffham, in October, 1861. Mr. Ellis, bird preserver, of that town, of whom I purchased them in the following spring, informed me that they were brought to him alive about the same time, one of them having been taken in a barley field, at Marham, hidden under the swathe, the other in a field at Cressingham, near a run of water. They were both very fat, and one was kept alive for some days." On the 15th August, 1878, one, also said to have been very fat, was picked up alive at Shottisham, near Norwich. Mr. Frank Norgate tells me that on October 10th, 1882, Mr. Newby, bird-stuffer, of Thetford, showed him a stuffed shearwater, "Manx, I believe," which had been killed near that place; and he also states that Mr. E. P. Mackenzie, of Santon Downham, has another, "apparently Manx," which was picked up on a neighbouring warren even still further inland. Mr. Charles Candler tells me that a Manx shearwater was taken alive in a harvest field, at Pulham, near Harleston, on the 10th September, 1888. Dr. Babington, in the " Birds of Suffolk," records several

---

* According to Edwards ("Gleanings of Natural History," iii., p. 315), this "old draught" is still preserved in the British Museum.

similar instances of the capture of this species far inland in that county. It is probable that the occurrence of these ocean-loving birds in such unaccustomed localities is involuntary on their part, and due to atmospheric disturbances (possibly also to the prevalence of fogs), which drive them from the element they rarely quit except during the brief period devoted to nesting.

PUFFINUS OBSCURUS (Gmelin).

DUSKY SHEARWATER.

Mr. Stevenson contributed to the "Trans. of the Norfolk and Norwich Nat. Soc.," vol. iii., pp. 467—473, a full account of the occurrence of this interesting bird in the county of Norfolk, from which the following narrative of the event is extracted:—

"In the 'Zoologist' for 1858 (p. 6096) I recorded the appearance, far inland in this county, of a petrel, which I felt little doubt at the time was an example of this rare species (rare, at least, on the shores of Great Britain), and which on examination recently, by the best authorities on these oceanic wanderers, has proved to be what I first described it.

"My original notes on this interesting bird may be thus summarized. About the 10th of April of the above year it was found dead by a gamekeeper on the Earsham estate, situated close to the south-eastern boundary of Norfolk, and within a mile of the well-known town of Bungay in Suffolk.* Captain Meade, who at that time hired the hall and the shooting, brought the bird, in the flesh, to the late Mr. John Sayer, birdstuffer, of St. Giles', Norwich, who at once observed its marked difference in size from any Manx shearwaters he had ever seen. Being from home myself at the time, I did not examine the bird in a fresh state; but I saw it within a

---

* Its flight inland, therefore, from the coast would probably have been between Lowestoft and Southwold.

week of its being stuffed, and its resemblance to the figure of the dusky petrel in the third edition of Yarrell's 'British Birds,' and in the supplement to the second edition (1856), struck me forcibly at first sight; confirmed, to a great extent, by the comparison of its measurements (though a mounted specimen) with the description given of the species by that author.

"It proved, on dissection, to be a male in very poor condition, and probably had been driven so far inland by a gale, and met its death through coming in contact, at night, with a tree or some other object, having a wound on one side of the head, as if from a violent blow. It showed no appearance of having been shot at; and the feathers, except on the spot mentioned, were clean and unruffled; but the inner web of one foot was partially nibbled away, as though a mouse or some other vermin had been at it.* Fortunately I noted these injuries at the time, which have enabled me to identify the specimen again, beyond any doubt, though lost sight of for the last thirteen years. Having been brought to the birdstuffer by Captain Meade, and returned to him when mounted and cased, I naturally inferred that the petrel belonged to him; and hearing some time after that he had left England, and all his effects at Earsham had been sold off, I presumed that this rarity was lost to us altogether. In the absence of the bird itself, I was unable to support my previous conviction as to the species; whilst subsequent accounts of extremely small Manx shearwaters being occasionally met with, made me question my own judgment in the first instance; more especially as my acquaintance with that class of marine birds was somewhat limited at that time. I specially mention this, because it will explain why I did not bring the fact of the dusky petrel having occurred in Norfolk under the notice of either the late Mr. Gould, when publishing his 'Birds of Great Britain,' or of Mr. Dresser for his 'Birds of Europe,' neither of which authors has

---

* This was my impression at the time; but the examination of a large number of pomatorhine and other skuas, killed on our coast in 1877, showed that the webs of the feet, in this class of birds, are frequently mutilated.

included this species in the above-named publications. The re-discovery of the Norfolk specimen was quite accidental. Early in the present year, Mr. J. H. Gurney, jun., and myself, being separately engaged in working out a complete list of the 'Birds of Norfolk,' and comparing notes on the subject, the right of this species to rank with other local rarities was questioned, and, 'drawing a bow at a venture,' Mr. Gurney put himself in communication with Mr. Hartcup, of Bungay, who proved to be a trustee for the family of the late Sir W. W. Dalling, Bart., and the Earsham estate. From him it was soon elicited that a good many birds killed on the estate were preserved at the Hall, and amongst these, most fortunately, was found the dusky petrel of 1858. The thanks of this society, and of naturalists generally, are due to Mr. Hartcup for the opportunities he has afforded for a thorough inspection (with permission to photograph it) of this unique specimen; and, having, myself, first obtained the confirmatory opinions of Professor Newton and Mr. Osbert Salvin, it was exhibited by the latter at a meeting of the Zoological Society on the 16th of May, 1882."*

PROCELLARIA† LEUCORRHOA, Vieillot.

FORK-TAILED PETREL.

The following note on the occurrence of this species in Norfolk was contributed by Mr. Stevenson to Mr. Dresser's "Birds of Europe:"—

"The earliest record of this species in Norfolk is that recorded by Messrs. Paget as picked up dead on the beach at Yarmouth, on December 5th, 1823;‡ and

---

* See "Proc. Zool. Soc.," 1882, p. 421.

† It is now known that the name *Thalassidroma*, formerly in pretty general use for the smaller petrels, is strictly synonymous with *Procellaria*, which, as the older term, has to take its place.

‡ This bird is said by the Messrs. Paget to be in "Mrs. Baker's collection;" but in Hunt's List he states that it was "in the

it is also worthy of note that in the edition of Bewick's 'British Birds,' published in 1826 (vol. ii., p. 244), this species is first figured and described by that author from a specimen and notes supplied him by the late Mr. Yarrell, the bird itself having been bought alive in Leadenhall Market, on the 3rd November, 1823, and said to have been caught on the Essex coast.

"Selby records one in the late Rev. R. Hamond's collection, as 'picked up dead upon a warren in Norfolk,' the date not given; but Selby's 'British Ornithology' was published in 1833. In 1849, Mr. J. H. Gurney recorded in the 'Zoologist' the occurrence of a male and female of this petrel at Yarmouth, in the months of October and December, 1849; and in Mr. Gurney's collection is also a specimen from Yarmouth, but the date uncertain. On the 17th November, 1862, an example, in my own collection, was shot on the coast at Salthouse; and another was picked up dead, inland, after a gale, at East Bradenham, on the 21st November, 1864. Since that date my notes show the following repeated occurrences :—

"1867. One at Yarmouth, on the 6th of July, as recorded at the time (July 13th) in the 'Field'—a strange date for such a bird. One November 14th; one December 2nd, male [on the sea bank]; one December 2nd, male [in the harbour]; one December 7th, male; one December 14th, female; one December 9th; all shot at or near King's Lynn, with many of the storm petrels at the same time. Records in the 'Field' of one at Colchester, and one at Spalding, about same date.

"1868. One at Babingley, near King's Lynn, December 19th.

"1869. One shot on the river Bure, near Yarmouth, October 26th.

"1870. One at Gooderstone, near Lynn, the first week in January.

---

possession of [C. S.] Girdlestone, Esq." Mr. Lubbock, also, in the MS. notes in his copy of Bewick, states that it "was bought by Girdlestone." Mr. Dawson Turner, also referring to the same bird, remarks that so stormy an autumn as that of 1823 "was never known."

"1871. One, on good authority [the late E. L. King], said to have been seen at Lynn, in February."

To these may be added one mentioned in Salmon's Diary, in a note copied from the "Norfolk Chronicle" for the 31st January, 1835, said to have been killed by Rev. George Hogg "while skimming over the waves near Holme," about the 25th January. Mr. Stevenson records one "said to have been obtained at Yarmouth" in April, 1876. One was seen on Durrant's stall, at Yarmouth, on the 18th November, 1883; and, in the autumn of the same year, Mr. J. H. Gurney, jun., had a fork-tailed petrel's wing sent him by the master of the Happisburgh floating light.

The above is a list of all the recorded occurrences of this bird in Norfolk with which I am acquainted; it has generally been met with under the same circumstances, and frequently associated with the storm petrel, and like it has probably always been an involuntary visitor.

PROCELLARIA PELAGICA, Linnæus.

STORM PETREL.

This pretty little wanderer may be found out at sea off the Norfolk coast every autumn and early winter, and is frequently observed from the light vessels. It is only after severe weather that it frequents the estuaries and harbours, and is sometimes, as will be seen, found far inland, dead or dying from exhaustion.

Hunt, in his "British Ornithology" (1822), states that Mr. Lombe, "kept a stormy petrel alive for a few days, and that in removing itself from one part of its place of confinement to another, it made use of its beak, in a manner similar to the parrot." (Cf. also fulmar, p. 360). In his "List" the same author says, "For the last two years [1826 and 1827?] this bird has been numerous off the Yarmouth coast, where numbers have been taken. Mr. J. Harvey, of that town, had, at one time, upwards of fifty specimens. In the winter before the last (1826) a poor man caught one of these birds alive, in the Rose Lane, in Norwich," the street in which Hunt lived. The Messrs. Paget say that "a few are generally shot every winter. In November, 1824, between two and three hundred shot after severe gales." In the Hooker MS., initialled by Dawson Turner, occurs the following entry:—" Storm petrel. So abundant was this bird at Yarmouth after two or three days of strong north-easterly wind, October 27th, 1834, that one man in a boat off the Jetty took twenty-five specimens swimming on the sea, and brought seven of them to Jacob Harvey alive. They appeared a remarkably tame and gentle bird."

Mr. Stevenson mentions in his notes six of these birds in one case, in the Dennis collection, at Bury, all killed on the 9th December, 1849, by flying against the floating light off Yarmouth beach; and again, "Several were killed at Yarmouth, in the autumn of 1855, during very severe gales, and many were knocked down with sticks in Lynn Harbour, where they flew, to use the words of an eye-witness, 'as thick as sand-martins,'" a circumstance of which the present writer was also an eye-witness. At the same time they were found inland as far as Newmarket and Cambridge. Again, early in October, 1867, after several days' stormy weather, many were found dead on the coast and inland at Beeston Regis, Little Fransham, and Lexham (Gunn, "Zoologist," 1867, p. 992; Stevenson, l. c., p. 1012). Between the 18th October and the 1st of November, 1869, many of these birds visited the coast, and one was picked up dead near Foundry Bridge, in this city (Gunn, "Zoologist," 1870, p. 1983); others at Hickling, Stalham, Walton, Hempton Green, near Fakenham, and other places. Others occurred during the terrific gale and snowstorm which visited this county on the 10th of February, 1871. In November, 1872, another visitation of these birds took place, and many were off the coast, although the only one recorded that year was obtained at Cromer on 15th of that month. Two of these birds, taken at Harford Bridges, near Norwich, after a north-west gale on the previous night, were brought to Mr. Cole on the 1st November, 1880. These are some of the principal occurrences of the storm petrel in Norfolk; many others are recorded, and probably not an autumn passes without some being either killed or seen, as happened early in December of the present year, 1889.

Mr. Booth, in his "Rough Notes," gives an interesting account of the appearance of this bird off the Norfolk coast, in November, 1872. He says he passed the greater part of that month at Yarmouth. On the 11th a terrible gale began to blow, with squalls of rain, from the north-north-east; large bodies of fowl were flying continually all day towards the north, and immense flocks of dunlins passing in rapid succession. On the 12th the gale was from the east-north-east; and a few fowl still flying north, but the number of dunlins not to be compared with those observed the previous day. On the 13th, it was still blowing with fearful gusts from the east. "Three or four stormy petrels remained feeding just off the harbour-mouth, and occasionally flying a short distance between the piers, where they were enabled to find shelter from the cutting blasts." On the 14th, blowing still harder from the east-north-east, "A dozen or so of stormy petrels were flitting here and there between the piers, near the harbour-mouth, and a few made their way up the river as far as the wharfs where the vessels and fishing-boats were moored." On the 15th "there was a slight lull in the morning, and all the common species of gulls were seen in hundreds just off the harbour-mouth during the whole of the day, several stormy petrels flying with them. At day-break, on the 16th, the wind had freshened, and a gale was blowing hard from the east-south-east. The unfortunate little petrels proved to be much exhausted by the long continuation of the storms, and were even driven right up the harbour, several being seen on Breydon mud flats. Off the denes to the north of the harbour they were hovering in dozens over the rolling breakers, and were frequently carried by the irresistible squalls of wind right across the carriage-drive. The surface of the North Sea on Monday morning, the 18th, was nearly as smooth as glass, the wind having completely dropped; and, leaving the harbour early in the 'Reliance,' we steamed out through the roads and round the sands. Immense numbers of stormy petrels were resting on the water, most of them asleep, with their heads buried under the feathers of their backs, strongly resembling small round balls as they floated about with their plumage puffed out in the tideway. They were passed during the whole of the day along the coast from Winterton to Lowestoft, many hundreds being noticed. No forked-tailed petrels were observed, though I carefully examined with the glasses all that were within view, as we steamed slowly past them. These poor little birds had evidently been completely tired out by the long continuation of the gale, and were now endeavouring to recruit their strength by seeking repose on the placid surface of the water." On the 20th "the stormy petrels, so numerous close inshore two days previous, had now almost entirely disappeared, and only two or three were noticed flitting in an apparently worn-out condition outside the Cross Sands."

Writing of the storm petrel at Yarmouth, Mr. A. Patterson tells me that twenty or thirty years ago this bird used to be fairly frequent off Yarmouth during the herring season. At that time the fish was "ferried" ashore and landed on the beach, and as it did not pay to land the "offal," as is now done for manure, it was all thrown overboard. This, of course, offered a great attraction for both fish and fowl, which were both much more frequent immediately off the shore at that time than at present, and amongst them came the storm petrels. "A worn out gunner," says Mr. Patterson, "told me that he had seen what he described as a 'body of petrels' during the early winters, and that the 'carriers' used to trail soft roes (milts) of herrings behind their boats, and the ravenous little creatures would come right under the stern, and were knocked down by osier wands carried for the purpose. What induced this destruction I did not learn—whether it was merely love of slaughter, mere pastime, or whether the game dealers offered any inducement, I cannot say, but probably all three combined."

There is at Northrepps an albino storm petrel, said to have been killed at Yarmouth, and formerly in the collection of the late Mr. Stephen Miller, at the sale of whose birds it was purchased by Mr. Gurney.

APPENDIX A.

SPECIES TO BE ADDED.

AQUILA CHRYSAETUS (Linnæus).

GOLDEN EAGLE.

Mr. Stevenson pointed out when speaking of the white-tailed eagle (vol. i., p. 4) that although frequently recorded, no authentic instance of the occurrence of the golden eagle in the county of Norfolk had to his knowledge ever come under the notice of any competent authority; the so-called golden eagles, as has also frequently been the case since, invariably proving on examination to have been immature examples of the white-tailed eagle.

The claim, and the only claim, so far as the present writer knows, of this species to a place in the list of Norfolk birds, rests upon the finding of the remains of an undoubted specimen on the salt marshes at Stiffkey, as first recorded in the "Field," and subsequently by Mr. Stevenson in the "Zoologist" for October, 1869, p. 1863, as follows:—

"*Golden Eagle in Norfolk.*—A correspondent in the 'Field,' of December 19, 1868, announced that a golden eagle had been found dead in the Stiffkey Marshes, in this county, and, though in a state of decomposition when discovered, that the sternum and feet were preserved. I have since had an opportunity of examining one of these feet, by which the species is clearly identified; and I am now enabled, for the first time, to record the occurrence of this eagle in Norfolk. The history of this specimen appears to be as follows:— It was first seen in November lying dead in the marshes, on the property of Mr. P. Bell, of Stiffkey, by a fisherman named Green, who mentioned the fact to Mr. T. J. Mann (the 'Field' correspondent), who was at that time shooting in the neighbourhood: he immediately visited the spot and secured such parts as were most likely to identify the species, the carcase being then too far gone for preservation; but, from his description of the tail-feathers, 'chesnut brown, shading off to a perfect black at the tips,' it was no doubt an adult bird. The foot sent to me in January last, by Mr. Mann, had the toes still supple, as if taken from a recently killed specimen; and from its small size, though having formidable talons, I have no doubt, on comparing it with the fine series of golden eagles in the Norwich Museum, that it belonged to a male bird. I could not ascertain at the time, either locally or from the Journal, that any eagle of this kind had escaped from confinement; and I suspect, therefore, from the locality in which it was found, close to the sea, that it was the victim of some random shot off the coast, and died almost as soon as it reached the shore."

As mentioned by Mr. Stevenson, the sternum was preserved, and is with the feet still in the possession of Mr. T. J. Mann, of Hyde Hall, Sawbridgeworth. The determination of the species must therefore be considered satisfactory, but of course there remains the possibility of the bird having been an "escape."

LANIUS MINOR, J. F. Gmelin.

LESSER GREY SHRIKE.

Mr. Murray Mathew, writing in February, 1870, records ("Zoologist," s. s., p. 2060) his having lately received a specimen of this species, "with a black band on the forehead and rose-tinted under parts," which had recently been obtained " in the immediate neighbourhood of Yarmouth;" and further information enables me to state that the bird had been killed in a garden at the north end of that town in the spring of 1869. A second example was caught in a greenhouse in the same locality in the last week of May, 1875. It lived in confinement a few days, and was then brought to Norwich to be preserved; upon dissection it proved to be a male. The specimen is now in Mr. Gurney's collection at Keswick.

Mr. J. H. Gurney, jun., in his "Catalogue of Norfolk Birds," contributed to Mason's "History of Norfolk," remarks as follows with regard to the form of the grey shrike known as *Lanius major*, Pallas: " Whether this variety be ultimately considered a distinct species from the great grey shrike or not, it is entitled to a place in the Norfolk List as a not unfrequent visitor."

With regard to the specific value of this form, see an exhaustive paper by Mr. Robert Collett, in the "Ibis" for 1886 (pp. 30-40).

TURDUS VARIUS, Pallas.

WHITE'S THRUSH.

An example of this beautiful thrush was killed on the 10th October, 1871, by Mr. Borrett, in the parish of Hickling. It was presented by him to the late Mr. Sotherton N. Micklethwait, and preserved by Mr. Gunn, who recorded the occurrence ("Zoologist," s. s., p. 2848). This specimen, the seventh known to have been killed in Britain, proved by dissection to be a male, and after Mr. Micklethwait's death was acquired for the Norwich Museum, where it now is. Mr. Borrett confirms the statement of other fortunate observers of this rare bird, that its appearance on the wing was so like that of a woodcock as to lead them to mistake it for one.

HYPOLAIS ICTERINA (Vieillot).

ICTERINE WARBLER.

We are indebted to the brothers F. D. and G. E. Power for the recognition of this rare visitant in Norfolk; it was killed at Blakeney, on September 11th, 1884. Full particulars of the occurrence will be found in a paper by Mr. F. D. Power, " Trans. Norfolk and Norwich Nat. Soc,," iv., pp. 36-43.

SYLVIA NISORIA, Bechstein.

BARRED WARBLER.

On the 4th September, 1884, a female of this species, a bird of the year, was killed by Mr. F. D. Power, at Blakeney, during the visit which produced the icterine warbler, and will be found recorded in the paper referred to above. On the 10th September, 1888, Mr. Power was fortunate enough to secure a second example of this rare visitor; it, also, was killed at Blakeney after easterly winds, and near the spot where the previous specimen was shot. This proved on dissection to be a male, and its food was found to consist almost entirely of earwigs. (Cf. " Zoologist," 1889, p. 135).

MOTACILLA ALBA, Linnæus.

WHITE WAGTAIL.

(Vol. i., p. 163). So lately as 1886, Mr. J. H. Gurney, jun., and myself were unable to include this species with certainty in our list of " Norfolk Birds," contributed to the " Trans. of the Norfolk and Norwich Nat. Soc." (iv. p. 275), but on the 14th April, 1888, its claim to be admitted was placed beyond doubt by the occurrence on the golf ground, at Great Yarmouth, of two males, which came into the possession of Mr. George Smith, of that town; while on the 1st May of the same year a third example was obtained by Mr. Pycraft from about the same locality.

EMBERIZA HORTULANA, Linnæus.

ORTOLAN.

Mr. Stevenson decided not to include this species in the "Birds of Norfolk," (cf. i., p. 199), as the genuineness of the only recorded Norfolk example at that time, was open to doubt. Subsequently, however (not to mention one netted at Yarmouth in April, 1866, and six others also said to have been netted in the same locality, on 5th May, 1871, which may or may not have escaped from captivity), an ortolan in very immature plumage was shot at Blakeney by Mr. Power, on the 12th September, 1884, and a second example, in the same locality, by Mr. J. H. Gurney, jun., on 5th September, 1889; these latter could hardly have been other than genuine immigrants.

SERINUS HORTULANUS, K. L. Koch.

SERIN.

A male serin finch was shot at Yarmouth on 13th June, 1885, which, in the opinion of Mr. W. R. Chase (in whose possession the specimen now is), showed no sign of having escaped from confinement. Mr. Stevenson also records ("Zoologist," 1888, p. 84) the capture by a net of one of these birds, on the North Denes, at Yarmouth, on the 5th February, 1887. It was brought alive to Mr. Cole, of Norwich, on the 8th, and seen in the flesh by Mr. Stevenson. It proved on dissection to be a male, and was preserved by Mr. Cole for Mr. Connop. Previous to either of these occurrences Mr. H. A. Macpherson, in April, 1877, as he informed Mr. J. H. Gurney, jun., purchased a pair of these birds of a London bird-catcher, which were said to have been recently caught at Yarmouth. The male died at once, and the female was so wild that after a few weeks she was suffered to fly away. The repeated observations of the occurrence of this species, and the fact that its range on the continent of Europe is known to have been of late years extending, appear to be evidence in favour of the examples being genuine wild birds; and, although the month of February seems an unlikely season for it to appear, this is not the only example which has been met with in that month. See Yarrell, ed. 4, ii., p. 112.

TICHODROMA MURARIA (Linnæus).

WALL-CREEPER.

The record of the occurrence of this species in the county of Norfolk is not a little singular. In the year 1875 Mr. H. P. Marsham, of Rippon Hall, near Norwich, was good enough to place in my hands, as secretary of the Norfolk and Norwich Naturalists' Society, a series of letters written to his great grandfather, Robert Marsham, F.R.S., of Stratton, by Gilbert White, and to this correspondence we owe the knowledge of the fact that on the 30th October, 1792, a wall-creeper was shot at Stratton Strawless, near Mr. Marsham's house. The corresponding letters from Marsham to White were in the possession of the late Professor Bell, and by a mutual exchange the complete series was published both in the "Trans. of the Norfolk and Norwich Nat. Soc." (ii., pp. 133-195) and in Professor Bell's edition of the "Natural History of Selborne." The first mention of the bird occurs in a letter from Marsham to White, dated Otober 30, 1792, in which he says,

"My man has just now shot me a bird, which was flying about my house; i am confident i have never seen its likeness before. But on application to Mr. Willughby,

i conclude it is a wall creeper or spider catcher. I find he had not seen it in England."*

White was much interested by the information, and sent him a translation of Scopoli's description of the bird, which induced Marsham to send in return a coloured drawing of two quill-feathers, of which he says,

"A young lady drew them for me, and they appear to me to be very exact copies and charmingly executed."

Professor Bell was kind enough to lend this drawing to Professor Newton, who found it to represent probably the fifth and seventh primaries of the wing of a female or young male of *Tichodroma muraria*, "leaving no doubt," as he remarks in a foot-note to the published letters, "as to the correctness of the determination of the specimen by Marsham and White." In his reply White remarks,

"You will have the satisfaction of introducing a new bird of which future ornithologists will say—'found at Stratton, in Norfolk, by that painful and accurate Naturalist, Robert Marsham, Esq.'"

I may be excused for adding the concluding remarks with regard to the coloured sketch sent him, which will be found reproduced on plate v., as they are so characteristic of this estimable man, and were written only five months before his death:

"I am much delighted with the exact copies [of the two quills] sent me in the frank, and so charmingly executed by the fair unknown, whose soft hand has directed her pencil in a most elegant manner, and given the specimens a truly delicate and feathery appearance. Had she condescended to have drawn the whole bird, I should have been doubly gratified! It is natural to young ladies to wish to captivate men; but she will smile to find her present conquest is a very old man." To which Marsham replies, "I am glad you liked the drawing of the two feathers; I hinted my wish for the

---

* This is quoted from Ray's translation. The actual words are, "They say it is to be found in England; but we have not as yet had the hap to meet with it" (p. 143). Professor Newton tells me that the author referred to is Merrett, and that in his "Pinax" (1667) the wall creeper is included as a British bird.

whole bird; but she lent a deaf ear: and in that manner, all young Women have treated me (when i ask favours) since i was turned of 40."

ÆGIALITIS ASIATICA (Pallas).

CASPIAN PLOVER.

The true home of this beautiful plover appears to be Western Asia, more especially the shores of the Caspian Sea, whence in winter it passes by the Red Sea shore and Abyssinia, to South and South-west Africa. In the western palæarctic region Mr. Dresser speaks of it as a rare straggler, but it has occurred twice in Heligoland, and strange to say at very opposite periods of the year, namely, a young bird in November, 1850; and an adult male, in full summer plumage, on the 19th May, 1859. These, until the occurrence of which I am about to speak, are the most westerly examples recorded, but Mr. Harting, in an excellent article "On rare and little known Limicolæ," contributed to the "Ibis" in 1870 (second series, vi., pp. 201—213), concludes some remarks on the distribution and routes of migration of this species with the observation that it is quite possible on some future occasion the bird may occur in England. It was with no little pleasure, therefore, that on the evening of the 23rd of May, 1890, I received from Mr. Lowne, of Yarmouth, the fresh skin of a handsome full plumaged male of this species which he sent for determination, as the bird was unknown to him.

I subsequently learnt the following particulars with regard to this interesting occurrence. During the morning of the 22nd of May, a date very nearly coinciding with the second appearance of the Caspian plover in Heligoland, two strange birds were seen in a large market garden, known as Sacret's piece, bordering on the North Denes, at Yarmouth, which attracted the attention of a man named Samuel Smith, who works the garden for a Mr. Bracey, but he had no opportunity of a shot. About 5.30 p.m., when they were on the golf ground, which forms a portion of the denes, Smith's step-son, Arthur Bensley, saw them, and having a gun with him, tried to get both birds in line for a double shot, but being unsuccessful, selected the brighter of the two, its companion being at the time about six yards distant from it; when he fired, the paler bird, presumably the female, flew off in a westerly direction, and was no more seen. Very shortly after, the bird was purchased of Smith by Mr. H. C. Knights, by whom it was shown in the flesh to Mr. G. F. D. Preston, and taken the next morning to Mr. Lowne for preservation; he, as before stated, forwarded the skin to me the same evening. The weather being very warm at the time, Mr. Lowne would not risk sending the bird in the flesh, hence it was that I saw only the skin, but I may mention that it had all the appearance of having been very recently removed, and that there were still many living parasites remaining on the feathers. Mr. Gurney also saw the skin while it was in my possession. The sternum Mr. Lowne sent to Professor Newton. The total length of the bird when in the flesh, Mr. Lowne tells me, was eight inches, and its weight two and a quarter ounces. Mr. Knights was good enough to give me the first offer of the bird, and through the liberality of some friends of the Norwich Museum I was enabled to purchase it for that institution, and to send this first British example of the Caspian plover for exhibition at the meeting of the Zoological Society, on the 17th of June.

In Mr. Dresser's figure of this species the tints of the plumage are not quite so bright as in the freshly killed bird before me, and the conspicuous black border below the chestnut pectoral band is wanting, although it is mentioned in the verbal description; the legs also are coloured and described as "ochreous yellow;" but in the Yarmouth bird they were, when fresh, undoubtedly "greenish ochreous," as described in the "Ibis" by Mr. Harting, a hue which Mr. Dresser states "is certainly an error," but which in this case I can confirm. Mr. Harting's otherwise excellent figure is much darker in colour than the Yarmouth specimen. In neither of the figures referred to has the bird the appearance of standing so high on the legs as in the example now recorded.

I submitted the parasites before mentioned to Dr. E. Piaget, who was kind enough to inform me that they were a new species, for which he proposed the name of *Nirmus assimilis*.

SOMATERIA SPECTABILIS (Linnæus.)

KING-DUCK.

In a foot-note at p. 192 of the present volume, I gave my reason for omitting this species from the "Birds of Norfolk," a step which I had already taken with Mr. Stevenson's full concurrence in 1879, in the second edition of Lubbock's "Fauna of Norfolk." Mr. Julian G. Tuck favoured me in the autumn of 1888 with some interesting notes on the birds observed by him at Hunstanton, and mentioned a young male "eider" seen by him cased at a fish-shop in that town, as apparently differing from other eiders that he had seen. As he had no books of reference with him he made a mental note of it as "a rather dark and small eider," and suggested that it might possibly be an example of the king-duck. I had an opportunity of examining this specimen in the last week of July, 1889—unfortunately after my article on the eider had been printed off—and was delighted to find it a young male *Somateria spectabilis*. The history of the bird is as follows:—It was seen alive off Hunstanton by several of the shore gunners, in January, 1888; among others, by Mr. Tuck's correspondent referred to in the "Zoologist," 1888, p. 148, and was eventually shot about the middle of that month. Mr. Osborne, fishmonger, of Hunstanton, bought it, and for him it was preserved by Mr. Clark, of Snettisham. The specimen did not leave Mr. Osborne's possession until I bought it of him, and transferred it to the Norfolk and Norwich Museum, where it now is. (See also "Zoologist," 1889, pp. 383-85, and "Trans. of the Norfolk and Norwich Nat. Soc.," v., pp. 105-106).

APPENDIX B.

Without attempting to bring the whole of the information contained in the first and second volumes of this work up to our present knowledge, there are a few species to which, from their great rarity or other causes, it seems desirable to refer. They are as follows:—

Noctua tengmalmi, TENGMALM'S OWL (vol. i., p. 60). In addition to that recorded by Mr. Stevenson two other examples of this bird are now known to have occurred in Norfolk. One in Mr. Dowell's collection was taken alive by a labouring man, at Beechamwell, near Swaffham, on the 27th June, 1849. It was found sitting in a bush, apparently dazzled by the bright sunlight, and "mobbed" by jays, blackbirds, and a flock of smaller tormentors. On Sunday, 30th October, 1881, a Tengmalm's owl, now in the possession of Mr. J. H. Gurney, was taken alive at Cromer Light-house at 2 a.m., the weather at the time being misty, and the wind south-south-east.

Ruticilla suecica, BLUETHROAT (vol. i., p. 96). At the time Mr. Stevenson wrote he could only claim this species for Norfolk from a single example; it has since proved to be a frequent, probably a regular, autumn migrant, occurring on certain parts of our coast, usually in September, and sometimes in considerable numbers. As many as eighty were observed by Dr. Power, at Cley, in September, 1884; and on various occasions since that time many have been seen and not a few shot; so that, as just suggested, this species may visit us every autumn in greater or fewer numbers, the vast majority being immature birds. Their stay is very brief, probably only

3 A

till they have rested from their flight over the sea, when they continue the journey to their winter quarters without further delay.

Salicaria luscinioides, Savi's Warbler (vol. i., p. 110). I am indebted to Prof. Newton for the following extract from a letter written by Mr. Gurney to Mr. Heysham, of Carlisle, bearing date "January 15, 1841," which clearly proves that the pair of birds recorded (vol. i., p. 112) as having been shot at South Walsham "in the summer of 1843,"* were killed in May, 1840. Mr. Gurney wrote, "I am extremely glad that the bird I sent thee turns out to be the *S. luscinoides* (what is the English name of the bird for I hate Latin?) & I beg thou wilt accept it as such. It is the male bird of a pair that were killed last May at South Walsham, a marshy parish about ten miles from here. The female I placed at the time in our Norwich Museum. The only differences which I noticed when they were sent to me (I had them in the flesh) between them and the specimens of the Reed Wren which I had previously seen, were that they were somewhat larger, brown instead of olive, and somewhat more rounded or rather more lanceolate in the shape of the end of the tail, which by the way was, I think, a little larger than in the common sort. Both kinds had a head like a Nightingale in point of shape." It may here be observed that it was not until October, 1840, that notice of the species as a British bird was published. As previously mentioned (vol. i., p. 112), Mr. Heysham's bird was subsequently procured for the Norwich Museum, which, by the transfer of the Lombe collection, now contains the two oldest known pairs of British specimens.

Of this apparently exterminated British bird it may be desirable to put on record that in 1848 the Rev. W. Allen (as he informed Professor Newton), in company with the late Mr. Brook Hunter, of Downham Market, obtained a nest in Feltwell Fen; and also that

---

* This statement was no doubt based upon that in the "Zoologist" for 1846 (p. 1308), and the error, arising probably from an accidental misprint, has hitherto escaped observation.

the specimen formerly in Mr. Stevenson's collection is now in the possession of Mr. J. J. Colman, M.P.

Meliozophilus dartfordiensis. DARTFORD WARBLER (vol. i., p. 133). As Mr. Stevenson always considered Lothingland to be geographically part of Norfolk, I shall be justified in stating that I learn from Sir Edward Newton that in the early spring of 1884 he saw a Dartford warbler in a patch of furze on a small remnant of heathland, in the parish of Lowestoft, on the south side of Lake Lothing, and not 300 yards from the houses in that town. But even this patch was stubbed up shortly afterwards, and though he frequently visited the spot he never saw the bird again. In the autumn of 1889, he observed a pair in another locality in Suffolk, and returning the following spring found a nest containing four eggs; in a third locality, in the same county a fortnight afterwards, he saw a pair feeding their young which had just left their nest. These afford the first evidence of the breeding of this species in the Eastern Counties, where it hitherto had always been regarded merely as an accidental visitor. It seems highly probable that this species occurred, although perhaps very locally, on the furze-covered hills which formerly skirted the shores of the counties of Suffolk and Norfolk, in support of which it may be mentioned that the Thirtle MS. recorded their occurrence "near Lowestoft, but rare;" and Mr. James Worthington, who was with Sir Edward Newton when he saw the pair, accompanied by their young, in the past spring, at once recognised them as the same species of bird he had seen on the hills at Corton and Gunton as a boy thirty years ago, and they were, he said, by no means uncommon there when that district was much less frequented than at present. It is quite possible, therefore, that the birds recorded by Mr. Stevenson from Yarmouth Denes may have been residents, and even that so extremely local a species, and one so difficult of observation, may still frequent our furze-covered heaths and escape detection.

Anthus obscurus, ROCK PIPIT (vol. i., p. 169). Mr. Stevenson's remark that this species "is a rare bird in Norfolk," has not been confirmed by subsequent observations. It had probably escaped notice, as he suggested may have been the case, but in autumn it certainly is at times quite abundant on some parts of the coast. In 1880 Mr. F. D. Power suspected this species of having bred at Blakeney. (Cf. "Trans. of Norfolk and Norwich Nat. Soc.," iii., p. 346.) The Scandinavian form of this bird, known as *Anthus rupestris* (Yarrell, edition 4, i., p. 590), has more than once been detected in Norfolk.

Otocorys alpestris, SHORE LARK (vol. i., p. 171). This species, which twenty years ago was regarded as one of our Norfolk rarities, visits us almost every autumn or winter with great regularity, showing considerable partiality to certain spots, where it may constantly be found, and, if not killed off, continues to frequent the same places perhaps for several months. Since 1880 Colonel Feilden has found a flock frequenting a particular locality every winter that he has had an opportunity of visiting the spot, and a very intelligent shore gunner there told him that so long as he could remember there had always been plenty of these birds there in winter, but he formerly believed them to be snow buntings. They are not restricted to one spot, but are found in parties varying in number all round the coast. Mr. Gätke informed Mr. J. H. Gurney, jun., that this species is now much commoner in Heligoland than formerly, and he thinks it is spreading westward.

Emberiza cirlus, CIRL BUNTING (vol. i. p. 198). Mr. Stevenson was unable to record more than two occurrences of this species, namely, in November, 1849, and December, 1855. The latter specimen is in Lord Leicester's collection, at Holkham; but since that time its claim to a place in the Norfolk list has been unquestionably established. In the autumn of 1875, Mr. E. T. Booth obtained two female examples of this bird at Hickling, which were not preserved; two others, now in Mr. E. Connop's collection, were netted on Breydon marshes, on the 29th of January, 1888. Mr. Alexander Napier also shot one at Holkham, which was not preserved.

Coccothraustes vulgaris, HAWFINCH (vol. i., p. 214). This species, regarded in 1846 by Messrs. Gurney and Fisher as "a rare bird in Norfolk," and, in their belief, only occurring as "an irregular migrant," has since proved to be a regular winter visitor, sometimes occurring in considerable numbers. In the winter of 1859-60 large numbers occurred; again, in 1872, about sixty were killed near Diss alone; in November, 1876, thirty were killed in one garden, at Diss, before the slaughter was stopped, the attraction being the numerous berries on some fine old yew trees; many others were met with in November and December, 1878, and in the winter of 1880-1, more especially in the month of January ("Zoologist," 1883, p. 313). Mr. Stevenson remarks that a young bird, shot in the latter end of June, 1856, at Kimberley, where it had been observed with the old birds and two other young ones in a garden, was "the first Norfolk bred Hawfinch he had ever seen." It now nests regularly in various localities in this county, as well as in Suffolk, and the numbers breeding here seem to be on the increase.

Loxia curvirostra, CROSSBILL (vol. i., p. 235). Although recorded by Miss Anna Gurney ("Trans. of the Norfolk and Norwich Nat. Soc.," ii., p. 19) as having bred in this county so long since as 1829, in April of which year a crossbill's nest was found at Sheringham, I am not aware that any other instance has been detected till quite recently; but it is possible, and even probable, that a pair or two may have from time to time remained to breed here; it is, however, very unlikely that had they done so in such numbers as have been met with of late, they would have escaped notice. Certainly for the past five years this species has bred in several widely distant localities in Norfolk and Suffolk, and sometimes in rather considerable numbers.

Professor Newton, in a foot-note to his account of this species (Yarrell's "British Birds," edition 4, ii., p. 190), mentions that Lord Lilford had stated to him his belief that it had several times bred in West Norfolk prior to the year 1877, and in the same locality to which I believe Lord Lilford referred they have since been known to nest several times. Mr. Frank Norgate, who had excellent opportunities of observing this species during the breeding season, tells me that he first discovered it nesting in Suffolk in 1885.

In 1887, Mr. Norgate wrote me with regard to the Norfolk and Suffolk border as follows:—"Eleven crossbills have been slain in the past spring, and I saw three others surviving. I have no doubt there were many nests, but it is difficult to find even the birds, more difficult to identify them even with a glass, and impossible to follow them through the big Scotch fir plantations." In the spring of 1888 Mr. Cole tells me ten of these birds were seen in the Norwich Cemetery, five of which were killed. In the spring of 1889 they are said to have bred freely in a certain locality in this county, and very recently they have again been seen in the Norwich Cemetery, where they remained till the last week in February, 1890.

Mr. Norgate has kindly given me permission to quote from his interesting notes on the breeding of this species, and I shall confine myself to those which relate to this county only, merely remarking that he found these birds equally numerous in certain parts of the adjoining county of Suffolk.

In the year 1889 Mr. Norgate saw many crossbills in the month of January, on Scotch firs. On the 5th March, he says "a man offered me three crossbills' nests of three, three, and three eggs, but his price was exorbitant, and I did not even see them. In North Norfolk, on March 23rd, two crossbills were obtained; they were feeding on fir seeds, but I did not see them. On March 26th I took a nest of four crossbills' eggs from a Scotch fir; the hen bird objected to leave the nest even after it was brought down from the tree, when three or four other crossbills came and fluttered about close to our heads, uttering their peculiar cry and showing their hooked beaks.

"On April 1st I brought home three more crossbills' nests of four, five, and four eggs, from Scotch firs, and had excellent views of many of these birds; they are very tame, allowing one to touch their nests ere they leave them. The red cocks have a pleasing little song, besides the usual call notes.

"On April 13th I took from a very small Scotch fir (little more than a bush in size) a nest of four sat-on eggs; after examining the greenish back, crossed mandibles, and forked tail of the bird, I touched the nest with my hand before she would leave it. I also obtained another nest of four crossbills' eggs taken from a Scotch fir not many yards off, on the 6th of April. I heard of several other crossbills' nests of eggs lately taken by boys. All the crossbills' nests and eggs which I have seen, except that of March 23rd, 1885,* resemble those of the greenfinch, the nests being composed of Scotch and other fir-twigs and dry grass, and lined with rabbit's felt, occasionally a feather or two; most of the greenfinches and other small birds' nests which happen to be built in the fir belts in the open rabbit warrens are of the same materials.

"Crossbills are not so much restricted in range in the nesting season as I had supposed. They or their torn fir cones may be found in most years in March in several widely separated localities in Norfolk and Suffolk, probably nesting in some numbers in most of the Scotch fir belts, and plantations, wherever that tree forms the chief timber crop, and especially in open and uncultivated lands."

I regret to add, in the words of Mr. Norgate, "wealthy collectors and dealers have caused great destruction of crossbills; last year (1889) probably one hundred have been killed!"

A nest which Mr. Norgate sent me is a very compact structure, composed externally of a few fir twigs, next to which are dried roots of grasses, succeeded by fine blades of dry grass closely matted together and thickly lined with rabbit's fur and a few feathers; the whole structure is about five and a-half inches across, and the interior a trifle over two inches; it was taken on the 6th April, 1889, and contained four eggs. The hen bird sat so close as to allow herself to be touched on the nest, and a red cock bird joined the hen in fluttering round.

Loxia pityopsittacus, PARROT CROSSBILL (vol i., p. 239). Mr. J. H. Gurney, jun., and Colonel Butler, on

---

* This nest was in an oak; it was composed chiefly of the inner bark of lime twigs, and was nearly as big as the nest of a mistletoe thrush.

April 16th, 1888, identified at Mr. Gunn's shop, where I also saw them shortly after, two parrot crossbills which had been killed a few days previously near Norwich. There can be no question as to the accuracy of Mr. Gurney's determination of the species, but it is remarkable that, although several other crossbills were killed in the same locality about that time, two only proved to be *L. pityopsittacus*.

The Rev. Julian Tuck was kind enough to send me for examination a very large example of *L. curvirostra*, which he thought might belong to this species; the sternum in all its measurements was exactly intermediate between those of the two species figured at the foot of page 210, vol. ii., of the 4th edition of Yarrell's "British Birds."

Syrrhaptes paradoxus, PALLAS'S SAND GROUSE. The remarkable irruption of this species into Great Britain, in 1863, was so far as Norfolk is concerned very fully dealt with by Mr. Stevenson; first in the "Zoologist" for 1863, pp. 8708 and 8849, and subsequently in the present work (i., p. 376); but a much more extensive influx occurred in the year 1888, of which I shall now have to speak very briefly. Between the years 1863 and 1888 I am not aware that any occurrence in England was recorded, but while investigating the various instances in the latter year, with a view of ascertaining the earliest date on which examples were seen in this county, I was surprised to learn from Mr. E. J. Boult, of Potter Heigham, that on the 27th May, 1876, he had seen fifteen or twenty of these birds rise from the south sand hills, at Winterton, and go away at a great pace to the northward. Although diligently searched for they were not seen again. I have not the slightest hesitation in accepting Mr. Boult's statement as perfectly accurate; and it is further borne out by the fact that in the same year examples were obtained both in Ireland and Modena, a circumstance of which Mr. Boult was not aware.

In 1888 the sand grouse arrived in this county on the 13th May, and soon spread in every direction, certain districts, however, appearing to offer superior attractions, and to these they seemed to be remarkably attached, for a long time making them their head quarters. I communicated a somewhat detailed account of this visitation to the "Zoologist," which will be found in the volume of that journal for 1888 (pp. 442-456), bringing my notes down to the 31st of October of that year, up to which time it seems probable that the large number of between 1100 and 1200 of these birds had been seen in Norfolk, and in addition not fewer than 186 killed; to this account I must refer the reader, and in the brief record which follows I shall confine myself to events subsequent to that date, merely repeating as being a matter of special interest the main facts that bear upon the supposed instances of the breeding of the species in this county.

Although many reports were circulated as to the finding of nests in various parts of the county, in no instance could they be substantiated by the production of either eggs or young. In one instance, however, it seems highly probable that eggs were really produced. A Mr. Tolman, of South Pickenham, near Swaffham, in the last week of June, shot what he states to have been a sand grouse, as it rose from its nest containing three eggs; these he described as "much like those of a water hen in colour, but rather darker, largest in the middle, and tapering off towards each end." Mr. Tolman broke two of these eggs and gave the third to his landlord, Colonel Applewhaite, who in attempting to blow his specimen, unfortunately broke it also; he thinks it was within three or four days of hatching. There can be no doubt of the good faith of Colonel Applewhaite and his tenant, Mr. Tolman, but it is most unfortunate that no fragment of the eggs was preserved.

Mr. Alexander Napier, whilst shooting with Lord Leicester and party on the Holkham sand hills, on the 13th October, met with a flock of thirty, all of which went away with the exception of a single bird, which he tells me he "felt convinced must have had either a nest or young. When first I saw it, it fluttered along in front of me just like a partridge with young. It was so tame that I called Lord Leicester and the others up to see it, and it did not fly up until we had approached to within three or four yards of it . . . it flew away very strongly, calling out most lustily." No further search was made for eggs or young, probably under the impression that the date was a very unlikely one for the bird to be accompanied by its young, but with so "paradoxical" a species, which has since been found with newly hatched young on the 8th of August, such a search would certainly have been desirable. The strange way in which some of these birds allowed themselves to be approached and even to be taken by hand when apparently uninjured is very remarkable, and it is possible that Mr. Napier's case may simply have been another similar instance.

It must not be supposed that the following list is by any means so complete as that of which it is the continuation, as from a variety of circumstances I was unable to give the time requisite to continue the investigation so carefully as I had done up to that date, but I think it is sufficiently complete for the present purpose, and I may add that I am indebted for information to the same sources to which I endeavoured to acknowledge my obligation on the previous occasion.

The birds which had so long made their home at Morston all disappeared from thence late in the autumn, and the only entries I have for November, 1888, are a solitary male killed at Blakeney, on the 4th of the month, which had frequented that neighbourhood for some three weeks previous to its being shot, and two others seen at the same place on the 20th.

In December a male and female were seen on the denes at Yarmouth on the 2nd; on the 14th four were seen by a party shooting at Thetford; on the 28th Mr. Hamond's keeper saw two or three near Beechamwell, and on the same day one was killed at Brancaster; on the 29th Mr. A. Napier informed me they were still to be found on the Warham Hills.

In January, 1889, between the 2nd and 9th, three males and three females were killed near Brancaster; and a single bird heard by Sir E. Newton and seen by Mr. Francis Newcome on the former date when shooting at Wilton; on the 7th a live female caught at Southrepps was brought to Mr. Gurney, at Northrepps;* on the 14th Mr. B. Bowen wrote to the Rev. J. G. Tuck that there were still sand grouse to be seen between Thornham and Ringstead, and that one had been killed on the marshes, at Thornham, in the first week of the previous December; on the same day (the 14th of January) a male and a female were killed near Wells; on the 28th three were seen, as I am informed by Mr. F. Newcome, during a partridge drive at Downham (Suffolk); and on the 29th the same gentleman tells me six others were seen at Wilton, in Norfolk.

On February the 16th one was shot at Buxton; and on the 17th two were seen at Cley-next-the-Sea. From this date to the 17th of the following October I have no record of their having been again met with,† but on the latter day I am informed that Mr. George Hunt saw a flock which rose at his feet on the sand hills, near Hunstanton, and that on the 4th of December eleven were seen by Sir William ffolkes when shooting at Docking, twenty were seen on the same day in the next parish by another person, possibly the same flock. Colonel Feilden tells me that on the 22nd of April, 1890, three of these birds were seen by Cringle and another shore-gunner to whom the species is well known, to come inland from Holkham Bay, disappearing in the sandhills. On the 9th of May, as I am informed by Sir Edward Newton, five were observed to fly from the land and drop on the beach at Lowestoft just above highwater mark; and on the 10th another flock of nine were seen flying southward at the same place. About the 15th of May a flock of fourteen were seen from the sea near Aldeburgh, Suffolk, as reported by Mr. Edward Neave in "Science Gossip" for 1890, p. 187.‡

---

* This bird had probably been injured, and is now dead, but a male, also taken by hand some time previously, is still alive (August, 1890) in the aviary at Northrepps.

† Professor Newton was informed on what seems to be good authority that during the summer of 1889 a small number frequented some open brecks at Eriswell, in the adjoining county, not very far from our own boundary; and it is quite possible that they bred there, though no eggs or young were observed.

‡ On the 30th May, 1890, Mr. Cordeaux writes me, "Sand grouse came direct in from the sea to Spurn Point last week, and

It will be observed that no sand grouse are recorded in Norfolk between the 16th of February and the 17th October, 1889; this seems significant of the departure of the birds which had wintered with us, and a return migration in the autumn; and, again, there seems little doubt that fresh arrivals took place in the spring of the present year 1890, at various points on the east coast, but there is no reason to believe that there was a fresh irruption from Tartary.

In the above list thirteen are mentioned as killed and some seventy as seen, not inclusive of three instances in which flocks were observed but the numbers not recorded. A few others were to my knowledge killed, though of them I have no precise information.

Since my last notes on the subject some seeds taken from the crops of birds killed in this county and sown, but at that time not satisfactorily determined, have produced plants that came into flower with Mr. H. D. Geldart, who was kind enough to grow them for me, and proved to be *Suæda maritima*, *Trifolium striatum*, *T. minus*, and *Anthyllis vulneraria*.

Otis tarda, GREAT BUSTARD* (vol. ii., p. 1.) On a subject so superlatively interesting as the history of the native race of Norfolk and Suffolk bustards, it may well be expected that a considerable amount of additional information has accrued since Mr. Stevenson completed his account of them (full as that was) in the autumn of 1868. Indeed, he himself, in the preface to the second volume of this work, written two years later (in 1870), stated that he had even then obtained further particulars, which he looked forward to giving in an appendix to the present third volume. These particulars, and others which have come to my knowledge, I purpose putting on record as briefly as the importance of the subject will permit, following so far as is possible the order of topics adopted by Mr. Stevenson.

I know of four couple here in North Lincolnshire. A flight of thirty is reported near Driffield, in Yorkshire, on the high wolds. I fancy all these are fresh arrivals."

* I wish specially to acknowledge my obligation to Professor Newton for his kind assistance in compiling this account.

The entry in the le Strange Household Book of 1527 has been cited (p. 2) as the earliest record of the bustard in Norfolk; but Mrs. Herbert Jones in her "Sandringham Past and Present" (London, 1883), quotes (p. 23) from the Chamberlain's accounts of the Borough of King's Lynn, preserved among the Corporation documents of that time, to the following effect:—"In 1371 the 44th of Edward III., 39s. 8d. was paid for wine, *bustards*, herons, and oats, presented to John Nevile, Admiral."*

Passing from this remote period to the end of the last century, among the papers left by the late Mr. Lubbock are some notes that are worth preserving. One is as follows:—

"Mr. Browne, the Rector of Blo Norton, tells me that in 1783 his grandfather, whilst riding on a Sunday from Eccles to Wretham, to perform divine service, passed seven bustards, which were walking about close to the road. They did not retreat or appear much alarmed. Mr. B. was fond of coursing, but notorious as a bad hare finder. 'To show you how near they were to me,' said he, '*I* could find a hare sitting amongst them.'"

Not long after we find the poet Cowper writing, on the 31st of January, 1793, to his cousin John Johnson (father of a rector of Yaxham and Welborne, in this county, of the same name, who was editor of Cowper's correspondence); and, after mentioning that he had been prevented from asking their friends the Courtenays to dine with them since their marriage owing to the illness of his friend, Mrs. Unwin, with whose family he was staying, continuing:—

"But this is no objection to the arrival here of a bustard; rather it is a cause for which we shall be

---

* Mrs. Jones observes that the word bustard does not occur in the list of supplies sent to Queen Isabella at Castle Rising, nor did bustards form a dish at the grand dinner given at Lynn to the Lord Chancellor in the reign of Henry IV., at which dinner, curlews, plovers, ducks, and so on abounded, and hence argues that bustards could hardly have been very common at that time. This negative evidence, I think, does not warrant the lady's conclusion.

particularly glad to see the monster. It will be a handsome present to *them*. So let the bustard come, as the Lord Mayor of London said of the hare, when he was hunting—let her come, a' God's name; I am not afraid of her."*

The "Sporting Magazine" for May, 1794, has a note to the effect that "a man, in company with the Rev. Mr. Graves, of Lackford, being in pursuit of some bustards, near Thetford," was so much injured by the bursting of his gun that he died in the workhouse there a few days after.

The sale "Catalogue of the Leverian Museum" shews (p. 63) that lot 1520, sold on the 19th May, 1806, was "a noble specimen of the male bustard, *otis tarda*, shot in Norfolk, its weight was 29 lbs.;" and, according to the annotated copy in possession of Professor Newton, it was bought by "Donovan ℔ Latham" at the price of four guineas.

Graves, in his "British Ornithology" (vol. iii.) published in 1821, states of this species—"In the spring of 1814 we saw five birds on the extensive plains between Thetford and Brandon, in Norfolk; from which neighbourhood, in 1819, we received a single egg, which had been found in [on?] an extensive warren."

Mr. Whitear in his diary, under date of June 17th, 1816, wrote, "I saw a fine male bustard in the open fields, between Burnham and Choseley; he suffered me to approach him to the distance of about a hundred yards; he then walked a few paces and took wing. He moved his wings like a heron."†

The dwindling of the Swaffham drove of bustards already mentioned by Mr. Stevenson (vol. ii., p. 7) receives further illustration from a letter dated 28th April, 1824, written by Robert Hamond to Selby, who had apparently asked him for a specimen:—

"I am sorry I cannot let a bustard accompany the stone curlew, but they are becoming exceeding scarce. I saw three about a month ago, & one female yester-

---

* "Private Correspondence of William Cowper," &c., ed. 2, London: 1824, vol ii., pp. 319, 320.

† "Trans. Norf. and Norw. Nat. Soc.," iii., p. 242.

day, & generally on the same spot; & I have found from observations that a female will frequent a certain field, undisturbed, for near a month before she deposits her eggs, which is generally about the 12th of May. There was a nest destroyed by the weeders last year near the spot I can generally see the one in question. I wish you would pay me a visit, & I think I could show you some. I am certain in winter, if not in summer; & I am now confident in asserting that the males come & visit the females, as they are never seen together at this season of the year." ("Trans. Norf. and Norw. Nat. Soc.," ii., pp. 400, 401.)

Nine years later (1833) Mr. Lubbock has a note:—
"At present four bustards known about Congham and Westacre, supposed all females; no increase among them."

I find the following note, dated November 30, 1870, in Mr. Stevenson's handwriting:—

"Mr. Drake, late of Billingford, told me to-day he had recently seen Tom Saul (late coachman to Cromer in Windham's time), and talking of bustards, he said somewhere about 1830 he remembered riding with the late A. Hamond and Mr. Fountaine of Narford, near Walton Bottom, below Westacre Field, where they saw something at a distance he could not make out, but took for sheep troughs." Mr. Hamond offered to make a bet that the objects were bustards. "Saul rode on, the birds rose and fled, and Mr. Hamond remarked, 'did you ever see sheep troughs take wing before.' Saul was of age in 1831, and this was about that time. . . . Mr. Drake says he has somewhere a bustard's egg given to his father by the late Mr. Downes, of Gunton, and his father always understood it was found on Beechamwell Common."

To Colonel Feilden I am indebted for the following extracts, copied by him from the Holkham game book, and given to me with the Earl of Leicester's approval:—

"1814. Oct. 7. Mr. Butcher killed a bustard on his farm and sent it here.

"1816. Nov. 11. [after a list of the birds killed by the different guns]. Wild day. Lord Spencer saw three bustards.

"1820. Sept. 7. Mr. Patterson shot 14 part., 1 hare, and 1 rabbit, two bustards. Gave away to Mr. Reeve 3 partridges."*

In regard to the matchless series of Norfolk bustards preserved at Congham House (vol. ii., pp. 33, 34), the details furnished by the late Mr. John Scales, an intimate friend of their former owner Mr. Robert Hamond, have now been published ("Trans. Norf. and Norw. Nat. Soc.," iv., pp. 113, 114), and the discrepancies, such as they are, between them and those supplied by Mr. E. J. Moor ("Zoologist," 1870, p. 2024) may be compared. At most they are not very important. Mr. Stevenson, from a paper which he left, seems to have leant towards Mr. Moor's account, which was written *forty-six* years after he had it from Mr. Hamond; but that of Mr. Scales was penned *thirty-five* years after the events recorded, in which he himself took an active part —a fact that, letting alone the shorter period during which memory could prove faithless, rather inclines one to put the greater trust in his version.† That there may be slight inaccuracies in each it is more than possible, and Mr. Stevenson has left a note as follows:—

"I have recently conversed with an old keeper, named Cater, now in his 79th year, who was formerly in the service of the Rev. R. Hamond, and remembers Mr. Scales assisting his master to stuff a very fine male bustard, but his impression seems to be that the male bird in the large case was one shot at and wounded by his brother, when flying over Narford field, and which, having crossed the river, was picked up dead on West-

---

* This Mr. Patterson was an American gentleman, and a guest at Holkham. As it was the custom to give game to the tenant on whose farm the shooting was, the name of Reeve shows the locality in which the bustards were killed.

† The age of the two witnesses at the time of giving their evidence also has to be considered. Scales wrote his testimony in 1855, and having been born in 1784, was then sixty-one years old. Moor in 1870 describes himself as having been a boy some "fifty-five years ago." Supposing him then to have been fifteen he would be seventy at the time he gave his testimony in 1870. Besides this, Moor only heard the particulars from Hamond in 1824, and does not say that he wrote them down at the time; whereas Scales must have been thoroughly acquainted with them.

acre field. We have, however, Mr. Hamond's own statement in writing that all the bustards represented in the lithograph [vol. ii., p. 34] were shot by him."

In 1872 Mr. Stevenson succeeded in tracing the history of two more specimens which had been in Mr. Robert Hamond's hands. In August, 1869, Mr. Richard Enfield, of Nottingham, wrote to Mr. J. W. Dowson, who was interested in the Norwich Museum, informing him that at the recent sale of Sir Robert Clifton's effects, at Clifton, near Nottingham, a fine male bustard, in good preservation, and said to have been obtained in Norfolk, had been bought by the Rev. H. Bell, who would have no objection to sell it to the Norwich Museum. Mr. Reeve, the curator of the Museum, tells me he well remembers this offer being brought before the committee, who did not, however, accept it, and nothing more was heard of the specimen until some three years after, when Mr. Stevenson, on enquiry, found it had passed into the possession of Mr. James H. Lee, of Linton Field, Nottingham, who confirmed the information previously obtained, and subsequently sent Mr. Stevenson a copy of the inscription at the back of the case, in the handwriting of its former owner (the uncle to the Sir Robert Clifton just named), as follows :—

"Male and female. Presented by the late Rev. Robt. Hamond, Swaffham, Norfolk. I believe shot there A great friend of mine."

As the case from its size could never have held more than this one specimen, and it was clear from the inscription that two birds had been sent, Mr. Stevenson set himself to find the missing female, and soon after learned from the late Dr. Beverley R. Morris that it was in his possession, having been bought at the Clifton sale by a Mr. Warwick, of Nottingham, who, in the spring of 1872, sold it to a Mr. Varley, from whom Dr. Morris had it. This specimen was in the possession of Dr. Morris's widow, who is very recently deceased, while the male remains in that of the family of Mr. Lee, who is also dead.*

---

* Mr. Stevenson's enquiries about other reputed Norfolk specimens produced, in 1869, a letter from Mr. Thomas Stubbs, of Alma

In 1888 Colonel Feilden was so fortunate as to obtain a specimen of a female bustard which there is good reason to believe was of the original Norfolk race. From the account which he published ("Trans. Norf. and Nor. Nat. Soc.," iv., p. 519) it had been given more than twenty years before as a plaything to the children of a labouring man living at Holkham, by the wife of a neighbour whose relations were inhabitants of Sedgeford, near Heacham. How it came into her possession cannot be ascertained; but the late Mr. Samuel Bone, deer-keeper at Holkham, himself brought up at Westacre,* recognized its value, and, protesting against the use to which it was being put, prevailed on the father of the children to have it put in a case. Here Colonel Feilden discovered it, and after having it re-mounted by Mr. Roberts, of Norwich, who found the body to be made of layers of cork, presented it to Lord Leicester, in whose collection at Holkham it now is.

Mr. Lubbock's papers also supply a contemporary, and therefore important, record of the supposed bustard seen near Harling (vol. ii., p. 8), and mentioned in his "Fauna of Norfolk" (ed. 2, p. 65), about which much doubt has been expressed:—

"About Christmas, 1840, Mr. G. Montgomerie, walking over Roudham Heath, came within fourteen or fifteen yards of a bustard—his dog stood at it in astonishment, and he did not know in the least what it was. On its taking wing, which it did slowly, he struck it with both barrels, but did not bring it down. The bustard in question took a long flight to the other side of the parish, where Mr. T. Montgomerie [owner of Garboldisham Hall] happened to be shooting, who

---

House, St. Agnes Gate, Ripon, who stated that he had a pair in his possession which he believed had been obtained at least thirty-five years before, but being then eighty-four years of age, he was unable to give any particulars of them. Mr. Stevenson could not have attached much importance to this statement, as, although he must have been aware of it when he wrote his article, he does not refer to it therein.

* Bone, in 1880, being then sixty-nine years old, told Colonel Feilden of his having in his youth found a bustard's nest with two eggs in it at Westacre.

chased it about a long time without being able to get a shot at all. The bird has since been seen in the parish of Larling. (Jan. 22d, 1841.)"

This story has been variously told and often disbelieved, but the above note, written at the time and probably from information obtained at first hand, makes it more credible than had hitherto appeared, and it will be observed that Mr. Lubbock speaks unhesitatingly of the bird as a bustard—which many people were inclined to doubt. Whether, however, it was the last of the Norfolk race or an immigrant is, of course, uncertain, though the presumption is certainly in favour of the latter view. More than two years had passed since the Lexham bird (vol. ii., p. 7)—itself quite possibly a stranger, though generally regarded as a native—had occurred, and many years since real Norfolk bustards had been seen anywhere near Roudham. Moreover, experience had not then shown that it is by no means uncommon for bustards to visit England in midwinter, which is now known to be a fact, and that when they do so occur it is most often in places remote from the ancient haunts of the native bird. That this should happen is not surprising when the limited extent of those haunts, compared with the rest of England, is considered.

Through the liberality of the representatives of the late Miss Postle, the bustard shot at Horsey in 1820 (vol. ii., pp. 30, 37) has found a home in the Norwich Museum, as also has the specimen killed at Lexham in May, 1838 (*tom. cit.*, p. 7)—by some supposed to be the last of the old Norfolk race—which was presented to the Museum in 1877 by the Rev. W. A. W. Keppel, and is there placed in the same case as the other specimens from this county.

More precise information concerning the specimen formerly in the Norwich Museum, and now in the collection of the Rev. C. S. Lucas, of Burgh (vol. ii., pp. 35, 36), is furnished by a letter dated Watton, 24th May, 1829, written by Mr. J. S. Futter to the late Mr. Samuel Woodward, and now in the possession of his son, Mr. Horace B. Woodward, as follows:—

"I was the other day at Eriswell [in Suffolk] where I saw a hen bustard which the keeper had taken in a rabbit trap alive. The bird has lost one foot and an eye, but will, I think, make a good specimen should your Museum directors wish to purchase it." (See "Trans. of the Norfolk and Norwich Nat. Soc.," ii., p. 612.)

It is probable from the above that the writer had seen the example in the flesh, and if so the date of its occurrence,* which hitherto has been doubtful, is fixed, and accords with that given by Bishop Stanley.

Another Eriswell specimen seems to exist, as Mr. Stevenson was informed by Mr. Harting, who saw it at East Acton in 1869, and subsequently learnt from its then possessor, the Rev. J. N. Ouvry-North, that it had been bought by his brother at the sale of the effects of the late Mr. James Gibson, formerly treasurer of the New England Company, which owned the chief part of the parish of Eriswell, and that it was believed to have been shot there by the gamekeeper, Thomas Rutherford, about the year 1818.

Of the two eggs mentioned (vol. ii., p. 39) as having belonged to Mr. Salmon, one given by him to Mr. Wolley remains, as before stated, in Professor Newton's possession; but the second, instead of having disappeared, as believed by Mr. Stevenson, is fortunately still safe in the Salmon collection, at the rooms of the Linnæan Society, in Burlington House, where Mr. Edward Bidwell found it, as he has reported ("Trans. of the Norfolk and Norwich Nat. Soc.," iv., p. 142). That gentleman has also recorded (*tom. cit.*, p. 141) the existence of an egg in Mr. Crowley's possession, which, at the sale of Mr. Heysham's collection (16th May, 1859), was bought by Canon Tristram, and is doubtless a Norfolk specimen sent by Hoy to Heysham, as a label upon it in the latter's handwriting testifies.

---

* Two misprints were made in Mr. Stevenson's notice of this example (vol. ii., p. 36). In line 10, " p. 3 " should be p. 111, and in line 11 " Cresswell " should be Cresswall.

The specimen formerly possessed by Mr. Scales (vol. ii., p. 40) perished in the fire that destroyed the rest of his collection, as elsewhere recorded ("Trans. of the Norfolk and Norwich Nat. Soc.," iv., p. 96), and that stated (*tom. cit.*) to belong to Mr. Coldham, of Anmer, was given, as Captain Mason informs me, to his sister, Miss Mason, after whose death in Scotland many year ago it was missing, and his efforts to find it have failed. He thinks it must have been taken not later than 1820.*

The Norwich Museum contains two specimens, both of which may be of Norfolk origin, though beyond the fact that one of them, mentioned before (vol. ii., p. 40), is believed to have been added by Mr. Salmon in the year 1835, when he arranged the eggs in the Museum, and that the other, which is unblown, was received in 1881 with the Whitear collection, nothing is known of their history. In 1885 Col. Feilden discovered in the possession of the late Mr. John Clarke, of Wigston Hall, Leicestershire, but then living at Great Yarmouth, another unrecorded specimen inscribed "Norfolk, 1824," which Mr. Clarke said was given to him a few years after that date by the late Rev. W. W. Turner. Mr. Turner had two more in his collection, which is now in the possession of his daughter, as Professor Newton subsequently learned from that lady.

One occurrence of the bustard in this county† remains to be noticed, and that was attended by some circumstances so remarkable as to render it almost unique in the annals of British ornithology. Though on the birds first appearance hostile steps were taken against it, they were fortunately ineffectual until better counsels prevailed, as they speedily did, and the visitor then received the utmost consideration from Mr. H. M. Upcher, on

---

* Captain Mason informs me also that in the time of James Coldham, Esq., nearly one hundred years ago, a bustard was shot on Anmer Field by a painter from Lynn, named Raven, who was doing some work on the house.

† According to Mr. Howlett, of Newmarket ("Field," 16th August, 1873), a bustard had been recently seen on the warrens at Wangford and Lakenheath, in Suffolk. The report may well have been true, but nothing more is known of the bird.

whose land at Hockwold it took up its abode, while every inducement for it to prolong its stay was offered. The arrival of the bird was made known to Mr. Upcher on the 24th January, 1876, and on his going to the place, a field of coleseed, where he was told he should see it, he found to his astonishment that it was a fine male. He at once gave orders that it should not be disturbed, and the situation it had selected, being in the heart of Blackdyke Fen, bounded by wide ditches, was a pretty safe one. Here it was seen for several days in succession, though it occasionally shifted its position at the approach of any person, always returning to the coleseed, however, at nightfall. On the 2nd and 3rd of February it was observed by Mr. (now Sir) Edward Newton, and Messrs. Francis and Edward Newcome —on the latter of these days Professor Newton being also of the party. On the 8th Mr. J. H. Gurney, jun., went to the place and had a good view it. The same day Mr. Upcher received by telegram an offer from Lord Lilford, who had been informed of the occurrence, of a hen bustard to turn out in the hope that her presence would induce the cock to remain. The offer was joyfully accepted: the bird arrived the next day, and on the 10th of February Mr. Upcher assembled a party of wedding guests, including the Messrs. Newcome, Sir Edward and Professor Newton, Mr. Salvin, and Mr. Harting, who had come from London expressly to be present. The hen bustard was accordingly released, the greatest care being taken not to alarm the cock in so doing. He, however, took wing, and though he soon returned, yet a snowstorm, which unhappily came on, had caused her to move away from the spot where he alighted. That night there was a sharp frost followed next day by a fog, so that it could not be seen whether the two birds had joined company; but on the 12th and 13th both were observed by the watcher, whom Mr. Upcher employed for the purpose, to be walking about together—and indeed the male was reported as strutting and "traping his wings like a turkey-cock." The next night, however, there unfortunately fell a deep snow: the male was seen close to the shelter of hurdles and straw which had been prepared for the female, who was supposed to be inside it, but she was not visible. The weather again changed: in the evening it blew a gale, and rained in torrents. On the 15th the snow had all gone, but the rain continued. The hen bustard was reported missing, and on search being made she was found dead in one of the ditches, having apparently crept for shelter into the grass by its side. On hearing of this disaster Lord Lilford liberally despatched a second female, which arrived on the 21st, but the weather was so bad that Mr. Upcher was afraid to turn her loose, and left her in the hut with a hurdle across the front. The cock bird was then on the ground, and seemed to have become accustomed to the presence of man. The next day, seeing him not very far from the hut the keeper went to liberate the hen; but, while so engaged, the cock unfortunately took wing. He returned in the evening, but settled on another piece of ground, where he was observed on the 23rd; but in the afternoon he was seen flying towards Brandon, and he never again, so far as is known, returned to Hockwold. On the 24th he was seen at Eriswell, and on the 25th at Elveden, after which nothing was heard of him, though enquiries were made in all the surrounding districts by Mr. Upcher, who was at much trouble to ensure his neighbours taking an interest in the bird, and to him and to Lord Lilford the thanks of all true naturalists are due.* The story of this unparalleled incident, which but for the unfavourable weather might have been followed by a very different result, here briefly told, will be found more fully narrated by Mr. Upcher in "The Field" of 8th April, 1876, whence it has been reprinted by Mr. Stevenson ("Trans. of the Norfolk and Norwich Nat. Soc.," ii., pp. 307-311), and by Mr. Harting in his "Essays on Sport and Natural History" (pp. 332-340).

Glareola torquata, PRATINCOLE. The extraordinary statement attributed to Hoy that this species bred an-

---

* Lord Lilford announced in the newspapers his willingness, in case of the appearance of the bird in any other part of the country suitable for the propagation of his species, and an assurance of good faith from the owner of the locality, to forward at once one or more hen bustards by way of inducing him to settle there.

nually in the fens of Norfolk, a locality quite unsuited to its habits, and commented upon by Mr. Stevenson in a foot-note, at p. 64, vol. ii., of the "Birds of Norfolk," has been corrected by the publication of a more accurate copy of the same letter, with others from celebrated Norfolk naturalists, in the "Trans. of the Norfolk and Norwich Nat. Soc." (ii., p. 404), where the following note by Professor Newton explains the origin of the error:—

"A copy of this letter, which is in an injured condition, was printed by Dr. Bree in the 'Field' newspaper for November 9th, 1867, from a transcript in the handwriting of, and lent to him by, Mrs. Lescher [sister of Mr. Hoy]. Some part of that portion which is now missing must then have been in existence, but the letter had probably already sustained some damage. . . . An attempt has been made here to restore the missing portion with the help of the older copy, the words supplied being enclosed by square brackets." The portion of the letter as restored by Professor Newton reads as follows:—"On the [20*th May*, 1827, *a pair of the pra*]tincole *Glareola austriaca*, were shot nea[*r Yarmouth. A specime*]n of the *Gallinula minuta*, of Montague, was [*shot near Yarmouth last*] May.

"I have received several sp[*ecimens of godwits, Scalopax œgocephala*] & *Limosa*, at different times from Ya[*rmouth. A few of them breed an*]nually in the fens near Yarmouth, i[*n the same locality as the Ruffs, wit*]h wh[*ic*]h they are sometimes taken." It will thus be seen that the accidental omission of a portion of the letter in the transcript completely altered its sense, and made a remark appear to refer to the pratincole, which in the original applied, and correctly so, to the godwit.

Professor Newton has very kindly forwarded me a series of letters with reference to a pair of these birds said to have been killed in Norfolk, in 1810, now in the collection of Mr. Thomas Boynton, of Ulrome Grange, Lowthorpe, Hull. It appears that Mr. Boynton purchased these birds of Dr. W. W. Boulton, of Beverley, who stated that he was informed they were "shot off the breakwater, near Yarmouth, November, 1810, by Mr.

Hunt, schoolmaster, Yarmouth," but Dr. Boulton has "quite forgotten the name of the old man from whom he purchased them."

Whether these are the pair of pratincoles referred to by Mr. Stevenson (vol ii., p. 64) as killed in May, 1827, on Breydon wall, and which he, with the assistance of Captain Longe, was unsuccessful in tracing, it is now impossible to say. Hunt, in his list of Norfolk Birds, so often before referred to, says these birds were killed in the "autumn" of 1827, but the entry in the Hooker MS. is very precise, and settles the date beyond question. It states in an entry dated "May 21st, 1827," that they were "shot on the marsh near Breydon, Yarmouth, *to-day*, are now in possession of Mr. Harvey;" the Pagets say they were killed on "Breydon wall," which may well be the "breakwater near Yarmouth," and we have only to supply the forgotten name of Harvey to Dr. Boulton's account of the purchase, and to suppose the date "1810" (in support of which there is no contemporary evidence) to have been a slip of the memory, to identify these birds with the pair killed on Breydon in 1827.

Mr. J. H. Gurney, jun., has seen these birds, and is of opinion that their appearance is indicative of their having been set up from skins. Supposing, however, that they are the pair referred to in the "Birds of Norfolk," the rough treatment they were admitted to have received (l. c., p. 65) would amply account for this appearance.

In the "Zoologist" for 1869 (p. 1492) Mr. Stevenson records the occurrence of a pratincole at Feltwell, where it was shot in the first week in June, 1868, and is now in the Newcome collection at that place.

APPENDIX C.

The following five species, I think, cannot be unreservedly admitted to the "Birds of Norfolk." I have, therefore, thought it best to refer to them in an appendix by themselves, stating, as fully as the evidence enables me, the claims in each individual case.

Calandrella brachydactyla, SHORT-TOED LARK. Mr. George Smith, of Yarmouth, records the occurrence of an example of this species in the "Zoologist" for 1890, p. 77, stated by him to have been shot near South Breydon Wall, Yarmouth, on the 7th November, 1889. It proved upon dissection to be a male. There can be no question as to the bird obtained by Mr. Smith being an example of this species, which is doubtless very likely to occur here in a state of nature, but it may be well to note that about the end of October, 1889, short-toed larks are known to have been imported into London with skylarks, and that two are recorded as presented to the Zoological Gardens on the 28th of the same month by Commander Latham; it is, therefore, not altogether impossible that the bird in question may have been an "escape."

Pyrrhula enuncleator, PINE-GROSBEAK (vol. ii., p. 234). Mr. J. H. Gurney, jun., "Zoologist," 1877, p. 845, considers the title of this species to a place in the Norfolk list as at least doubtful, but I follow Professor Newton in the 4th edition of Yarrell's "British Birds," who, with evident reluctance, allowed the bird to remain, simply rejecting the report with regard to its having nested in this county.

Charadrius fulvus, EASTERN GOLDEN PLOVER. This species has been admitted to the list of British birds on the strength of a specimen found in Leadenhall Market, in December, 1874, said to have been received from Norfolk (Dresser, "Ibis," 1875, p. 513). It is also included in the 2nd edition of Lubbock's "Fauna of Norfolk," 1879; and in Mr. Stevenson's list of Norfolk Birds in White's "Directory of Norfolk," dated 1883, but in most subsequent lists it has been regarded as doubtful, and Mr. Howard Saunders, in the 4th edition of Yarrell's "British Birds," observes that "although there is nothing improbable" in the statement that it was sent from Norfolk, "the evidence appears to be hardly strong enough to justify the admission of this species as a British bird," and, therefore, as a Norfolk bird also. Mr. J. H. Gurney, jun., remarks, "Catalogue of Birds of Norfolk" (in Mason's "History of Norfolk"), with regard to this occurrence, "Without wishing to throw doubt on this particular bird, I must say that my experience of London markets leads me to consider 'localities' worth next to nothing from that quarter," a sentiment with which my experience of some other markets has led me to perfectly agree.

Ardea alba, GREAT WHITE HERON (vol. ii., p. 149). Mr. Stevenson was evidently of opinion that this species had some claim to be regarded as a Norfolk species, and has left the following note on the subject:—"Thurtell's bird probably doubtful, but I cannot help thinking that Miller's bird might be relied on, though Lubbock does not mention it. His friend Girdlestone's birds many of them came into Miller's hands." Under the circumstance I do not feel justified in rejecting the species. Professor Newton possesses a letter written from Yarmouth in 1831 by Dawson Turner to P. J. Selby, in which reference is made to this species as a Norfolk bird. ("Trans. of the Norfolk and Norwich Nat. Soc.," ii., p. 413).

Porphyrio smaragdonotus, GREEN-BACKED GALLINULE. This species, with respect to the validity of the claim of which to be regarded as an occasional straggler to the British shores ornithologists differ, was not included by Mr. Saunders in the 4th edition of "Yarrell."

It has been met with five times in Norfolk, and only one of these examples can with any certainty be regarded as an "escape;" the others all occurred in localities and under circumstances which are by many considered compatible with their being voluntary immigrants. The examples referred to occurred—the first at Tatterford, near Fakenham, on the 10th October, 1876; the second near Hickling Broad, on the 7th September, 1877 (as a bird of this species was missing from Northrepps Hall about the same time, it is probable this may have been an "escape"); the third was shot on some swampy ground between Stalham and Barton Broads, on the 1st November, 1877; the fourth, also at Barton Broad, on the 23rd August, 1879; and the fifth was shot at Horning, on the 16th October, 1885. Mr. J. H. Gurney, jun., has given his reasons for regarding this species as an occasional migrant to our shores, in an article which appeared in the "Zoologist" for 1886, p. 71.

APPENDIX D.

SPECIES DISCARDED FROM THE NORFOLK LIST.

Scops asio, AMERICAN MOTTLED OWL (vol. i., p. 44). Mr. Stevenson omitted this species from his list in White's "Directory of Norfolk," 1883; and Mr. J. H. Gurney, jun., followed his example in 1884. There is no fresh evidence on which to reinstate it, but it is right to say that the late Mr. Gurney always maintained his belief in it.

Loxia bifasciata, EUROPEAN WHITE-WINGED CROSSBILL (vol. i., p. 242). Professor Newton has shown (Yarrell's "British Birds," 4th edition, ii., p. 213) that the specimen of this bird, on which Mr. Stevenson relied for its claim as a Norfolk species, was really killed at Drinkstone, near Bury St. Edmund's, in Suffolk, and a crossbill, purchased by Mr. J. H. Gurney on the 9th October, 1872, as having been taken alive on the rigging of a vessel which arrived at Yarmouth in October, 1870, proved to be the American form *L. leucoptera.* This bird lived in Mr. Stevenson's aviary till December, 1874. On the 1st September, 1889, an example of *L. bifasciata*, now in Mr. W. W. Spelman's collection, was killed at Burgh Castle, which is on the Suffolk shore of Breydon water, and another was seen at the same time in company with it which was not procured.

Sturnella ludoviciana, AMERICAN MEADOW STARLING (vol. i., p. 245). Although this species undoubtedly occurred (probably as an "escape") at Thrandeston, in Suffolk, the report of its having been seen at South Walsham, is, I think, not sufficiently conclusive to entitle it to a place in this work.

Ardea garzetta, LITTLE EGRET (vol. ii., p. 150). This bird, now in the Norwich Museum, proved to be the American *Ardea candidissima*, and has every appearance of having been mounted from a skin. See J. H. Gurney, jun., "Trans. of the Norfolk and Norwich Nat. Soc.," vol. iii., p. 565.

Ardea russata, BUFF-BACKED HERON (vol. ii., p. 151). This species was admitted into the Norfolk list in 1870 by Mr. Stevenson on the faith of a specimen in Saffron Walden Museum, said to have been killed at Martham in 1827; this bird is not, however, entered as British in the museum catalogue published in 1845. Mr. Stevenson never examined it, and it was destroyed in 1873 in consequence of an attack of moth.

Scolopax sabinii, SABINE'S SNIPE (vol. ii., p. 343). This being now generally regarded as a melanism of *S. gallinago*, must be omitted as a species. The variety, however, has been met with in this county once since the publication of Mr. Stevenson's article, viz., at Hockwold, on August 5th, 1889, by Mr. H. E. S. Upcher.

INDEX.

A

Adams's Diver, 268
Adriatic Black-headed Gull, 333
Ægialitis asiatica, Appendix A, 382
Alca torda, 286
American Meadow Starling, Appendix D, 413
American Mottled Owl, Appendix D, 413
Anas acuta, 163
  ,, *boscas*, 165
  ,, *circia*, 177
  ,, *clypeata*, 134
  ,, *crecca*, 182
  ,, *mergoides*, 221
  ,, *penelope*, 185
  ,, *strepera*, 157
Anser ægyptiacus, 41
  ,, *albifrons*, 32
  ,, *brachyrhynchus*, 14
  ,, *canadensis*, 43
  ,, *ferus*, 1
  ,, *leucopsis*, 34
  ,, *paludosus*, 27
  ,, *ruficollis*, 39
  ,, *segetum*, 21
  ,, *torquatus*, 36
Anthus obscurus, Appendix B, 387
Aquila chrysaetus, Appendix A, 375
Arctic Skua, 353
Arctic Tern, 305
Ardea alba, Appendix C, 411
Ardea garzetta, Appendix D, 414
  ,, *russata*, Appendix D, 414
Auk Little, 281
  ,,   ,, in summer plumage, 283

B

Bargander, 121
Barred Warbler, Appendix A, 378
Bean Goose, 21
Bernicla leucopsis, 34
Bernacle Goose, 34

Bewick's Swan, 53
Bimaculated Duck, 184
Birds found dead on the shore, 275, 278, 285, 287, 294, 334
Black-backed Gull, Greater, 341
  ,,   ,, Lesser, 337
Black Guillemot, 280
Black-headed Gull, 322
  ,,   ,, Adriatic, 333
  ,,   ,, at Hockwold, 326
  ,,   ,, ,, Horsey, 328
  ,,   ,, ,, Hoveton, 330
  ,,   ,, ,, Rollesby, 329
  ,,   ,, ,, Scoulton, 327
  ,,   ,, ,, Stanford, 323
  ,,   ,, ,, Wretham, 326
  ,,   ,, Food of, 332
Black-necked or Eared Grebe, 250
Black Tern, 312
Black-throated Diver, 269
Bluethroat, Appendix B, 385
Brent Goose, 36
  ,, "Stranger," 39
Buff-backed Heron, Appendix D, 414
Buffle-headed Duck, 222
Buffon's Skua, 357
Bunting, Cirl, Appendix B, 388
  ,, Ortolan, Appendix A, 379
Burrow Duck, 121
Bustard, Appendix B, 396

C

Calandrella brachydactyla, Appendix C, 410
Canada Goose, 43
Capped Petrel, 361
Caspian Plover, Appendix A, 382
Caspian Tern, 296
Charadrius fulvus, Appendix C, 411
Cirl Bunting, Appendix B, 388
Coccothraustes vulgaris, Appendix B, 388

INDEX.

Colymbus adamsi, 268
    ,, *arcticus*, 269
    ,, *glacialis*, 265
    ,, *septentrionalis*, 272
Common Gull, 335
Common Tern, 301
Cormorant, 287
    ,, nesting in Norfolk, 288
Close Time Committee of British Association, origin of, 237, Note
Crossbill, Appendix B, 389
    ,, parrot, Appendix B, 391
    ,, white-winged, Appendix D, 413
Cygnus bewicki, 53
    ,, *ferus*, 45
    ,, *immutabilis*, 80, 111
    ,, *minor*, 56
    ,, *olor*, 58

D

Dabchick, 256
Dartford Warbler, Appendix B, 387
Decoys in Norfolk, 170
Diver, Adams's, 268
    ,, Black throated, 269
    ,, Great Northern, 265
    ,, Red throated, 272
    ,, Red throated, fossil, 275
Divers, mode of submergence, 267
Domestic Goose, origin of, 7
    ,, Swan, weight of, 81
Duck Bimaculated, 184
    ,, Buffle-headed, 222
    ,, Common Scoter, 197
    ,, Eider, 190
    ,, Ferruginous, 208
    ,, Gadwall, 157
    ,, Golden eye, 219
    ,, Harlequin, 219, Note
    ,, King, Note, 192, Appendix A, 384
    ,, Longtailed, 216
    ,, ,, in summer, 219
    ,, Paget's Pochard, 207
    ,, Pintail, 163
    ,, Pochard, 201
    ,, Red Crested, 199
    ,, Scaup, 210
    ,, Shoveller, 134
    ,, Steller's, 192
    ,, Surf Scoter, 197, Note
    ,, Tufted, 211
    ,, Velvet Scoter, 195
    ,, Wigeon, 185

Duck, Wild, 165
Ducks, Hybrid, 184, 207, 221
    ,, males assuming female plumage, 153
Dusky Shearwater, 367

E

Eagle, Golden, Appendix A, 375
Eared or Black-necked Grebe, 250
Eastern Golden Plover, Appendix C, 411
Egret, Little, Appendix D, 414
Egyptian Goose, 41
Eider Duck, 190
    ,, King, Note, 192, Appendix A, 384
Emberiza cirlus, Appendix B, 388
    ,, *hortulana*, Appendix A, 379
European White-winged Crossbill, Appendix D, 413

F

Fen Goose, 1
Ferruginous Duck, 208
    ,, ,, aged, 209
Fork-tailed Petrel, 369
Fratercula arctica, 284
Fuligula albeola, 222
    ,, *clangula*, 219
    ,, *cristata*, 211
    ,, *ferina*, 201
    ,, *ferinoides*, 207
    ,, *glacialis*, 216
    ,, *histrionica*, 219, Note
    ,, *homeyeri*, 208, Note
    ,, *marila*, 210
    ,, *nyroca*, 208
    ,, *rufina*, 199
Fulmar, 359
Fulmarus glacialis, 359

G

Gadwall, 157, 161
    ,, breeding of, 160, 161
Gallinule, Green-backed, Appendix C, 411
Gannet, 292
Garganey Teal, 177
Glareola torquata, Appendix B, 407
Glaucous Gull, 343
Golden Eagle, Appendix A, 375
Golden-eye Duck, 219

INDEX.   417

Golden Plover, Eastern, Appendix C, 411
Goose, Bean, 21
　„　Brent, 36
　„　Canada, 43
　„　Egyptian, 41
　„　Grey lag, 1
　„　Pink-footed, 14
　„　Red-breasted, 39
　„　Spur-winged, 44
　„　White-fronted, 32
　„　Domestic, origin of, 7
　„　　„　rearing of, &c., 9
　„　Wild, taken at a gas lamp, 18
　„　"Wyld killed wt ye cros bowe," 39
Goosander, 229
Great Black-backed Gull, 341
Great Crested Grebe, 233
　„　　„　　„　nesting habits of, 239
　„　　„　　„　temperature of nests, 240
　„　　„　　„　winter resorts, 244
Great Northern Diver, 265
Great Skua, 346
Great White Heron, Appendix C, 411
Grebe, Black-necked or Eared, 250
　„　"furs," 234, 252
　„　Great Crested, 233
　„　Little, 256
　„　Red-necked, 245
　„　Sclavonian, 248
Green-backed Gallinule, Appendix C, 411
Grey Lag Goose, 1
Grosbeak, Pine, Appendix C, 410
Gull, Adriatic B. H., 333
　„　Black-headed, 322
　„　Common, 335
　„　Glaucous, 343
　„　Great Black-backed, 341
　„　Herring, 339
　„　Iceland, 336
　„　Kittiwake, 334
　„　Lesser, Black-backed, 337
　„　Little, 320
　„　　„　in summer plumage, 321
　„　Sabine's, 319
Gull-billed Tern, 307
Gulleries, 323, 326, 327, 328, 329, 330
Guillemot, Black, 280
　„　Common, 275
　„　　„　eggs found at Cromer, 277

Guillemot, Common, numbers dead on beach, 278
　„　　„　said to have bred on Hunstanton Cliff, 275, Note
　„　Ringed, 279

H

Harlequin Duck, 219, Note
Hawfinch, Appendix B, 388
Heron, Buff-backed, Appendix D, 414
　„　Great White, Appendix C, 411
Herring Gull, 339
Holkham Lake, 187, 225
Hooded Merganser, 228
Hooker Sir W., MS., 40, 42
Horsey Gullery, 328
Hoveton Gullery, 330
Hybrid between Bernacle Goose and Canada Gander, 35
Hybrid ducks, 184, 207, 221
Hypolais icterina, Appendix A, 378

I J

Iceland Gull, 336
Icterine Warbler, Appendix A, 378

K

Kittiwake, 334
King Duck, 192, Note
　„　　„　Appendix A, 384

L

Lanius minor, Appendix A, 376
Lark, Shore, Appendix B, 388
　„　Short-toed, Appendix C, 410
Larus argentatus, 339
　„　*canus*, 335
　„　*fuscus*, 337
　„　*glaucus*, 343
　„　*leucopterus*, 336
　„　*marinus*, 341
　„　*melanocephalus*, 333
　„　*minutus*, 320
　„　*ridibundus*, 322
　„　*sabinii*, 319
　„　*tridactylus*, 334
Lesser Black-backed Gull, 337
　„　Grey Shrike, Appendix A, 376
Lesser Tern, 309
le Strange Household Book, 39, 99, 135
Little Auk, 281

3 E

INDEX.

Little Egret, Appendix D, 414
" Grebe, 256
" " destroyed by pike, 261
" " flight of, 258
" " mode of progression of very young, 254
" Gull, 320
Long-tailed Duck, 216
Loxia bifasciata, Appendix D, 413
" *curcirostra*, Appendix B, 389
" *pityopsittacus*, Appendix B, 391

M

Manx Shearwater, 365
Meadow Starling, American, Appendix D, 413
Meliozophilus dartfordiensis, 387
Meres, Stanford and Thompson, 139
" Wretham, 142
Merganser, Hooded, 228
" Red-breasted, 226
Mergulus alle, 281
Mergus albellus, 224
" *anatarius*, 221
" *cucullatus*, 228
" *merganser*, 229
" *serrator*, 226
Motacilla alba, Appendix A, 378
Muscicapa parva, Errata
Mute Swan, 58
" " age attained by, 86
" " fatting, 97
" " marking, 92
" " marks, 93, 102
" " misnamed "Mute," 86
" " nesting habits of, 63
" " prolific, 69
" " to roast, 98
" " St. Helen's Swanpit, 96
" " service in destroying water weeds, 81
" " "upping," 88
" " weight of, 81

N

Noctua tengmalmi, Appendix B, 385
Norfolk decoys, 170
" swan marks, 102
Northumberland Household Book, 99
Nyroca Duck, 208

O

Œdemia fusca, 195
" *nigra*, 197

Œdemia perspicillata, 197, Note
Estrelata hæsitata, 361
Origin of "Close Time Committee" of the British Association, 237, Note
Ortolan, Appendix A, 379
Otis tarda, Appendix B, 396
Otocorys alpestris, Appendix B, 388
Owl, American Mottled, Appendix D, 413
Owl, Tengmalm's, Appendix B, 385

P

Paget's Pochard, 207
Pallas's Sand Grouse, Appendix B, 392
Parrot Crossbill, Appendix B, 391
Pelican, 295
" fossil, 296
Petrel, Capped, 361
" Fork-tailed, 369
" Fulmar, 359
" Storm, 371
Phalacrocorax carbo, 287
" *graculus*, 290
Pike destructive to Wild Fowl, 261, 327
Pine Grosbeak, Appendix C, 410
Pink-footed Goose, 14
Pintail Duck, 163
Pipit, Rock, Appendix B, 387
Plectropterus gambensis, 44
Plover, Caspian, Appendix A, 382
Plover, Eastern Golden, Appendix C, 411
Pochard, Common, 201
" " nests of, 204
" Paget's, 207
" Red Crested, 199
Pochard, Grass, 203
Podicipes auritus, 248
" *cristatus*, 233
" *fluviatilis*, 256
" *griseigena*, 245
" *nigricollis*, 250
Polish Swan, 111
"Popeler," 135, Note
Pomatorhine Skua, 348
Porphyrio smaragdonotus, Appendix C, 411
Pratincole, Appendix B, 407
Procellaria leucorrhoa, 369
" *pelagica*, 371
Puffin, 284
" Small-billed, 284, Note
Puffinus anglorum, 365
" *griseus*, 364
" *obscurus*, 367

INDEX.

Pyrrhula enuncleator, Appendix C, 410

R

Razorbill, 286
Red-breasted Flycatcher, Errata
" Goose, 39
" Merganser, 226
Red-crested Pochard, 199
Red-necked Grebe, 245
Red-throated Diver, 272
Ringed Guillemot, 279
Rock Pipit, Appendix B, 387
Rollesby Gullery, 329
Roseate Tern, 300
Ruddy Sheld Drake, 133
Ruticilla suecica, Appendix B, 385

S

Sabine's Gull, 319
" Snipe, Appendix D, 414
Salicaria luscinioides, Appendix B, 386
Salmon's MS. Diary, 138, 323
Sand Grouse, Pallas's, 392
Sandwich Tern, 298
Savi's Warbler, Appendix B, 386
Scaup Duck, 210
Sclavonian Grebe, 248
Scolopax sabinii, Appendix D, 414
Scops asio, Appendix D, 413
Scoter, Common, 197
" Velvet, 195
Scoulton Gullery, 327
Sea birds found dead on the shore, 275, 278, 285, 287, 294, 334
Serin, Appendix A, 379
Serinus hortulanus, Appendix A, 379
Shag, 290
Shearwater, Dusky, 367
" Manx, 365
" Sooty, 364
Sheld Drake, 121
" " Ruddy, 133
Shore Lark, Appendix B, 388
Short-toed Lark, Appendix C, 410
Shoveller, 134
" nesting, 138
Shrike, Lesser Grey, Appendix A, 376
Skua, Arctic, 353
" Buffon's, 357
" Great, 346
" Pomatorhine, 348
Smew, 224
Snipe, Sabine's, Appendix D, 414
Somateria molissima, 190

Somateria spectabilis, 192, Note, Appendix A, 384
" *stelleri*, 192
Sooty Shearwater, 364
Spur-winged Goose, 44
Stanford, large bag of wild birds, 177
" and Thompson Meres, 139
Starling, American Meadow, Appendix D, 413
Steller's Duck, 192
Stercorarius catarrhactes, 346
" *crepidatus*, 353
" *parasiticus*, 357
" *pomatorhinus*, 348
Sterna anglica, 307
" *caspia*, 296
" *dougalli*, 300
" *fluviatilis*, 301
" *hybrida*, 306
" *leucoptera*, 316
" *macrura*, 305
" *minuta*, 309
" *nigra*, 312
" *sandvicensis*, 298
Storm Petrel, 371
Sturnella ludoviciana, Appendix D, 413
Sula bassana, 292
Sylvia nisoria, Appendix A, 378
Syrrhaptes paradoxus, Appendix B, 392
Swan, Bewick's, 53
" Hooper, 45
" Mute, 58
" Polish, 111
" Spanish, 57
" fatting, 97
" ferruginous tint of head and neck, 76
" - marking, 92
" receipt for roasting, 98
" "upping" on the Yare, 88
Swan-marks, 93, 102
Swan-pit, St. Helen's, 75, 88, 96

T

Tadorna rutila, 133
" *vulpanser*, 121
Teal, Common, 182
" Garganey, 177
" hybrids of, 184
Tengmalm's Owl, Appendix B, 385
Tern, Arctic, 305
" Black, 312
" Caspian, 296

Tern, Common, 301
- " Gullbilled, 307
- " Lesser, 309
- " Roseate, 300
- " Sandwich, 298
- " Whiskered, 306
- " White-winged Black, 316

Thompson and Stanford Meres, 139
Thrush, White's, Appendix A, 377
Tichodroma muraria, Appendix A, 380
Tufted Duck, 211
- "   "   nesting of, 214

Turdus varius, Appendix A, 377

U

Uria troile, 275
- " *grylle*, 280

V

Velvet Scoter, 195

W

Wagtail, White, Appendix A, 378
Walsingham Lord, mixed bag of wild birds, 177
Wall-creeper, Appendix A, 380
Warbler, Barred, Appendix A, 378
- " Dartford, Appendix B, 387
- " Icterine, Appendix A, 378
- " Savi's, Appendix B, 386

Way-goose, origin of, 11
Western Duck (Steller's), 192
Whiskered Tern, 306
White-eyed Duck, 208
White-fronted Goose, 32
White-winged Black Tern, 316
- "   " Crossbill, European, Appendix D, 413

White Wagtail, Appendix A, 378
White's Thrush, Appendix A, 377
Whooper, 45
Wigeon, 185
- " aged, 190
- " in summer, 188
- " on Holkham Lake, 186, 188

Wild Duck, 165
- "   " aged, 170
- "   " dates of nesting, 167
- "   " nests in trees, 166
- "   " varieties of, 168

Wild fowl, large number sent to Yarmouth in 1878, 175
Wretham Meres, 142

GENERAL INDEX.

A

	VOL.	PAGE
Accentor, Alpine	i	90
,, Hedge	i	88
Accentor alpinus	i	90
,, *modularis*	i	88
Accipiter nisus	i	24
Adams's Diver	iii	268
Adriatic Black-headed Gull	iii	333
Ægialitis asiatica	iii	382
Agelaius phœniceus	i	244
Alauda alpestris	i	171
,, *arborea*	i	179
,, *arvensis*	i	175
Alca torda	iii	286
Alcedo ispida	i	314
Alpine Accentor	i	90
,, Swift	i	346
American Bittern	ii	174
,, Little Stint	ii	366
American Meadow Starling	i	245
,, ,, ,,	iii	413
,, Mottled Owl	i	44
,, ,, ,,	iii	413
Ammer, Yellow	i	196
Anas acuta	iii	163
,, *boscas*	iii	165
,, *circia*	iii	177
,, *clypeata*	iii	134
,, *crecca*	iii	182
,, *mergoides*	iii	221
,, *penelope*	iii	185
,, *strepera*	iii	157
Anser ægyptiacus	iii	41
,, *albifrons*	iii	32
,, *brachyrhynchus*	iii	14
,, *canadensis*	iii	43
,, *ferus*	iii	1
,, *leucopsis*	iii	34
,, *paludosus*	iii	27
,, *ruficollis*	iii	39
,, *segetum*	iii	21
,, *torquatus*	iii	36

	VOL.	PAGE
Anthus arboreus	i	166
,, *obscurus*	i	169
,, ,,	iii	387
,, *pratensis*	i	167
,, *ricardi*	i	168
,, *rupestris*	i	171
Aquila chrysaëtus	i	4
,, ,,	iii	375
Archibuteo lagopus	i	29
Arctic Skua	iii	353
,, Tern	iii	305
Ardea alba	ii	149
,, ,,	iii	411
,, *cinerea*	ii	130
,, *comata*	ii	151
,, *garzetta*	ii	150
,, ,,	iii	414
,, *purpurea*	ii	145
,, *russata*	ii	151
,, ,,	iii	414
Ash-coloured Harrier	i	35
,, ,,	i	39
Aster palumbarius	i	23
Auk, Little	iii	281
Avocet	ii	237

B

	VOL.	PAGE
Baillon's Crake	ii	401
Barbary Partridge	i	405
Bargander	iii	121
Barn Owl	i	51
Barred Warbler	iii	378
Bar-tailed Godwit	ii	253
Bean Goose	iii	21
Bearded Titmouse	i	150
Bee-eater	i	313
Bernicla leucopsis	iii	34
Bernacle Goose	iii	34
Bewick's Swan	iii	53
Bidcock—Water Rail	ii	404
Bimaculated Duck	iii	184

	VOL.	PAGE
Bittern, American	ii	174
,, Common	ii	159
,, Little	ii	154
Bittour or Bittern	ii	173
Blackbird	i	83
Blackcap	i	125
Black Curlew or Ibis	ii	191
,, Grouse	i	374
Black Guillemot	iii	280
Black-headed Gull	iii	322
,, ,, Adriatic	iii	333
Black-backed Gull, Greater	iii	341
,, ,, Lesser	iii	337
Black-headed Bunting	i	185
Black-necked or Eared Grebe	iii	250
Black Redstart	i	99
,, Stork	ii	182
Black-tailed Godwit	ii	248
Black Tern	iii	312
Black-throated Diver	iii	269
Black-winged Stilt	ii	244
Black Woodpecker, Great	i	291
Blood Olf	i	234
Blue-throated Warbler	i	96
,, ,,	iii	385
Blue Titmouse	i	142
Bohemian Pheasant, var.,	i	367
,, Waxwing	i	154
Bombycilla garrulus	i	154
Botaurus lentiginosus	ii	174
,, *minutus*	ii	154
,, *stellaris*	ii	159
Bottlebump or Bittern	ii	166
Brambling	i	202
Brent Goose	iii	36
,, "Stranger"	iii	39
Broad-billed Sandpiper	ii	359
Brown Linnet	i	227
,, Snipe	ii	348
Bubo maximus	i	47
Buff-backed Heron	ii	151
,, ,,	iii	414
Buff-breasted Sandpiper	ii	358
Buffle-headed Duck	iii	222
Buffon's Skua	iii	337
Bullfinch	i	233
,, Pine or Grosbeak	i	234
Bunting, Black-headed	i	185
,, Cirl	i	198
,, ,,	iii	388
,, Common	i	184
,, Lapland	i	181
,, Ortolan	i	199
,, ,,	iii	379

	VOL.	PAGE
Bunting, Snow	i	182
,, Yellow	i	196
Burrow Duck	iii	121
Bustard, Great	ii	1
,, ,,	iii	396
,, Little	ii	42
Butcherbird or Shrike	i	62
Buteo vulgaris	i	27
Buzzard, Common	i	27
,, Rough-legged	i	29
,, Honey	i	32

C

	VOL.	PAGE
Calandrella brachydactyla	iii	410
Calamophilus biarmicus	i	150
Calidris arenaria	ii	116
Californian Quail	i	437
Cambridge Godwit	ii	204
Canada Goose	iii	43
Capped Petrel	iii	361
Caprimulgus europæus	i	348
Carduelis elegans	i	222
Carolina Crake	ii	392
Carrion Crow	i	258
Caspian Plover	iii	382
Caspian Tern	iii	296
Cayenne Night Heron, supposed specimen of	ii	175
Certhia familiaris	i	295
Chaffinch	i	199
Charadrius cantianus	ii	97
,, *fulvus*	iii	411
,, *hiaticula*	ii	84
,, *intermedius*	ii	95
,, *minor*	ii	96
,, *morinellus*	ii	76
,, *pluvialis*	ii	66
Chiffchaff	i	133
Ciconia alba	ii	177
,, *nigra*	ii	182
Cinclus aquaticus	i	68
,, *melanogaster*	i	69
Cirl Bunting	i	198
,, ,,	iii	388
Circus æruginosus	i	35
,, *cineraceus*	i	39
,, *cyaneus*	i	37
Clinkers or Avocets	ii	240
Coal Titmouse	i	146
Coccothraustes chloris	i	218
,, *vulgaris*	i	214
,, ,,	iii	388
Coffin bird	i	300

	Vol.	Page
Colin Virginian	i	436
Collared Pratincole	ii	64
Columba œnas	i	355
,, *palumbus*	i	351
,, *risoria*	i	360
,, *turtur*	i	359
Colymbus adamsi	iii	268
,, *arcticus*	iii	269
,, *glacialis*	iii	265
,, *septentrionalis*	iii	272
Common Bittern	ii	159
,, Bunting	i	184
,, Buzzard	i	27
,, Creeper	i	295
,, Crossbill	i	235
,, ,,	iii	389
,, Curlew	ii	194
,, Gull	iii	335
,, Heron	ii	130
,, Linnet	i	227
,, Partridge	i	420
,, Pheasant	i	361
,, Redshank	ii	207
,, Redstart	i	98
,, Sandpiper	ii	230
,, Snipe	ii	305
,, Starling	i	247
,, Swift	i	343
,, Tern	iii	301
,, Whitethroat	i	128
,, Wren	i	296
Coot	ii	425
Coracias garrula	i	310
Cormorant	iii	287
Corn Crake	ii	387
Corvus cornix	i	260
,, *corax*	i	256
,, *corone*	i	258
,, *frugilegus*	i	264
,, *monedula*	i	277
Coturnix vulgaris	i	429
Coucou-roux	i	309
Courser, Cream-coloured	ii	48
Crake, Baillon's	ii	401
,, Carolina	ii	392
,, Corn	ii	387
,, Little	ii	396
,, Spotted	ii	393
Crane	ii	125
Creeper, Common or Tree	i	295
Crex bailloni	ii	401
,, *porzana*	ii	393
,, *pratensis*	ii	387
,, *pusilla*	ii	396

	Vol.	Page
Crossbill, Common	i	235
,, ,,	iii	389
,, European white-winged	i	242
,, Parrot	i	239
,, ,,	iii	391
,, White-winged	iii	413
Crow, Carrion	i	258
,, Hooded, grey-backed, Danish, or Royston	i	260
Cuckoo	i	303
Cuculus canorus	i	303
,, *hepaticus*	i	309
Curlew, Black	ii	191
,, Common	ii	194
,, Great Harvest	ii	198
,, Pigmy	ii	350
,, Stone	ii	51
,, Sandpiper	ii	350
Curruca atricapilla	i	125
,, *cinerea*	i	128
,, *hortensis*	i	126
,, *sylviella*	i	129
Cursorius europæus	i	48
Cygnus bewicki	iii	53
,, *ferus*	iii	45
,, *immutabilis*	iii	80
,, ,,	iii	111
,, *minor*	iii	56
,, *olor*	iii	58
Cypselus alpinus	i	346
,, *apus*	i	343

D

	Vol.	Page
Dabchick	iii	256
Danish Crow	i	260
Dartford Warbler	i	133
,, ,,	iii	387
Deviliu or Swift	i	343
Dipper	i	68
Diver, Adams's	iii	268
,, Black throated	iii	269
,, Great Northern	iii	265
,, Red throated	iii	272
Dotterel	ii	76
Dove, Ring or Woodpigeon	i	351
,, Rock	i	358
,, Stock	i	355
,, Turtle	i	359
Double Snipe	ii	299
"Drain" Dunlins	ii	382
Draw-water	i	222
Duck, Bimaculated	iii	184
,, Buffle-headed	iii	222

424 GENERAL INDEX.

	VOL.	PAGE
Duck, Common Scoter	iii	197
„ Eider	iii	190
„ Ferruginous	iii	208
„ Gadwall	iii	157
„ Golden eye	iii	219
„ Harlequin (Note)	iii	219
„ King (Note)	iii	192
„ „	iii	384
„ Longtailed	iii	216
„ Paget's Pochard	iii	207
„ Pintail	iii	163
„ Pochard	iii	201
„ Red Crested	iii	199
„ Scaup	iii	210
„ Shoveller	iii	134
„ Steller's	iii	192
„ Surf Scoter (Note)	iii	197
„ Tufted	iii	211
„ Velvet Scoter	iii	195
„ Wigeon	iii	185
„ Wild	iii	165
Dunlin	ii	371
Dusky Shearwater	iii	367

E

	VOL.	PAGE
Eagle, Golden	i	4
„ „	iii	375
„ White-tailed or Cinereous Sea	i	1
Eared, or Black-necked Grebe	iii	250
Eastern Golden Plover	iii	411
Egret, Little	ii	150
„ „	iii	414
Egyptian Goose	iii	41
Eider Duck	iii	190
„ „ King	iii	192
„ „ „	iii	384
Emberiza cirlus	i	198
„ „	iii	388
„ *citrinella*	i	196
„ *hortulana*	i	199
„ „	iii	379
„ *miliaria*	i	184
„ *schœniclus*	i	185
Erythaca rubecula	i	90
European White-winged Crossbill	iii	413

F

	VOL.	PAGE
Falco æsalon	i	21
„ *candicans*	i	7
„ *gyrfalco*	i	8

	VOL.	PAGE
Falco islandicus	i	8
„ *peregrinus*	i	9
„ *rufipes*	i	19
„ *subbuteo*	i	18
„ *tinnunclus*	i	21
Falcon, Greenland	i	7
„ Iceland	i	8
„ Peregrine	i	9
„ Red-footed	i	19
Fen Goose	iii	1
Fern Owl, or Nightjar	i	348
Ferruginous Duck	iii	208
Fieldfare	i	75
Fire-crested Wren or Regulus	i	138
Fishing hawk	i	5
Flycatcher, Pied	i	66
„ Red-breasted	Errata	
„ Spotted	i	63
Fork-tailed Petrel	iii	369
Fratercula arctica	iii	284
French Partridge	i	404
Fringilla cœlebs	i	199
„ *montifringilla*	i	202
Fulica atra	ii	425
Fuligula albeola	iii	222
„ *clangula*	iii	219
„ *cristata*	iii	211
„ *ferina*	iii	201
„ *ferinoides*	iii	207
„ *glacialis*	iii	216
„ *histrionica* (Note)	iii	219
„ *homeyeri* (Note)	iii	208
„ *marila*	iii	210
„ *nyroca*	iii	208
„ *rufina*	iii	199
Fulmar	iii	359
Fulmarus glacialis	iii	359

G

	VOL.	PAGE
Gadwall	iii	157
„	iii	161
Gallinula chloropus	ii	411
Gallinule, Green-backed	iii	411
Gannet	iii	292
Garden Warbler	i	126
Garganey Teal	iii	177
Garrulus glandarius	i	280
Glareola torquata	ii	64
„	iii	407
Glaucous Gull	iii	343
Glossy Ibis	ii	191
Goatsucker, or Nightjar	i	348
Godwit, Bar-tailed	ii	253

	VOL.	PAGE		VOL.	PAGE
Godwit, Black-tailed	ii	248	Grey Linnet	i	227
,, Cambridge	ii	204	,, Partridge	i	420
Gold-crested Wren, or Regulus	i	134	,, Phalarope	ii	436
			,, Plover	ii	101
Golden Eagle	i	4	,, Shrike, Great	i	61
,, ,,	iii	375	,, ,, Lesser	iii	376
Golden-eye Duck	iii	219	,, Snipe	ii	318
Golden Oriole	i	86	,, Wagtail	i	163
,, Plover	ii	66	Grosbeak, or Hawfinch	i	214
,, ,, Eastern	iii	411	,, ,,	iii	388
Goldfinch	i	222	,, Green, or Greenfinch	i	218
Goose, Bean	iii	21	,, Pine, or Bullfinch	i	234
,, Brent	iii	36	,, ,,	iii	410
,, Canada	iii	43	Grouse, Black	i	374
,, Egyptian	iii	41	,, Pallas's Sand	i	376
,, Grey lag	iii	1	,, ,,	iii	392
,, Pink-footed	iii	14	Grus cinerea	ii	125
,, Red-breasted	iii	39	Guillemot, Black	iii	280
,, Spur-winged	iii	44	,, Common	iii	275
Goosander	iii	229	,, Ringed	iii	279
Goshawk	i	23	Gull, Adriatic Black-headed	iii	333
Grakle, Minor	i	255	,, Black-headed	iii	322
Gracula religiosa	i	255	,, Common	iii	335
Grasshopper Warbler	i	104	,, Glaucous	iii	343
Great Black-backed Gull	iii	341	,, Great Black-backed	iii	341
,, Black Woodpecker	i	291	,, Herring	iii	339
,, Bustard	ii	1	,, Iceland	iii	336
,, Crested Grebe	iii	233	,, Kittiwake	iii	334
,, Grey Shrike	i	61	,, Lesser Black-backed	iii	337
,, Northern Diver	iii	265	,, Little	iii	320
,, Plover	ii	51	,, Sabine's	iii	319
,, Skua	iii	346	Gull-billed Tern	iii	307
,, Snipe	ii	299	Gyrfalcon	i	8
,, Spotted Woodpecker	i	288			
,, Titmouse	i	139	H		
,, White Heron	ii	149			
,, ,, ,,	iii	411	Hæmatopus ostralegus	ii	122
Grebe, Black-necked or Eared	iii	250	Half Snipe	ii	334
,, Great Crested	iii	233	Haliæetus albicilla	i	1
,, Little	iii	256	Harlequin Duck (Note)	iii	219
,, Red-necked	iii	245	Harnser or Heron	ii	130
,, Sclavonian	iii	241	Harrier, Ash-coloured	i	35
Green-backed Gallinule	iii	411	,, ,,	i	39
Greenfinch, or Green Linnet	i	218	,, Hen	i	37
Greenland Falcon	i	7	,, Marsh		35
Green-legged Shank or Knot	ii	357	,, Montagu's		39
Green Olf	i	221	Hawfinch	i	214
,, Plover or Lapwing	ii	103	,,	ii	388
,, Sandpiper	ii	215			
Greenshank	ii	234	Hedge Sparrow, or Accentor	i	88
Green Woodpecker	i	285	Hen Harrier	i	37
Grey-backed Crow	i	260	Heron Buff-backed	ii	151
Grey-headed Wagtail	i	164	,, ,,	iii	414
Grey Lag Goose	iii	1	,, Common	ii	130

	VOL.	PAGE
Heron Great White	ii	149
,, ,, ,,	iii	411
,, Night	ii	174
,, Purple	ii	145
,, Squacco	ii	151
Herring Gull	iii	339
Himantopus melanopterus	ii	244
Hirundo riparia	i	338
,, *rustica*	i	324
,, *urbica*	i	328
Hobby	i	18
,,	i	25
,, Orange-legged	i	20
Honey Buzzard	i	32
Hooded Crow	i	260
,, Merganser	iii	228
Hoopoe	i	298
House Martin	i	328
,, Sparrow	i	209
Hypolais icterina	iii	378

I

	VOL.	PAGE
Ibis falcinellus	ii	191
Ibis, Glossy	ii	191
Iceland Falcon	i	8
,, Gull	iii	336
Icterine Warbler	iii	378

J

	VOL.	PAGE
Jackdaw	i	277
Jack Snipe	ii	334
Jay	i	280

K

	VOL.	PAGE
Kentish Plover	ii	97
Kestrel	i	21
Kite	i	26
Kittiwake	iii	334
King Duck (Note)	iii	192
,, ,,	iii	384
Kingfisher	i	314
King Harry, black cap	i	224
,, ,, red cap	i	224
Knot	ii	354

L

	VOL.	PAGE
Landrail	ii	387
Lanius collurio	i	62
,, *excubitor*	i	61
,, *minor*	iii	376
,, *rutilus*	i	64

	VOL.	PAGE
Lapland Bunting	i	181
Lapwing	ii	103
Lark, Shore	i	171
,, ,,	iii	388
,, Short-toed	iii	410
,, Sky	i	175
,, Wood	i	179
Larus argentatus	iii	339
,, *canus*	iii	335
,, *fuscus*	iii	337
,, *glaucus*	iii	343
,, *leucopterus*	iii	336
,, *marinus*	iii	341
,, *melanocephalus*	iii	333
,, *minutus*	iii	320
,, *ridibundus*	iii	322
,, *sabinii*	iii	319
,, *tridactylus*	iii	334
Lesser Black-backed Gull	iii	337
,, Grey Shrike	iii	376
,, Redpole	i	230
,, Spotted Woodpecker	i	293
,, Tern	iii	309
,, White-throat	i	129
Linnet, Common, Brown, or Grey	i	227
Limosa melanura	ii	248
,, *rufa*	ii	253
Linota canescens	i	228
,, *cannabina*	i	227
,, *linaria*	i	230
,, *montium*	i	231
Little Auk	iii	281
,, Bittern	ii	154
,, Bustard	ii	42
,, Crake	ii	396
,, Egret	ii	150
,, ,,	iii	414
,, Gallinule	ii	396
,, Grebe	iii	256
,, Gull	iii	320
,, Owl	i	59
,, Ringed Plover	ii	96
,, Stint	ii	361
Long-eared Owl	i	44
Long-tailed Duck	iii	216
Long-tailed Titmouse	i	148
Lophortyx californicus	i	437
Loxia bifasciata	i	242
,, ,,	iii	413
,, *curvirostra*	i	235
,, ,,	iii	389
,, *pityopsittacus*	i	239
,, ,,	iii	391

GENERAL INDEX.

M

	VOL.	PAGE
Machetes pugnax	ii	261
Macrorhamphus griseus	ii	348
Magpie	i	280
Manx Shearwater	iii	365
Marsh Harrier	i	35
„ Titmouse	i	147
Martin, House	i	328
„ Sand	i	338
„ Snipe	ii	224
Maybird or Whimbrel	ii	199
Maychit	ii	233
Meadow Pipit	i	167
„ Starling, American	i	245
„ „ „	iii	413
Mealy Redpole	i	228
Meliozophilus dartfordiensis	i	133
„	iii	387
Merganser, Hooded	iii	228
„ Red-breasted	iii	226
Mergulus alle	iii	281
Mergus albellus	iii	224
„ *anatarius*	iii	221
„ *cucullatus*	iii	228
„ *merganser*	iii	229
„ *serrator*	iii	226
Merlin	i	21
Merops apiaster	i	313
Milvus ictinus	i	26
Minor Grackle	i	255
Missel Thrush	i	74
Moorhen	ii	411
Montagu's Harrier	i	39
Motacilla alba	iii	378
„ *boarula*	i	163
„ *flava*	i	164
„ *rayi*	i	165
„ *yarrelli*	i	160
Muscicapa atricapilla	i	66
„ *grisola*	i	65
„ *parva*	iii	xiii
Mute Swan	iii	58

N

	VOL.	PAGE
Night Heron	ii	174
Nightingale	i	123
Nightjar	i	348
Noctua passerina	i	59
„ *tengmalmi*	i	60
„	iii	385
Norfolk Plover	ii	51
Nucifraga caryocatactes	i	281
Numenius arquata	ii	194

	VOL.	PAGE
Numenius phæopus	ii	199
Nutcracker	i	281
Nuthatch	i	301
Nycticorax gardeni	ii	174
Nyroca Duck	iii	208

O

	VOL.	PAGE
Œdemia fusca	iii	195
„ *nigra*	iii	197
„ *perspicillata* (Note)	iii	197
Œdicnemus crepitans	ii	51
Œstrelata hæsitata	iii	361
Olf, Blood	i	234
„ Green	i	221
Olivaceous Crake	ii	396
Orange-legged Hobby	i	20
Oriole, Golden	i	86
Oriolus galbula	i	86
Ortolan Bunting	i	199
„ „	iii	379
Ortyx virginianus	i	436
Osprey	i	5
Otis tarda	ii	1
„ „	iii	396
„ *tetrax*	ii	42
Otocorys alpestris	i	171
„ „	iii	388
Otus brachyotus	i	50
„ *vulgaris*	i	44
Ouzel, Ring	i	84
Owl, American Mottled	i	44
„ „ „	iii	413
„ Barn	i	51
„ Eagle	i	47
„ Little	i	59
„ Long-eared	i	44
„ Scop's-eared	i	42
„ Short-eared	i	50
„ Snowy	i	57
„ Tawny	i	54
„ Tengmalm's	i	60
„ „	iii	385
Oxbird or Dunlin	ii	379
Oyster Catcher	ii	122

P

	VOL.	PAGE
Paget's Pochard	iii	207
Pallas's Sand Grouse	i	376
„ „	iii	392
Pandion haliæetus	i	5
Parrot Crossbill	i	239
„ „	iii	391

	VOL.	PAGE
Partridge, Barbary	i	405
" Common or Grey	i	420
" Red-legged or French	i	404
Parus ater	i	146
" *caudatus*	i	148
" *cæruleus*	i	142
" *major*	i	139
" *palustris*	i	147
Passer montanus	i	206
" *domesticus*	i	209
Pastor roseus	i	253
" Rose-coloured	i	253
Pectoral Sandpiper	ii	367
Peewit or Lapwing	ii	103
Pelican	iii	295
Perdix cinerea	i	420
" *petrosa*	i	405
" *rufa*	i	404
Perdrix de marais	i	419
Peregrine Falcon	i	9
Pernis apivorus	i	32
Petrel, Capped	iii	361
" Fork-tailed	iii	369
" Fulmar	iii	359
" Storm	iii	371
Phalacrocorax carbo	iii	287
" *graculus*	iii	290
Phalarope, Grey	ii	436
" Red-necked	ii	439
Phalaropus hyperboreus	ii	439
" *lobatus*	ii	436
Phasianus colchicus	i	361
" *torquatus*	i	365
" *versicolor*	i	365
Pheasant Bohemian, var.	i	367
" Common	i	361
Phœnicura ruticilla	i	98
" *suecica*	i	96
" *tithys*	i	99
Philomela luscinia	i	123
Pica caudata	i	280
Picus major	i	288
" *martius*	i	291
" *minor*	i	293
" *viridis*	i	285
Pickcheese	i	142
Pied Flycatcher	i	66
" Wagtail	i	160
Pigmy Curlew	ii	350
Pine Grosbeak	i	234
" "	iii	410
Pink-footed Goose	iii	14
Pintail Duck	iii	163

	VOL.	PAGE
Pipit, Meadow	i	167
" Richard's	i	168
" Rock	i	169
" "	iii	387
" Tree	i	166
Platalea leucorodia	ii	184
Plectrophanes lapponica	i	181
" *nivalis*	i	182
Plectropterus gambensis	iii	44
Plover, Caspian	iii	283
" Eastern Golden	iii	411
" Great	ii	51
" Green or Lapwing	ii	103
" Golden	ii	66
" Grey	ii	101
" Kentish	ii	97
" Little-ringed	ii	96
" Norfolk	ii	51
" Ringed	ii	84
Pochard, Common	iii	201
" Paget's	iii	207
" Red Crested	iii	199
Podiceps auritus	iii	248
" *cristatus*	iii	233
" *fluviatilis*	iii	256
" *griseigena*	iii	245
" *nigricollis*	iii	250
Polish Swan	iii	111
Pomatorhine Skua	iii	348
Porphyrio smaragdonotus	iii	411
Pratincole, Collared	ii	407
" "	iii	64
Procellaria leucorrhoa	iii	369
" *pelagica*	iii	371
Puffin	iii	284
Puffinus anglorum	iii	365
" *griseus*	iii	364
" *obscurus*	iii	367
Purple Heron	ii	145
" Sandpiper	ii	384
Pyrrhula enucleator	i	234
" "	iii	410
" *vulgaris*	i	233

Q

Quail, Common	i	429
" Californian	i	437

R

Rail, Land	ii	387
" Water	ii	404

	VOL.	PAGE
Rallus aquaticus	ii	404
Raven	i	256
Razorbill	iii	286
Recurvirostra avocetta	ii	237
Red-backed Shrike	i	62
Redbreast	i	90
Red-breasted Flycatcher	iii	xiii
" Goose	iii	39
" Merganser	iii	226
" Snipe	ii	348
Red-crested Pochard	iii	199
Red-footed Falcon	i	19
Red-legged Partridge	i	404
Redleg or Redshank	ii	213
Red-necked Grebe	iii	245
" Phalarope	ii	439
Redpole, Lesser	i	230
" Mealy	i	228
Redshank, Common	ii	207
" Spotted	ii	203
Redstart, Black	i	99
" Common	i	98
Red-throated Diver	iii	272
Redwing	i	82
Red-winged Starling	i	244
Reed Pheasant	i	153
" Warbler	i	115
Reeve	ii	261
Regulus cristatus	i	134
" Fire-crested	i	138
" *ignicapillus*	i	138
Richard's Pipit	i	168
Ringed Guillemot	iii	279
" Plover	ii	84
Ring-dove	i	351
" Ouzel	i	84
Rock Dove	i	358
" Pipit	i	169
" "	iii	387
Roller	i	310
Rook	i	264
Roseate Tern	iii	300
Rose-coloured Pastor	i	253
Rough-legged Buzzard	i	29
Royston Crow	i	260
Ruddy Sheld Drake	iii	133
Ruff	ii	261
Ruticilla suecica	iii	385

S

Sabine's Gull	iii	319
" Snipe	ii	343
" "	iii	414

	VOL.	PAGE
Salicaria locustella	i	104
" *luscinioides*	i	110
" "	iii	386
" *phragmitis*	i	108
" *strepera*	i	115
Sand Grouse, Pallas's	i	376
" "	iii	392
Sanderling	ii	116
Sand Martin	i	338
Sandpiper, Broad-billed	ii	359
" Buff-breasted	ii	358
" Common	ii	230
" Curlew	ii	350
" Green	ii	215
" Pectoral	ii	367
" Purple	ii	384
" Spotted	ii	233
" Wood	ii	226
" Yellow-shanked	ii	215
Sandwich Tern	iii	298
Savi's Warbler	i	110
" "	iii	386
Saxicola œnanthe	i	102
" *rubetra*	i	101
" *rubicola*	i	100
Scaup Duck	iii	210
Sclavonian Grebe	iii	248
Scolopax brehmi	ii	333
" *gallinago*	ii	305
" *gallinula*	ii	334
" *major*	ii	299
" *russata*	ii	332
" *rusticola*	ii	272
" *sabinii*	ii	343
" "	iii	414
Scops aldrovandi	i	42
" *asio*	i	44
" "	iii	413
Scops' Eared Owl	i	42
Scoter, Common	iii	197
" Velvet	iii	195
Sea Eagle	i	1
" Dotterel	ii	86
Sea-pie	ii	122
Sea Woodcock	ii	260
Sedge Warbler	i	108
Serin	iii	379
Serinus hortulanus	iii	379
Shag	iii	290
Shearwater, Dusky	iii	367
" Manx	iii	365
" Sooty	iii	364
Sheld Drake	iii	121
" " Ruddy	iii	133

GENERAL INDEX.

	VOL.	PAGE
"Shoeing-horn," or Avocet	ii	237
Shore Lark	i	171
„ „	iii	388
Short-eared Owl	i	50
Short-toed Lark	iii	410
Shovelard or Spoonbill	ii	184
Shoveller	iii	134
Shrike, Great Grey	i	61
„ Lesser Grey	iii	376
„ Red-backed	i	62
„ Woodchat	i	64
Siskin	i	224
Sitta europæa	i	301
„ *cæsia*	i	301
Skua, Arctic	iii	353
„ Buffon's	iii	357
„ Great	iii	346
„ Pomatorhine	iii	348
Skylark	i	175
Smew	iii	224
Snipe, Brown	ii	348
„ Common	ii	305
„ Great	ii	299
„ Grey	ii	348
„ Jack	ii	334
„ Martin	ii	224
„ Red-breasted	ii	348
„ Sabine's	ii	343
„ „	iii	414
„ Solitary	ii	299
„ Summer	ii	224
„ „	ii	230
Snow Bunting	i	182
Snowy Owl	i	57
Solitary Snipe	ii	299
Somateria molissima	iii	190
„ *spectabilis* (Note)	iii	192
„ „	iii	384
„ *stelleri*	iii	192
Song Thrush	i	78
Sooty Shearwater	iii	364
Spanish Swan	iii	57
Sparrow, Hedge	i	88
„ House	i	209
„ Tree	i	206
Sparrowhawk	i	24
Spink or Chaffinch	i	199
Spoonbill	ii	184
Spotted Crake	ii	393
„ Flycatcher	i	65
„ Redshank	ii	203
„ Sandpiper	ii	233
„ Woodpecker, Great	i	288
„ „ Lesser	i	293

	VOL.	PAGE
Spowes or Whimbrel	ii	201
Spur-winged Goose	iii	44
Squatarola cinerea	ii	101
Squacco Heron	ii	151
Starling, American Meadow	i	245
„ „	iii	413
„ Common	i	247
„ Red-winged	i	244
Starna cinerea	i	419
„ *palustris*	i	419
Steller's Duck	iii	192
Stercorarius catarrhactes	iii	346
„ *crepidatus*	iii	353
„ *parasiticus*	iii	357
„ *pomatorhinus*	iii	348
Sterna anglica	iii	307
„ *caspia*	iii	296
„ *dougalli*	iii	300
„ *fluviatilis*	iii	301
„ *hybrida*	iii	306
„ *leucoptera*	iii	316
„ *macrura*	iii	305
„ *minuta*	iii	309
„ *nigra*	iii	312
„ *sandvicensis*	iii	298
Stilt, Blackwinged	ii	244
Stints, or Dunlins	ii	373
Stint, American Little	ii	366
„ Little	ii	361
„ Temminck's	ii	363
Stock Dove	i	353
„ „	i	355
Stonechat	i	100
Stone Curlew	ii	51
Stonehatch	ii	85
Stork, Black	ii	182
„ White	ii	177
Storm Petrel	iii	371
Strepsilas interpres	ii	113
Strix flammea	i	51
Sturnella ludoviciana	i	245
„ „	iii	413
Sturnus vulgaris	i	247
Sula bassana	iii	292
Summer Snipe	ii	224
„ „	ii	230
Surnia nyctea	i	57
Swallow	i	324
Swan, Bewick's	iii	53
„ Hooper	iii	45
„ Mute	iii	58
„ Polish	iii	111
„ Spanish	iii	57
Swift, Alpine	i	346

GENERAL INDEX. 431

	Vol.	Page
Swift, Common	i	343
Sylvia hypolais	i	133
,, *nisoria*	iii	378
,, *rufa*	i	133
,, *sylvicola*	i	130
,, *trochilus*	i	132
Syrnium stridulum	i	54
Syrrhaptes paradoxus	i	376
,, ,,	iii	392

T

	Vol.	Page
Tadorna rutila	iii	133
,, *vulpanser*	iii	121
Tangle-picker	ii	114
Tawny Owl	i	54
Teal, Common	iii	182
,, Garganey	iii	177
Temminck's Stint	ii	363
Tengmalm's Owl	i	60
,, ,,	iii	385
Tern, Arctic	iii	305
,, Black	iii	312
,, Caspian	iii	296
,, Common	iii	301
,, Gullbilled	iii	307
,, Lesser	iii	309
,, Roseate	iii	300
,, Sandwich	iii	298
,, Whiskered	iii	306
,, White-winged Black	iii	316
Tetrao tetrix	i	374
Thrush, Missel	i	74
,, Song	i	78
,, White's	iii	377
Tichodroma muraria	iii	380
Titlark	i	167
Titmouse, Bearded	i	150
,, Blue	i	142
,, Coal	i	146
,, Great	i	139
,, Long-tailed	i	148
,, Marsh	i	147
Totanus calidris	ii	207
,, *flavipes*	ii	215
,, *fuscus*	ii	203
,, *glariola*	ii	226
,, *glottis*	ii	234
,, *hypoleucus*	ii	230
,, *macularius*	ii	233
,, *ochropus*	ii	215
Tree Creeper	i	295
,, Pipit	i	166
,, Sparrow	i	206

	Vol.	Page
Tringa canutus	ii	354
,, *maritima*	ii	384
,, *minuta*	ii	361
,, *pectoralis*	ii	367
,, *platyrhyncha*	ii	359
,, *pusilla*	ii	366
,, *rufescens*	ii	358
,, *subarquata*	ii	350
,, *temmincki*	ii	363
,, *torquata*	ii	381
,, *variabilis*	ii	371
Troglodytes vulgaris	i	296
Tufted Duck	iii	211
Turdus iliacus	i	82
,, *merula*	i	83
,, *musicus*	i	78
,, *pilaris*	i	75
,, *torquatus*	i	84
,, *varius*	iii	377
,, *viscivorus*	i	74
Turnstone	ii	113
Turtle Dove	i	359
Twite	i	231

U

	Vol.	Page
Upupa epops	i	298
Uria troile	iii	275
,, *grylle*	iii	280

V

	Vol.	Page
Velvet Scoter	iii	195
Virginian Colin	i	436

W

	Vol.	Page
Wagtail, Grey	i	163
,, Grey-headed	i	164
,, Pied	i	160
,, White	iii	378
,, Yellow	i	165
Wall-creeper	iii	380
Warbler, Barred	iii	378
,, Blackcap	i	125
,, Bluethroated	i	96
,, Dartford	i	133
,, ,,	iii	387
,, Garden	i	126
,, Grasshopper	i	104
,, Icterine	iii	378
,, Reed	i	115
,, Savi's	i	110
,, ,,	iii	386

GENERAL INDEX.

	Vol.	Page
Warbler, Sedge	i	168
,, Willow	i	132
,, Wood	i	130
Water-hen	ii	411
Water-rail	ii	404
Waxwing	i	154
Western Duck (Steller's)	iii	192
Wheatear	i	102
Whinchat	i	101
Whimbrel	ii	199
Whiskered Tern	iii	306
Whistling Plover	ii	66
White-eyed Duck	iii	208
White-fronted Goose	iii	32
White Spoonbill	ii	184
,, Stork	ii	177
White-tailed Eagle	i	1
White-throat, Common	i	128
,, Lesser	i	129
White-winged Black Tern	iii	316
,, ,, Crossbill, European	i	242
,, ,, ,,	iii	413
White's Thrush	iii	377
White Wagtail	iii	378
"Whole" Snipe	ii	334
Whooper	iii	45
Wigeon	iii	187

	Vol.	Page
Wild Duck	iii	165
Willow Warbler	i	132
Wind-hover, or Kestrel	i	21
Woodchat Shrike	i	64
Woodcock	ii	276
"Woodcock" Snipe	ii	304
Woodlark	i	179
Woodpecker, Great Black	i	291
,, Great Spotted	i	288
,, Green	i	285
,, Lesser Spotted	i	293
Woodpigeon	i	351
Wood Sandpiper	ii	226
Wood Warbler	i	130
Wren, Common	i	296
,, Fire-crested	i	138
,, Golden-crested	i	134
Wryneck	i	294
"Wypes," or Lapwings	ii	109

Y

	Vol.	Page
"Yarwhelp," or Godwit	ii	252
Yellow ammer, or Bunting	i	196
Yellow-shanked Sandpiper	ii	215
Yellow Wagtail	i	165
Yunx torquilla	i	294

www.ingramcontent.com/pod-product-compliance
Lightning Source LLC
Chambersburg PA
CBHW032009300426
44117CB00008B/957